THE
EVOLU[ ]Y
BIOLOGY OF
PLANTS

THE
EVOLUTIONARY BIOLOGY OF
PLANTS

KARL J. NIKLAS

THE UNIVERSITY OF CHICAGO PRESS
CHICAGO AND LONDON

Karl J. Niklas is professor of botany at Cornell University. His two previous books, the award-winning *Plant Biomechanics: An Engineering Approach to Plant Form and Function,* and *Plant Allometry: The Scaling of Form and Process,* are both published by the University of Chicago Press.

The University of Chicago Press, Chicago 60637
The University of Chicago Press, Ltd., London
© 1997 by The University of Chicago
All rights reserved. Published 1997
Printed in the United States of America
06 05 04 03 02 01 00 99 98 97   1 2 3 4 5

ISBN: 0-226-58082-2 (cloth)
ISBN: 0-226-58083-0 (paper)

Library of Congress Cataloging-in-Publication Data

Niklas, Karl J.
    The evolutionary biology of plants / Karl J. Niklas.
       p.   cm.
    Includes bibliographical references and index.
    ISBN 0-226-58082-2 (cl : alk. paper). — ISBN 0-226-58083-0 (pa : alk. paper)
    1. Plants—Evolution.   I. Title.
QK980.N55   1997
581.3'8—dc20   96-31060
                                                                       CIP

⊗ The paper used in this publication meets the minimum requirements of the American National Standard for Information Sciences—Permanence of Paper for Printed Library Materials, ANSI Z39.48-1984.

Contents

Preface

It is not enough to believe what you see. You must also understand what you see.

Leonardo da Vinci

No author can hope to capture the full scope of evolution between the covers of one book. Each can only bring a certain perspective to round out the full evolutionary pageantry of life. I wrote this work on evolution from the perspective of plant biology and in terms of what I know best—the plant fossil record. I was trained as a paleobotanist, so I am naturally drawn to the study of past life. Without the window on the past that paleontology gives us, biologists would be totally ignorant of the existence (let alone the appearance) of well over 90 percent of all the species that ever lived. Likewise, without the fossil record, we would be deprived of an invaluable biochronometer for measuring the tempo of long-term evolutionary changes. The study of fossils therefore neither requires nor deserves apology.

Nevertheless, as "a student of death" I am also aware that fossils, no matter how well preserved or well studied, cannot animate evolutionary mechanisms to our complete satisfaction. To discover and fully appreciate these mechanisms, biologists must understand life's genetics, development, and ecology as well as its necrology. One of my goals in writing this book was to give the living and the dead equal attention and to render as comprehensive a view as practical of the enlivened present and the now dead past.

That goal profoundly influenced the organization of this book. I believe evolution is best taught and understood by considering the various levels of biological organization as well as the different kinds of evidence for evolution, simultaneously rather than individually. Evolution is neither molecular change nor phenotypic change. It is both, and very much more because evolution is a continuing process: some of today's descendants will be tomorrow's ancestors. For this reason I wrote with my mind turned toward the elements of genetics, development, ecology and the fossil record whose juxtaposition best illuminates the topics at hand.

This book is divided into four parts, each with two chapters, following a short introduction in which I set the stage for reading the main text by presenting the concepts that reappear in various guises throughout. One of these—Sewall Wright's metaphor of adaptive evolution—deserves special mention. Like many others, I believe the metaphor of "walks on fitness landscapes" is exceptionally powerful because it illustrates how the combined effects of random and nonrandom evolutionary phenomena can shape life's history. Much of this book is devoted to this Wrightian metaphor. Each of the eight chapters employs or amplifies it in one way or another, but part 3 (chapters 5 and 6) is the most overt expression of this perspective on evolution.

Chapters 1 and 2 deal with basic evolutionary principles. Most introductory biology and advanced evolution textbooks cover this material, but the levels at which these topics are typically introduced to undergraduate and graduate students differ in both presentation and subtlety. Chapters 1 and 2 were written to bridge these two levels of instruction as well as to reiterate basic principles with a phytocentric bias. I believe students learn by repetition and that a botanical perspective gives new insight into concepts, no matter how well they are presented from a zoological perspective. Chapters 3 and 4 offer a short history of plant life, from its early origins to the evolution of flowering plants, presenting plant form and function relationships in a historical context different from that found in most paleobotany textbooks. Chapters 5 and 6 deal with the theory of plant form and function relationships. In these chapters I discuss computer simulated adaptive walks in complex hypothetical fitness landscapes along the lines first proposed by Sewall Wright. These two chapters especially reflect my research approach and interests. They are plainly idiosyncratic, but the topics covered are patently relevant to any general understanding of plant evolution. Chapters 7 and 8 discuss the tempos and modes of plant evolution. The plural is warranted because all the evidence we have shows that plants have evolved in many directions and at very different rates. In these chapters, molecular and morphological data drawn from living and fossil plants are used to illuminate the birth and death of species.

I wrote this book for today's students who will be tomorrow's teachers and scientists. I have tried to keep the style simple and to use jargon sparingly. Jargon is unquestionably necessary—as anyone who has read a basic mathematics, physics, or biology textbook can attest—because it affords an economical way of communicating complex ideas. But technical words glibly unleashed in the classroom or in print tend to sound more like explanations than descriptions, and jargon in any guise can easily alienate students. The study of evolution has more than its fair share of technical terms because it draws on every facet of the biological and physical sciences. Botany too is rich in terminology. For these reasons I have tried to use technical terms as little as practical and to define each when it is first used in the text.

Because this book is intended for classroom use, I drew the line drawings by hand (with the aid of a computer) in order to emphasize rather than codify plants or concepts. Each one intentionally emphasizes certain features over others, so they are best understood as cartoons rather than fastidious representations of reality. As a student, I learned the elements of biology more from my teachers' line drawings than from their words, so I hope the drawings in this book will be used in the classroom even if my words are not. If a diagram is worth a thousand words, then a word is a millidiagram!

I must thank as well as absolve from responsibility those who have directly or indirectly helped me write this book. At the most basic level, I am deeply indebted to all my students who asked those pesky questions that inspire fear and awe in any lecturer because they challenge us to think harder on subjects we think we knew well and to conclude that the teacher is really the student. I also owe a great debt to my colleagues at Cornell University—Harlan P. Banks, William L. Crepet, Jerrold L. Davis, Jeffrey J. Doyle, James L. Gulledge, Kevin Nixon, Thomas G. Owens, Dominick J. Paollilo Jr., Carol Reiss, Mark Staves, Robert Turgeon, and Randy Wayne—all of whom suffered my interminable questions with grace (and in some cases mild amusement) and many of whom read earlier drafts of this book and gave invaluable advice on how to improve the final version. I am also grateful to Otto Stein (University of Massachusetts, Amherst), who made invaluable suggestions to improve this book. I am particularly indebted to my teachers and dear friends—Sherwin Carlquist, who taught me the wonders of plant form-function relationships and island biology; Tom L. Phillips, who opened my eyes to paleobotany and taught me the full meaning of academic generosity; Lawrence J. Crockett, who taught me as well as so many others the wonders of botany with wit and brilliance; Bruce H. Tiffney, who serves as a role model in many ways and continues to instruct me on how to write (with the aid of limitless red ink); and finally Susan Abrams and Alice Bennett, who edited my books with charitable professionalism and unselfish friendship. Above all, I thank my parents, who gave me the opportunity to be myself, and Edward D. Cobb, who has given me more support than anyone truly deserves.

Introduction

They take in moisture and the carbon dioxide we breathe out and the energy of the sunlight, and they produce sugar and oxygen, and then we eat the plants and get the sugar . . . and that's the way the world goes around.

John Updike

Remembrance of a particular form is but regret for a particular moment.

Marcel Proust

Astronomers like to say we are fashioned from stardust—that the elements in us were formed in ancient, long vanished stars and differ not at all from those found in the outer reaches of space. If so, then it is also true that we are made from starlight—that our substance is predicated on the way plants convert the energy of sunlight into chemical energy. Because photosynthesis underwrites most of life on Earth, it is only fitting that this book is about living things that thrive without intention, build without blood or brain, move without muscle, summon without self-awareness, and feed the world without intent. In short, this book is about plants.

But it is also important to know how plants achieved the ability to fashion their living substance from sunlight, how they moved this photosynthetic machinery onto land, painted the continents green, fabricated the largest living things that ever existed, and continue to cope with the animals they unintentionally feed and shelter. This book therefore is also about evolution. The inaugural unicellular forms of life in Earth's ancient oceans could not harvest and use sunlight's energy, yet their multicellular descendants now live on land and grow many times larger than the largest animals that ever lived. Clearly plant life has changed over time, and organic change over time is evolution.

Evolution does not refer to the changes occurring in an *individual* organism. The alterations in the chemical composition, size, shape, and structure attending the growth and development of an individual plant or animal are trivial in an evolutionary sense because the individual cannot evolve. Evolution involves descent with modification from ancestral forms of life. The theory of descent with modification is the cornerstone of Charles Darwin's book *On the Origin of Species by Means of Natural Selection*. The principal propositions of this theory are:

1. All life evolved from one or a few simple kinds of organisms.
2. Species evolve from preexisting species.
3. The birth of species is gradual and of long duration.
4. Higher taxa (genera, families, and so forth) evolve by the same evolutionary mechanisms responsible for the evolution of species.
5. The greater the similarities between taxa, the more closely the taxa are related to one another and the shorter their divergence time.
6. Extinction is primarily the result of interspecific competition.
7. The geological record is incomplete; the absence of transitional forms between species is due to gaps in the geological record.

With the exception of the presumed cause of extinction, the theory of descent with modification is strictly phenomenological. That is, its propositions describe relationships without invoking mechanisms, and so no rationale for *why* or *how* new species evolve is embedded in the theory. To fill this intellectual void, Darwin proposed an ancillary but ultimately more far reaching theory because it explains *why* life evolves. This theory proposes that descendants differ from their ancestors because in each generation randomly generated variants within populations are differentially selected by the environment. Organisms survive and reproduce under diverse biological and physical conditions because random evolutionary forces obtain variations among related individuals while nonrandom evolutionary forces mindlessly sort these variants, leaving behind individuals better suited to their particular environment. This is the theory of natural selection; its theme and variations occupy the bulk of Charles Darwin's *Origin of Species*. The basic precepts of the theory of natural selection are:

1. The number of individuals in populations tends to increase at a geometric rate.
2. But the number of individuals in populations tends to remain the same because
3. The environment has limited resources, and so
4. Only a fraction of the offspring will survive and successfully reproduce.
5. Those offspring that do survive and reproduce differ from those that die because
6. The individuals in a population are not identical owing to heritable variation.

7. The struggle to survive and reproduce determines which variants in the population will perpetuate the species.
8. Natural selection results in the accumulation of favorable heritable traits by eliminating individuals bearing unfavorable heritable traits.

Darwin's theory of natural selection tells us that organisms have the means as well as the motivation to change over time. Here the means is natural variation among individuals in a population, while the motivation is the cruel reality that organisms must either evolve to survive the gauntlet of a changing environment or suffer and ultimately perish. One of the great insights of Darwin's theory is that correlated variables have meaning only in relation to one another—that one among many variables cannot be conceived of as cause or effect. This is a subtle but important insight. Organisms evolve, and by doing so they change their own environment. Reciprocally, when the environment changes, organisms are obliged to change if they are to survive and reproduce. Thus, although organisms are both the means and the end of evolution, its mechanisms involve the interaction between the environment and the organism.

Although it goes a long way to explain *why* organisms evolve, the theory of natural selection speaks little or not at all to the issue of *how* they evolve. Understanding the mechanisms of organic evolution requires a genetic theory, one that can explain and quantify how features are passed on from parent to progeny. Such a theory was first provided by Gregor Mendel and later amplified by the authors of the "modern synthesis" (so called because it represents the amalgamation of insights gained from population genetics, developmental biology, and paleontology). The postulates of the modern synthesis or the neo-Darwinian view of evolution are:

1. Evolution is the change of gene (allele) frequencies in the gene pool of a population over many generations.
2. The gene pools of species are isolated from one another, and the gene pool of each species is held together by gene flow.
3. Each individual of a sexually reproductive species has only a portion of the genes in the gene pool of its species because
4. The alleles and allelic combinations of the individual are contributed by two different parents and may be modified by chromosomal or genic mutations.
5. Mutations are the ultimate source of new genes.
6. Individuals favored by natural selection will contribute larger portions of their genes or gene combinations to the gene pool of the next generation.

7. Changes in allelic frequencies in populations come about primarily by natural selection, even though random genic and chromosomal variations frequently occur.

8. Barriers that restrict or eliminate gene flow between the subpopulations of a species are essential for genic and phenotypic divergence of the subpopulations of a species.

9. Speciation is complete when gene flow does not occur between a divergent population and the population of its parent species.

Because it is the consequence of randomly generated genetic variation and patterned environmental sorting (natural selection), evolution is a curious phenomenon indeed. On the one hand, populations of sexually reproductive organisms are capable of producing a veritable cosmic number of genetically different sperm or egg cells, while the number of genetically different offspring that even a single pair of parents can produce is staggeringly large. On the other hand, only the tiniest fraction of these genetic possibilities are ever produced by chance, and among these an even smaller number may survive, go on to reproduce, and contribute to the next generation. Thus, at a very basic level, evolution is the dynamic outcome of what "genetically is possible," "genetically happens," and "biologically works."

This small fraction of life's possibilities is ensured a time-tested viability because every living thing has a repeatedly challenged (and proven) lifeline extending through countless generations of ancestors. By preserving the partial products of past successes, evolution continues to build on organisms' ability to deal with life's exingencies despite the intervention of random events. This mechanism constitutes adaptive evolution. Yet the notion of adaptative evolution as an ineluctable process producing "perfect" organisms must be rejected, because every organism must work only well enough to survive and reproduce, and every organism is as much a consequence of random events as it is the product of ordered environmental sorting (Ayala 1988). Adaptive evolution, therefore, is as much the result of contingency as of patterned events.

Contingency is an important factor in the history of life (Gould 1995). If the evolutionary experiment were to begin anew, if all Earth's life were catastrophically destroyed and evolution had to start again from scratch, it would undoubtedly take a different course. Subtle differences in the starting conditions of any complex system tend to cascade and affect the outcome of subsequent events. The random death of one species or the birth by sheer chance of another early in a "replicate experiment" of evolution would likely have profound biological and physical consequences for following events. This conclusion follows from the statistical improbability of duplicating the exact sequence of events in any process governed even in small part by random events. This improbability is the basis of Dollo's law, which states that evolution is irreversible. Even though *general* trends in evolution may be reversed (indeed many, such as body size, are reversed in the history of individual lineages), the elementary laws of probability show that the *details* of an evolutionary trajectory have a vanishingly small probability of reappearing.

Yet even though future evolutionary events are contingent on past events, some of the themes of organic evolution would reappear in any replicate experiment because no form of life can avoid the consequences of physical and chemical processes and laws. The effect of gravity is invariably to compress or bend things. Gases expand when heated and contract when cooled. Moving fluids exert pressure against obstructions to their flow. Light's intensity decreases with distance from its source. Hydrogen bonds will develop and hold water molecules together. Even complex heredity molecules must subscribe to the laws of chemistry, just as the most structurally complex organism must obey the laws of physics and the principles of engineering. Fortunately, chemical and physical laws and processes are remarkably indifferent to time and space and lineage, and so every evolutionary experiment must comply with them. In this sense many properties of any evolutionary trajectory are predictable.

Nicholas Humphrey (1987) makes this point by posing a fascinating question. If you were forced to consign to oblivion one of the following masterpieces, which would it be: Newton's *Principia,* Chaucer's *Canterbury Tales,* Mozart's *Don Giovanni,* or Eiffel's Tower? Humphrey's answer is Newton's *Principia,* because it is the only achievement that follows from first principles and so would inevitably be found by another scientist. Darwin's *Origin of Species* is perhaps an even better example, because in 1859 both Alfred Wallace and Charles Darwin independently proposed the theory of evolution by natural selection.

The study of evolution requires a balanced understanding of past and present life. The fossil record tells us what happened, often in stirring detail, and long-term evolutionary rates and patterns can be adduced only from that record. Humans are too short-lived to witness the interplay of random and nonrandom evolutionary forces on the time scale required to fully appreciate the meaning of evolution. Nevertheless, the fossil record sheds little direct light on how and why things happened the way they did. The genetic and developmental mechanisms of evolution are hidden from paleontologists, who can only infer how these mechanisms affected past forms of life based on the genetics and development of living things. Thus, in addition to reviewing the fossil record, this book reviews the current signature of plant life to understand how the external appearance (morphology), internal structure (anatomy), and the behavior of plants are shaped by genes and sorted by the environment.

There are a number of ways to visualize adaptive evolution, but one of the most attractive was first proposed by Sewall Wright at the 1932 International Congress of Genetics in Ithaca, New York. Wright envisioned evolution as a "walk" over a fitness "landscape." In his original paper, the walk was a sequence of genotypes leading from regions of low to high fitness, while the landscape was rendered as a spatial domain depicting all conceivable genotypes, each assigned a relative fitness value. As originally cast, however, this model for adaptive evolution is largely impractical. The genotypes of fossil plants are unknown, just as the genotypes of living plants are largely unknown, but the landscape of plant evolution can be rendered in terms of phenotypes and their relative fitness, just as walks over the landscape can be seen in terms of the phenotypic alterations required to scale adaptive peaks. Regardless of how it is rendered, Wright's metaphor for adaptive evolution is a *thought experiment,* one that must be performed in full appreciation that evolution is as much a game of chance as it is structured by environmental sorting. The landscape over which plant life evolved changed unpredictably many times. Thus plant evolution reflects the consequences of many chance events. But the landscape is also ordered and predictable—ordered in the sense that physical laws and processes influencing fitness cannot be violated, and predictable in the sense that the physical environment will dispose of less fit phenotypes and leave behind the more fit. Thus every walk, whether cast in genotypic or in phenotypic terms, reflects the combined effects of random and nonrandom evolutionary forces.

A virtue of the phenotypic version of Wright's fitness landscape is that hypothetical adaptive walks can be compared with those preserved in the plant fossil record. Here the concern is not to rewrite evolutionary history but to discover why plants may have evolved as they did. If they evolved in accordance with physical and chemical laws and processes, then hypothetical adaptive walks based on these laws and principles should agree with the phenotypic trends seen in the fossil record. The cardinal sin threatening this test is the conceited notion that all possibilities and all effects have been anticipated. Plant life is extraordinarily complex, and manipulating a few simple shapes to construct the domain of all possible phenotypes is truly naive. Likewise, the ways plants have dealt with the requirements for survival or reproduction are demonstrably intricate and varied. Thus it is ingenuous to believe that we understand the complexity and intricacy of plant life well enough to simulate the precise course of evolution. But the tart response in the face of these pitfalls is "nothing ventured, nothing gained." By refusing the effort, we pass up the possibility of gaining some little insight into evolutionary history. Simulated adaptive walks may fail the test of the fossil record, but we may learn something along the way. Some preconceptions about plant evolution may be revealed as incorrect or irrelevant; the staggering diversity of plant form and structure and its relation to fitness may be quantitatively codified; and the experiment of plant life will be opened to further interpretation.

Part 1 Evolutionary Basics: A Review

Plate 1. Genetic variation, which fuels the fires of evolution, is subject to artificial as well as natural selection. For example, variations in fruit shape and color among *Cucurbita pepo* cultivars are the result of plant breeding selection programs focusing on intergenic interactions typically involving two loci. (Starting a the upper left and going in a clockwise direction, the squash cultivars are spaghetti, zucchini [middle], sweet dumpling, yellow crookneck, and acorn.) Although the genetics of crooknecked fruits are complex, two loci influence whether fruits are disk shaped, spherical, or oblong. Dominant alleles at both loci result in the expression of disk-shaped fruits (not shown); a dominant allele at one locus and a recessive allele at the other locus result in spherical fruit shapes (e.g., acorn); recessive alleles at both loci result in oblong fruit shapes (e.g., zucchini). Uniform external fruit color is also governed by two loci (*White* and *Yellow*). A dominant allele at the *White* locus prevents the expression of the dominant *Yellow* allele, resulting in a white fruit (not shown); two recessive alleles for *White* and one dominant allele at *Yellow* result in a yellow fruit (e.g., crookneck); and recessive alleles at both loci produce green fruit color (e.g., acorn). Two loci also influence whether fruits are deeply ribbed, but the genetic interactions between these loci and others are not well understood.

1   Adaptive Evolution

*The development of a predictive theory [of evolution]
depends on being able to specify when a population is
in [a] better or worse evolutionary state. For this purpose an objective definition of adaptedness is necessary.*
Lawrence Slobodkin

Charles Darwin and Alfred Wallace did not originate
the concept of adaptation, but they were the first to
offer a coherent explanation for how organisms evolve the ability to cope
with particular, often vastly different environments. Darwin's explanation, which was supported by far more data than Wallace's, was that differential survival and reproductive success were the result of the naturally
occurring variation among the individuals within a population. Because
populations produce more offspring than their environments can support, some variants are less capable than others of surviving or reproducing. These less fit individuals are eliminated by the environment. Over
the course of many generations, the selective sorting by the environment
leaves behind individuals that are better suited to their particular local
environments (Dobzhansky et al. 1977; Ginsburg 1983). Darwin's theory,
called natural selection, provides the cornerstone for modern evolutionary theory. Individuals may perish or reproduce, but only populations
can evolve over many generations by the accumulation of traits advantageous to survival or reproductive success. Darwin convincingly argued
that all forms of life, plants as well as animals, descend with modification
from previous forms. The first forms of unicellular life, which inau-
gurated this process billions of years ago, diversified in appearance and
structure and evolved adaptations to various habitats. Darwin envisaged
the pattern of evolution as a "tree of life" with multiple branches continually emerging from a common trunk and continually pruned by natural
selection.

Darwin was not successful at explaining how the tree of life extended
new branches. If a population evolves by the environmental sorting of
variants, then natural selection can only diminish or amplify preexisting
heritable variations. Because natural selection cannot engender variation,

without some mechanism capable of generating new variants within populations, evolution would wind down and eventually halt. Throughout his lifetime, Darwin's attempts to propose a reasonable mechanism capable of creating new variants were unsuccessful. Indeed, the major obstacle to the acceptance of his theory of natural selection was the lack of a corresponding theory of genetics that could explain how variations in traits first arise and how parental traits can be precisely transmitted to offspring. In this sense the theory of natural selection was something of a paradox: "Like begets like, but not exactly."

With the hindsight of history, we know that this paradox was resolved in Darwin's time by the botanist Gregor Mendel. But the implications of Mendel's work went largely unappreciated during the nineteenth century. Ironically, the rediscovery of Mendel's work in the early twentieth century brought greater resistance to Darwin's theory, because the traits examined by Mendel and geneticists of the early twentieth century were qualitative "either/or" characteristics (red or white flowers, wrinkled or smooth peas, and so forth) rather than the quantitative ones that vary over a continuum within a population and are central to Darwin's theory of natural selection. Additionally, during most of the 1920s, most geneticists focused on mutations and on the rapid leaps and radical changes in the appearance of organisms these random genetic changes can produce. The effects of mutations on the appearance of the organisms studied at the time contrasted sharply with the gradualistic evolutionary changes in populations Darwin had postulated.

Nevertheless, between 1918 and 1931 R. A. Fisher and J. B. S. Haldane in England, Sewall Wright in the United States, and S. S. Chetverikov in Russia proposed theoretical models of evolutionary processes that began to integrate Mendelian genetics with natural selection. Most of these models were far too complex for most biologists to follow, and it was not until the 1930s, with the publication of Theodosius Dobzhansky's *Genetics and the Origin of Species,* that Darwinism and Mendelism became integrated and accessible. Several important books were published in the following years that deepened this integration: Ernst Mayr's *Systematics and the Origin of Species* (1942), J. S. Huxley's *Evolution: The Modern Synthesis* (1942), George Gaylord Simpson's *Tempo and Mode in Evolution* (1944), and G. L. Stebbins's *Variation and Evolution in Plants* (1950). With their emphasis on the extensive genetic variation within populations and the importance of quantitative inheritance, these books blended the fields of paleontology, biogeography, developmental biology, and taxonomy with population genetics. The result was what Huxley called the "modern synthesis," a neo-Darwinism emphasizing the importance of populations as the fundamental units of evolution, the universal role of natural selection, and the gradual nature of evolutionary changes. Darwin's intellectual legacy in the modern synthesis is unmistakable (Stebbins and Ayala 1985).

The theory of evolution continues to evolve—this is the nature of scientific thinking. Some evolutionists have challenged whether evolution is a gradual process and whether natural selection is more central to it than are random evolutionary forces (Eldredge and Gould 1972; Gould and Eldredge 1977; Gould 1980). At its core, the debate is over adaptive evolution. Few biologists deny that plants and animals appear well suited to their particular environments, but no consensus exists on whether the features that permit organisms to cope with their physical and biological contexts are determined primarily by selection pressure or by random, albeit fortuitous, historical accident. The debate is not over whether evolution *exists,* but over contending evolutionary models for its tempo and mode. Arguing about these evolutionary models is like arguing about different theories for the birth of stars—the sun continues to shine even as astronomers debate how.

In this chapter I will review the concept of adaptive evolution, beginning with what is meant by "adaptation" and "adaptive evolution." This agenda involves exploring the roles of selection and random forces in plant evolution. The conclusion that emerges is that evolution is a dialectic—a tension between random and nonrandom forces—whose synthesis is adaptation. This conclusion serves as the basis for discussing Darwin's gradualistic view of evolution and the alternative "saltational" view that sees species (and adaptations) as evolving suddenly and non-gradualistically (see chapter 2).

Adaptation: State and Process

Plants transact their lives under the most exacting environmental conditions by means of very sophisticated "devices" or adaptations. But there is no doubt that each adaptation works only in a very particular habitat. Examples of plants well suited to their environments are as diverse as they are fascinating. The cactus has its bristling armament of spines that can radiate heat and defend its water-packed stem from herbivory; the insectivorous plant has its glistening sugar-baited leaves to lure, ensnare, and digest unwary insects; the silversword plant has its felty mantle of leaf hairs to scatter light and insulate against the piercing frost; and the vine may have its aerial roots growing within cisternlike leaves that trap and recycle water. The terrestrial landscape is resplendent with examples of plant adaptations. Likewise, aquatic plants appear well suited to the environmental circumstances of crashing waves and fluctuating tides. The giant kelp keeps afloat with gas-filled bladders and deflects the force of waves with long, flexible stipes; the seagrass pollinates its flowers and sets seed under cover of low tide; and the unicellular dinoflagellates living inside the cnidarian animal jointly construct a coral reef. Move the spiny cactus from the arid desert to a wet tropical forest, or flush the giant kelp or the flowering seagrass with freshwater, and each languishes and dies.

Four points must be kept in mind whenever adaptations are discussed:

1. Adaptations are features (or sets of correlated features) that allow organisms to survive or reproduce under very specific environmental conditions, and so an adaptation can be understood only in terms of its particular environmental context.
2. Each adaptation operates in the particular biological context of the organism possessing it.
3. The benefits conferred by a particular adaptation can be gauged only in relative, not absolute terms.
4. There is no reason a priori to assume that each adaptation evolved through natural selection for its current function.

Consider that the cactus spine would be an impossible adaptation unless the cactus stem or some other organ type carried on photosynthesis. The cactus spine is a highly modified leaf that functions in ways otherwise not normally entertained by green foliage leaves because the cactus stem is capable of assuming the tasks of typical photosynthetic leaves (light interception, gas exchange, etc.). We must also bear in mind that, however tempting it is to think a particular adaptation is a "perfect" solution to life, each adaptation need only suit an organism to its environment *well enough* to permit survival or reproductive success (Knoll and Niklas 1987). The cactus spine functions well, but it is neither perfect nor the "only" solution for survival in arid deserts. Many desert plants lack spiny leaves yet grow, survive, and reproduce in deserts as well as spine-bearing cacti (e.g., the resurrection plant, *Selaginella lepidophylla;* Washington palm, *Washingtonia filifera;* boojum tree, *Fouquiera columnaris;* agave, *Agave shawii;* Joshua tree, *Yucca brevifolia;* sagebrush, *Artemisia tridentata;* and creosote bush, *Larrea divaricata*). The Madagascar *Pachypodium lealii* and the African quiver tree, *Aloe dichotoma,* produce large foliage leaves when water is plentiful but amputate these organs during times of water deprivation (fig. 1.1). The quiver tree goes so far as to lose entire branches and terminal rosettes of leaves under extreme drought, keeping only enough apical buds to produce more leaves later in the season when water becomes available. Finally, although the cactus spine dissipates heat from sun-drenched stems that would otherwise stew, there is no reason to assume that the spine evolved as a consequence

Figure 1.1. Plants adapted to living in arid deserts. From left to right: a North American cactus; the Madagascar *Pachypodium lealii* (subspecies *saundersii*); and the African *Aloe dichotoma*. All three store water in succulent stems. The cactus retains its spiny leaves; *P. lealii* and *A. dichotoma* drop their large foliage leaves in times of drought. *A. dichotoma* amputates branches and terminal crowns of foliage leaves during times of extreme water deprivation.

of selection for this present-day function. It may have evolved as a consequence of a random genetic mutation that altered leaf development from a course otherwise normally producing a broad, photosynthetic leaf. In arid deserts this mutation may have benefited survival or reproductive success, but spiny leaves cannot be viewed as an adaptation per se to the exigencies of desert life. Indeed, numerous scenarios for the appearance of spiny desert plants can be based on the inextricable link between the development of the leaf and that of the stem. Perhaps spiny leaves were a developmental corollary of natural selection favoring succulent, water-packed stems. If so, the succulent stem is an adaptation, whereas the spine developmentally hitchhiked along as a useful device.

These four points emphasize that an adaptation must be thought of first and foremost as a state of being and may be defined as *any heritable feature (or set of associated features) whose presence increases the probability that the individual bearing it will survive or successfully reproduce under a given set of environmental conditions.* This definition necessarily speaks of survival or successful reproduction in relative, not absolute terms. It also recognizes that an adaptation must be heritable and therefore will evince natural variation among individuals within a population. Accordingly, the hypothesis that a feature is adaptive must be tested by comparing individuals drawn from the same population (or from different populations of the same species) and evaluating how far survival or reproductive success correlates with the feature presumed to be adaptive. If spines are truly beneficial to the survival or reproductive success of plants living in arid deserts, then the degree of spininess should correlate in some way with either survival or reproductive success and with the degree of aridity.

Testing whether a feature is adaptive is often difficult because it requires knowing which, if any, among many potential selection pressures underpins the functional significance of the presumed adaptation. Consider spines, thorns, and prickles. These three plant modifications share many morphological features. Each is sharp, stiff, and nasty, suggesting that each may be an adaptation to herbivory (fig. 1.2) This notion gains support when we see that the size and number of the thorns on hawthorns increase in proportion to how severely trees are browsed by deer or cattle. But the spines of desert-dwelling cacti are equally effective at radiating heat from stems, while the prickles of roses can act as grappling hooks for climbing and for binding stems together. Clearly, each of these plant modifications may serve in manifold capacities, and so each may be the historical consequence of many selection pressures. Spines, thorns, and prickles further illustrate the ambiguity that can result when comparisons are made across rather than within species to evaluate whether similar-

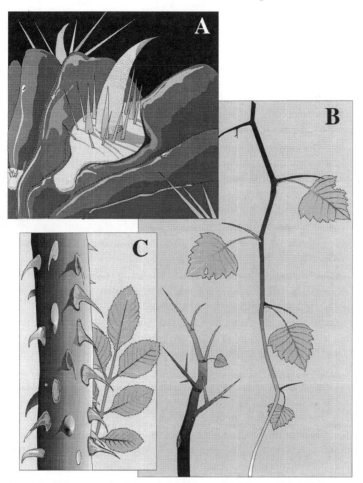

Figure 1.2. Spines, thorns, and prickles. Spines are modified leaves (A); thorns are modified branches (B); and prickles are modified outgrowths of the epidermis and cortex (C). All three modifications can dissipate heat from stems and deter herbivory, as they do for the cactus *Ferocactus* (A). The number and size of thorns on hawthorn trees (*Crataegus* spp.) increase with increasing herbivory (the slender, decumbent branch is from a tree protected from grazing deer; the larger, more branched thorn at lower left in B is from a short, intensively grazed tree). The prickles of multiflora roses (*Rosa multifloria*) help scrambling stems mechanically support one another (C).

appearing structures confer differential survival or reproductive success. Biologists often compare different species in terms of presumed adaptive features and may assume that the form that is numerically dominant among coexisting species is the most adaptive. But this assumption is only one among many possible inferences. An equally plausible assumption is that the most abundant species occupied the habitat first and simply had more time to become established. Alternative explanations exist, and none is clearly superior to another a priori without fastidious experiment and observing populations over many years.

Comparisons may also be made among individuals of different species that use different adaptations to acquire the same resources. All plants essentially require the same resources to survive and grow (light, water, minerals, space, etc.), and so in theory all plants may compete with one another. But comparing species with different adaptations can yield ambiguous conclusions. Isolating which adaptation, if any, is truly responsible for the differential survival or reproductive success of coexisting species is far from simple because these are comparisons between "apples and oranges" (e.g., spines, thorns, and prickles). In any circumstances, survival or reproductive success likely involves the total complement of differing biological features rather than individual features or small sets of correlated ones.

Adaptations may be spoken of as states of being—as features (or sets of associated features) that, through operational interaction with the environment, confer survival or reproductive advantage. But the term "adaptation" is often used in a very different context—specifically, in reference to the process of becoming adapted. Here adaptation is spoken of in terms of how natural selection molds a species to its environment so the likelihood of survival or reproductive success is increased.

The issue of becoming adapted lies at the heart of a vigorous scientific debate. To dispute that species appear suited to their particular habitats would flout the undeniable fact that heritable traits permit organisms to survive and reproduce in their specific environmental settings. But acknowledging this kind of adaptation does not affirm that inherited traits are gradually reworked and modified through natural selection so that the probability of survival or reproductive success increases over time. Heritable genetic mistakes influencing one or many interconnecting developmental pathways occur frequently enough that they cannot be discounted as an important route by which novel traits appear in a population. Logically, however, the mechanism by which new traits initially appear is not the salient issue. For a trait (or set of traits) to become the norm rather than remain a novelty or "sport," its carriers must have some survival or reproductive advantage over their "normal" kindred. That is, novelties introduced by chance must confer a competitive edge if the individuals bearing them are to accumulate in a population. Only in this way do novel individuals become evolutionarily meaningful. Thus, when we speak of the process of adaptation we are referring not to the mechanism by which features make their first appearance in a population, but rather to how and why the individuals carrying them increase in number when these features confer some advantage to survival or reproductive success.

The reality of differential survival is as brutal conceptually as it is evident from life table statistics for a stand of trees grown from a single cohort of seeds. For example, Lee Van Valen (1975) showed that among 170,000 palm seeds growing in an area of less than 6,000 m^2, fewer than 30% may survive to the seedling stage, only 1% may grow into saplings, and fewer than 0.01% of the original number may reach sexual maturity. The extreme mortality (or "differential survival") does not imply direct competition among the members of a population (intraspecific competition), even though this may occur. To survive and reproduce, organisms must cope with the physical environment as well as with other organisms. The death of so many palm seeds is attributable as much to their inability to withstand physical conditions as to competing with one another for sunlight, water, minerals, and space. The only inference we can draw from Van Valen's data like those shown in figure 1.3 is that the individuals within a population, novel and normal alike, will grow, survive, and reproduce with differing success in the same setting.

Nevertheless, Darwin's rhetoric about competition among individuals within the same population is very clear: "Individuals of the same species . . . require the same food, and are exposed to the same dangers. In the case of varieties of the same species, the struggle will generally be almost equally severe and the contest soon decided" (Darwin 1972, 60). Evidently Darwin believed that differential survival or reproductive success reflected the way intraspecific competition culled the least fit members of a population. Darwin also believed that even though a species' or variety's demands on the environment remain the same, its success will differ in different environments.

Perhaps the most extensive experiment illustrating these points was performed by H. V. Harlan and M. L. Martini (1938). In 1923 they mixed equal numbers of seeds of eleven varieties of barley, *Hordeum vulgare,* and sent the seeds to colleagues who planted them and harvested their progeny over successive plant generations to determine how the varieties competed in a variety of localities differing in climate and soil conditions. The experiment lasted more than twelve years. It gives considerable insight into the speed of natural selection and how the competitiveness of the barley varieties differed across environments. In each locality, one or a few varieties quickly came to dominate, but which ones did so clearly depended on environmental conditions (fig. 1.4; table 1.1). Additionally, because more varieties persisted in some localities than in others (e.g., compare Moccasin, Montana, with Ithaca, New York, in

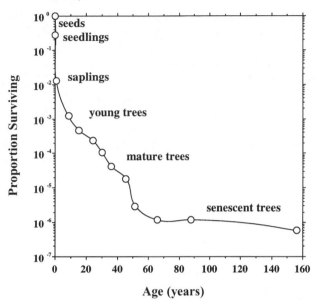

Figure 1.3. Survivorship curve for the palm *Euterpe globosa*. Of the cohort of 170,000 seeds planted in the first year, fewer than 30% survive to the seedling stage, fewer than 1% grow into young trees, and a miniscule fraction reach maturity. Data from Van Valen 1975.

table 1.1), some environments may preserve, at least for a time, a greater genetic diversity in populations than other environments.

The meaning of "differential survival or reproductive success" changes, but not radically so, when species compete for much the same resources (interspecific competition). Experiments indicate that even in these circumstances intraspecific competition, especially in close quarters, is more severe for plants than interspecific competition. For example, John Harper (1961) planted the seeds of two species of poppy, *Papaver rhoeas* and *P. lecoquii,* at three densities (540, 2,400, and 4,300 seeds/m^2). Seeds of the two types were mixed in equal proportions to produce the two populations at the extreme densities, while two ratios of seeds were used to produce populations at the intermediate density (one *P. rhoeas* seed to eight *P. lecoquii* seeds and eight *P. rhoeas* seeds to one *P. lecoquii* seed). In the latter case the predominant seed type could be considered the "native" species and the other the "exotic" species. Harper found that the probability of a seed's producing a mature plant decreased much more rapidly when it was the native (dominant) species than when it was the exotic species. Also, the probability of one of two species surviving decreased with increasing density of that species. These and other

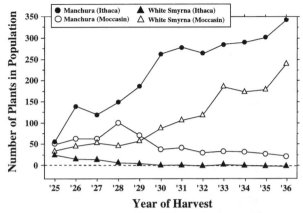

Year of Harvest

Figure 1.4. Changes in the number of plants of two barley varieties (Manchura and White Smyrna) in two populations growing in different environments (Ithaca, New York, and Moccasin, Montana). Equal numbers of seeds of both varieties were planted in 1923 to establish each population. The seeds produced each year were planted to carry each population into the next year. Within four to nine years, one variety dominated the other, apparently as a consequence of intraspecific competition, but the competitiveness of the varieties differed in the two populations, presumably because of environmental differences (Manchura was more successful than White Smyrna in Ithaca; White Smyrna was more successful than Manchura in Moccasin). Data from Harlan and Martini 1938, table 2.

Table 1.1   Percentages of Total Harvest of Six Barley Varieties from the Same Seed Mixture after Repeated (Ten or Twelve) Annual Harvestings from Populations Growing in Different Localities

Variety of Barley	Locality			
	Ithaca, New York (12 yrs)	St. Paul, Minnesota (10 yrs)	Moccasin, Montana (12 yrs)	Moro, Oregon (12 yrs)
Manchura	78	0	5	0
Hannchen	7	75	4	7
White Smyrna	0	1	56	14
Coast	13	20	20	73
Gatami	2	4	14	0
Meloy	0	0	1	6

Source: Data from Harlan and Martini 1938.

experimental results show that the members of a species tend to be their own worst enemies. Indeed, there is absolutely no evidence that any plant species has ever been driven to extinction by competition with another plant species. On the other hand, evidence abounds that animals can extinguish plant populations and entire plant species.

That the intensity of natural selection tends to increase with increasing population density also bears on individuals totally isolated from their parent population. These individuals are undoubtedly "exotic" in terms of their new biological and also physical environment, so "differential survival or reproductive success" may be more a matter of chance than of intense selection. That a few individuals, introduced by chance into a new and otherwise biologically vacant habitat, can survive and reproduce, and even flourish, despite low fitness is evident in the failure of varieties of many island species to successfully compete with their mainland counterparts. The effect of population density on the relative importance of natural selection and random forces such as genetic drift may be especially important for colonizing species, whose populations tend to undergo rapid and significant changes in size (see Thomas and Bazzaz 1993). Nevertheless, differential survival and reproductive success give new meaning to the old Latin precept *carpe diem*—seize the day.

The Hardy-Weinberg Equilibrium Equation

Here we will consider how differential reproductive success can alter the genetic composition of a population of sexually reproducing plants. The principal objective is to explore the arithmetic of reproductive success in order to understand the concept of Darwinian fitness, defined as any measure of an individual's relative contribution to the gene pool of the next generation.

For illustration, let us assume that reproductive success is governed by flower color and that flower color is dictated by a single gene (fig. 1.5). We will further assume that this gene has only two alleles: a dominant allele, denoted by *A*, that results in an individual bearing red flowers, and a recessive allele, denoted by *a*, that produces a white-flowered individual. (A gene is any one of many discrete units of hereditary information consisting of DNA located on a chromosome, and an allele is one of the alternative forms a gene can take.) Every plant in our hypothetical population carries two copies of each gene; that is, every plant has two of each kind of chromosome (every individual is a diploid organism). In these circumstances there are only three possible genetic combinations, or genotypes, in terms of flower color: *AA, Aa,* and *aa*. Because *A* is dominant and *a* is recessive, however, there are only two physical manifestations of flower color, or phenotypes: red-flowered individuals (*AA* or *Aa*) and white-flowered individuals (*aa*).

The mathematics of Mendelian genetics shows that in the absence of selection against flower color, the frequencies of the three genotypes in

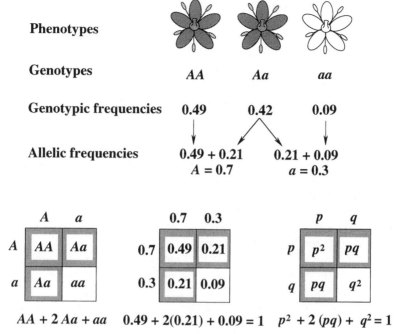

Figure 1.5. Hardy-Weinberg equilibrium equation showing that the gene pool of a non-evolving population remains constant over the generations. The illustration is for a population of flowering plants where flower color is dictated by one gene with two alleles (*A* and *a*). Three genotypes are possible (*AA, Aa,* and *aa*); two genotypes result in red flowers (*AA* and *Aa*), and one produces white flowers (*aa*). The allelic frequencies in the first generation (*A* = 0.7 and *a* = 0.3) remain the same in subsequent generations, provided all the conditions for a nonevolving population hold true (see text).

the population will not change from one generation to the next. That is, if the reproductive success of individuals bearing white flowers equals the reproductive success of individuals bearing red flowers, then the frequencies of the *A* and *a* alleles will not change generation by generation.

Imagine a population consisting of 700 individual plants, of which 343 bear the *AA* genotype, 294 the *Aa*, and 63 the *aa*. The frequencies of the three genotypes are *AA* = 343/700 = 0.49, *Aa* = 294/700 = 0.42, and *aa* = 63/700 = 0.09 (fig. 1.5). Because each individual is diploid, there are 1,400 genes for flower color in the population. The *A* and *a* alleles account for 70% and 30% of the entire gene population, respectively, because each of the 343 *AA* genotypes has two "doses" of the *A* allele, while each of the 294 *Aa* genotypes has one dose of the *A* allele and one dose of the *a* allele. Thus the total number of *A* alleles equals (343×2) + $(294 \times 1) = 686 + 294 = 980$, which is 70% of 1,400. Since there are only two alleles for the gene governing flower color, the remaining 30% of the gene population must consist of the *a* allele.

Now, during sexual reproduction each parent can transmit only one of its two alleles for flower color. Assuming that the union of sperm and egg is completely random, the probability of drawing the *A* or *a* allele is equal to the original frequency of occurrence of either allele in the entire population. That is, the probability of randomly drawing the *A* allele is 70% (or, expressed as a decimal fraction, 0.7), while that of randomly drawing the *a* allele is 30% (or 0.3). Consequently, in the absence of mutation or some phenomenon that biases the random draw of alleles, the frequencies of the three genotypes remain the same from generation to generation. For example, the probability of drawing the *A* allele from the total population of sperm cells and the probability of drawing the *A* allele from the total population of egg cells to produce an individual with the *AA* genotype is $0.7 \times 0.7 = 0.49$. Likewise, the probability of drawing two *a* alleles from the pool of sperm and egg to produce the *aa* genotype is $0.3 \times 0.3 = 0.09$.

The mathematics of Mendelian genetics can be generalized by assigning the probability *p* for drawing the *A* allele and the probability *q* for drawing the *a* allele (see fig. 1.5). Because the sum of any combination of probabilities for a given circumstance always equals one, it follows that $p + q = 1$. Thus the probability of producing the *AA* genotype is p^2, the probability of producing the *Aa* genotype is *2pq,* and the probability of producing the *aa* genotype is q^2. In other words, $p^2 + 2pq + q^2 = 1$ (fig. 1.5). This "sum of probabilities" equation is called the Hardy-Weinberg equilibrium equation, in honor of the two individuals who independently derived it from Gregor Mendel's laws of inheritance.

Incidentally, unless they are absolutely linked, genes, even those lying on the same chromosome, are expected to behave in populations as if they are independent. Here "independence" means that, in the absence of selection, the frequencies resulting from the fusion of sperm and egg for various combinations of alleles at any two loci equal the product of the frequencies of the two separate loci (i.e., alleles, whether linked or not, are expected to be associated at random within populations of randomly cross-fertilizing individuals). The independent assortment of genes, however, can be thwarted by absolute linkage or by differential selection preventing the attainment of equilibrium frequencies. Under these conditions, selection against a particular allele conferring low fitness reduces the allelic frequencies of other linked genes. In this sense some genes may "come along for the ride" that gradually reduces the frequency of less fit alleles.

The Hardy-Weinberg equilibrium equation is a formal mathematical statement showing that the frequency of occurrence of genotypes within a population will remain constant provided six conditions hold true:

1. The population consists of sexually reproductive organisms.
2. The population is (and remains) very large.
3. Individuals do not migrate into or out of the population.
4. Mating is completely random.
5. No mutation occurs.
6. All possible genotypes have an equal probability of reproductive success.

These conditions describe what will happen in the absence of mutation or natural selection. However, they never hold true for real populations of organisms, and so they proscribe the genotypic frequencies expected in a nonevolving population (the Hardy-Weinberg equilibrium equation provides a null hypothesis; that is, the condition obtained when evolution does not occur). Of equal theoretical importance, the equation shows that in the absence of selection pressures, genetic variation under Mendelian inheritance does not necessarily disappear generation by generation.

With this background, we can now explore the concept of Darwinian fitness—the measure of an individual's relative contribution to the gene pool of the next generation. The theory of natural selection posits that some naturally occurring factor or force biases the draw of different alleles, because some genotypes have lower rates of reproductive success than others. As a consequence of a biased draw, the frequencies of different genotypes within successive generations of a population will change. Suppose that, in our hypothetical population, a red-flowered plant is always reproductively successful—that is, each individual invariably passes its genes on to the next generation. Because Darwinian fitness is a relative term, each red-flowered individual is said to have 100% reproductive success. Denoting Darwinian fitness as $\bar{\omega}$, the fitness of all red-flowered individuals will be $\bar{\omega} = 1$. Now suppose that each white-flowered plant passes its genes on to the next generation at a lower rate because of some disadvantage. White flowers may fail to attract pollinators as well as pigmented flowers because they offer less desirable nectar rewards. White-flowered individuals may have lower survival rates or set fewer seeds than those bearing pigmented flowers (for examples, see Waser and Price 1983; Levin and Brack 1995). In these circumstances white-flowered individuals have a fitness less than 100% or, in notation, $\bar{\omega} < 1$. For the sake of argument, suppose each white-flowered plant produces 20% fewer offspring than a red-flowered plant. If so, its fitness relative to that of a red-flowered individual is $\bar{\omega} = 1 - 0.2$, or 0.8. The fitness of the red- and the white-flowered plants differs by 0.2 owing to some kind of selection pressure. This difference in fitness is called the se-lection coefficient; it is symbolized by s. The selection coefficient is a *rel-ative measure* of selection *against* reproductively less superior genotypes. Thus, s ranges in value from 0 (for a genotype experiencing no selection) to 1 (for a genotype that is lethal to its carriers).

In the absence of selection pressures or some other disturbing force, the Hardy-Weinberg equilibrium equation shows that sexual reproduc-tion will keep the amount of genetic variation within a population con-stant from generation to generation. Conversely, when natural selection occurs, the genotypic variant with the highest Darwinian fitness is ex-pected to gradually prevail at the expense of other genotypes. This can be seen by comparing the allelic frequencies in our hypothetical population before and after selection against white-flowered individuals:

| | Genotype | | |
	AA	Aa	aa
Before selection			
Darwinian fitness, $\bar{\omega}$	1	1	1
Frequency	p^2 +	$2pq$ +	q^2
After selection			
Darwinian fitness, $\bar{\omega}$	1	1	0.8
Frequency	p^2 +	$2pq$ +	$0.8q^2$

This comparison shows that the A allele will increase in one generation. The increment of change is given by the formula $\Delta p = spq^2/(1 - sq^2)$. In our example, the change in the A allele in one generation is $(0.2)(0.7) \times (0.3)^2/[1 - (0.2)(0.3)^2] = 0.0126/0.982 = 0.0128$. Noting that the fre-quency of the A allele in the second generation equals the sum of the original frequency and the increment, we see that $p = 0.7 + 0.0128 = 0.7128$. Because $p + q = 1$, the frequency of the a allele must concur-rently decline from 0.3 to 0.2872 (i.e., $q = 1 - 0.7128 = 0.2872$) after selection.

The change in the frequencies of the A and a alleles may seem trivial, but it reflects the passage of just one generation. If the selection pressure persists, the difference in the frequencies of the two alleles is amplified generation by generation. Assuming $s = 0.2$, the frequency of the A allele will increase from 70% to roughly 80% in ten generations. With a higher selection coefficient, say $s = 0.4$, the frequency will increase to 85% in ten generations (fig. 1.6). Note that the *rate* at which the genetic compo-sition of a population changes in absolute time depends on how long it

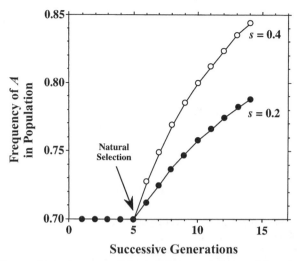

Figure 1.6. Changes in the frequency of the A allele in a population of flowering plants plotted against successive generations. The frequency of the A allele in the population remains constant in the absence of selection pressure on flower color (first five generations) and then changes as a consequence of natural selection on flower color (beginning with generation six). Changes in the A allele are plotted for two selection coefficients ($s = 0.4$ and $s = 0.2$).

takes a particular type of organism to complete its reproductive cycle. The generation time of small organisms is, on the average, much shorter than that of large organisms. Thus changes in genotype frequencies are expected to occur more rapidly for small organisms, such as unicellular plants or bacteria.

The Struggle to Measure Fitness

From the previous discussion of Darwinian fitness, it should be apparent that Herbert Spencer's description of natural selection as the "survival of the fittest" is misleading. The fitness of an individual can be gauged only in relative, not absolute terms. Those individuals that survive in a generation are not the best; they are merely those that are more reproductively successful than all their contemporaries. For this reason it is more appropriate to speak of the "survival of the tolerably more fit."

Although relative individual fitness poses little conceptual difficulty, Darwinian fitness is virtually impossible to measure directly. Many biological properties contribute to an individual's ability to contribute to the gene pool of the next generation relative to that of other individuals. For flowering plants, the individual seed must germinate, grow quickly, and establish a healthy, sexually mature plant. Flowers must be successfully pollinated and then produce and support viable seeds. All other things being equal, seed number is also important in the game of natural selection. Consequently biologists have measured Darwinian fitness by comparing individuals in terms of their relative viability, longevity, developmental speed, fertility, or fecundity. Clearly each of these biological properties reflects a separate component of Darwinian fitness, but for practical reasons each is rarely measured individually.

It is generally assumed that all the components of Darwinian fitness are correlated with the one measurement actually taken. This assumption may be true, especially in terms of the relation between survival and reproductive success. We may expect that the longer a plant survives, the larger it becomes and the more reproductive organs it can bear and metabolically foster. The relation between the ability to survive and the ability to successfully reproduce can be remarkably intimate. Every individual invests some physical measure of itself in the production of its offspring. The formation of spores, seeds, and fruits requires energy supplied by the plant bearing them. Nevertheless, the assumption that different measurements of the components of an organism's biology give equivalent measures of Darwinian fitness is problematic at best. A sterile plant that lives for hundreds or perhaps thousands of years is very fit in terms of survival but evidently is not equipped to perpetuate its own kind. The converse is true for a plant that lives for a few days or weeks but leaves behind thousands of offspring during its brief lifetime. This comparison between a "sterile but essentially deathless plant" and a "fecund but ephemeral plant" is purely hypothetical, but it suffices to show that there is no a priori reason to assume that survival and reproductive success are correlated. The only practical justification for assuming that Darwinian fitness can be measured in terms of one among many important components contributing to an individual's fitness is that it is largely impractical to simultaneously measure all the components.

Another consideration is how to measure and compare the fitnesses of different populations. How do we calculate the "adaptiveness" of one population compared with that of another? Perhaps the most obvious measure is the average population fitness. But what does the average value of many relative fitnesses mean? Suppose we have two populations of the same kind of flowering plant. If in one population 90% of the plants produce 70 seeds each, while the remaining 10% produce 35 seeds each, the average population fitness, $\overline{W}_1$, equals $[(0.9 \times 70) + (0.1 \times 35)]/105 = 0.63$. If, in the second population, 50% of the plants produce 90 seeds each, while the remaining 50% produce 50 seeds each, the average population fitness, $\overline{W}_2$, equals $[0.5 \times 90) + (0.5 \times 50)]/140 = 0.50$. Even though W_1 exceeds W_2, the average number of seeds produced by the first population is smaller than that produced by the second population (66.5 and 70, respectively).

Other measures of population fitness have been proposed. Among these are a population's degree of polymorphism, persistence in time, susceptibility to deleterious mutations, ability to increase in size, interspecific competitiveness, and production of developmentally anomalous individuals. Each of these measurements of adaptiveness has merit, but each poses practical and theoretical problems. The struggle to find an inclusive measure of population fitness continues, just as does the search for an inclusive measure of Darwinian fitness uniformly applicable to all species. In many respects the two efforts are inextricably connected, because the various properties contributing to an individual's fitness are very often those required to maintain a population or an entire species (see Wallace 1981).

Variation among Individuals

Contrary to the expectations our previous discussion of the Hardy-Weinberg equation may have inspired, genetic variation in real populations decreases at a fairly steady rate depending on population size. This was shown by R. A. Fisher (1958), who demonstrated that the genetic variance in a population of N individuals breeding at random will be halved by the effect of random survival acting alone in only $1.4N$ generations. Because genetic variation fuels evolution, continued evolution requires a continuous renewal of genetic variation within a population. Here I will outline the three ways genetic variation is replenished: mutation of genes and chromosomes, genetic recombination, and the immigration of genes from one population to another. Genic and chromosomal mutations are thought to be the most important source of genetic variation, so I will consider them first.

The simplest form of genetic mutation is a single base-pair substitution in DNA that alters the sequence of amino acids in a protein (fig. 1.7). Each amino acid in a protein is specified by a triplet sequence of DNA nucleotides called a codon. During the process called transcription, the information specified by DNA codons is transferred to a messenger RNA molecule in the form of mRNA codons. The corresponding DNA and the mRNA codons specify the same information but may differ whenever a thymine is specified, because thymine in a DNA codon is replaced by uracil in an mRNA codon. The nucleotide triplet in a DNA codon may be altered by genetic mutation so that one amino acid is replaced by

DNA

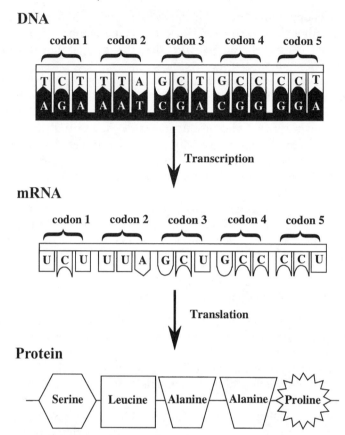

Figure 1.7. Diagram of DNA transcription and messenger RNA translation for the synthesis of a portion of a protein. Three successive nucleotide base pairs (one codon) in either the DNA or mRNA molecule are required to specify a particular amino acid. Some codons are synonymous codons (e.g., GCT and GCC encode for the same amino acid, alanine). Base-pair substitutions replacing one synonymous codon with another will not change the amino acid sequence.

another in a protein. A mutation resulting in the replacement of one DNA nucleotide by another is called a base-pair substitution (because replacing a nucleotide in one strand of DNA requires replacing its corresponding partner on the adjoining DNA strand (see fig. 3.3 for details about the structure of DNA). Base-pair substitutions can adversely affect the function of a protein or may have little or no effect on its ability to perform its metabolic task, because there are nonsynonymous and synonymous DNA codons. Nonsynonymous codons consist of a sequence of nucleotides that if changed encodes for a different amino acid. A base-pair substitution in a nonsynonymous codon, therefore, will change the sequence of amino acids in a protein. A synonymous codon encodes for the same amino acid despite differences in the triplet sequence of nucleotides. The DNA codons GCT, GCC, GCA, and GCG (and the corresponding mRNA codons GCU, GCC, GCA, and GCG) are synonymous (they all specify the same amino acid, alanine). A base-pair substitution changing GCT to GCC will not alter the amino acid sequence and so may have little or no effect on the function of the protein.

Under most conditions, mutation rates are too low to produce rapid evolution in populations of eukaryotic organisms. For example, let μ represent the mutation rate of one allele into another, say A to a in the previous example treating red- and white-flowered plants. If p_t represents the frequency of the A allele in a population at time t and q_t represents the frequency of the a allele in the same population at time t, then in the absence of natural selection the change in the A allele, represented by Δp, equals $p_t - q_t = -\mu p_{(t-1)}$. This formula shows that the frequency of A steadily decreases in proportion to the mutation rate. It also shows that the decrease is proportional to the frequency of all the nonmutated genes in the population. Thus, even at high mutation rates, the change in the frequency of the A allele will progressively decline because there are fewer and fewer A alleles to mutate into the a allele. The frequency of the A allele after N generations is roughly approximated by the formula $p_t = p_o\, e^{-\mu N}$, where p_o is the original frequency of A (Wallace 1981). To find the number of generations required to reduce the original frequency of A by half, we can set $e^{-\mu N} = 0.5$. Thus, $\mu N = -\ln 0.5 = 0.7$, and so $p_t = 0.5$ when $N = 0.7/\mu$. The mutation rate for plants ranges between 10^{-4} and 10^{-6}. Assuming that $\mu = 10^{-5}$, which lies in the middle of this range, the allelic frequency of A will drop to 0.50 after 70,000 generations (fig. 1.8).

Naturally the *time* required for this change depends on the duration of each generation. Even if the duration is less than one year, a decrease in the frequency of A from 1.0 to 0.905 (a decrease of less than 10%) takes roughly 10,000 years. And mutations can also be reversed in populations. In theory, the result can be a stable equilibrium between the normal and mutated alleles of the same gene (see fig. 1.8). However, the mutational forces that tend to maintain this equilibrium are weak compared with chance events (see Wallace 1981).

Mobile elements may provide yet another important source of mutation that can create evolutionary novelty, although we need much more information on these genetic phenomena. Mobile elements are DNA sequences that can replicate and insert themselves into different regions of the eukaryotic nuclear genome. Some mobile elements are preferentially

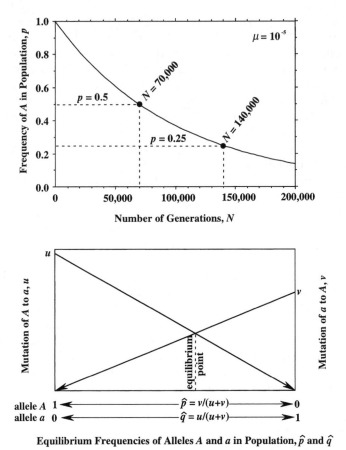

Equilibrium Frequencies of Alleles A and a in Population, $\hat{p}$ and $\hat{q}$

Figure 1.8. Changes in the frequency of the A allele in a hypothetical population experiencing "forward" mutation of allele A to allele a (upper graph) and "forward and reverse" mutations for the same alleles (lower graph). Upper graph: Assuming that the rate of mutation μ of allele A to allele a equals 10^{-5}, the allelic frequency of A drops to half its original value after 70,000 generations (and to a quarter after approximately 140,000 generations). Lower graph (adapted from Wallace 1981): Forward and reverse mutations occurring between A and a can achieve a stable equilibrium in allelic frequencies (equilibrium point indicated by dashed vertical line). Denoting the rate of mutation of A to a per generation as u and the rate of mutation of a to A per generation as v, equilibrium occurs when $up = vq$, where p is the frequency of A and q is the frequency of a. The frequency of A at equilibrium $\hat{p}$ equals $v/(u + v)$; the frequency of a at equilibrium $\hat{q}$ equals $u/(u + v)$.

transposed into sites that regulate genes. These elements, therefore, have the ability to mutate the timing or expression of developmental pathways, as in the homeotic conversion of floral sex organs (Bradley et al. 1993). Other mobile elements can insert themselves directly into gene loci and, when imprecisely excised from the locus at a later time, can result in novel gene products (e.g., a protein with new enzymatic properties in corn; see Wessler et al. 1986). Although most of the mutations created by mobile elements are neutral or deleterious, they may provide an important avenue for increasing the genomic variation among the members of a population, which in turn may confer some long-term evolutionary advantage. For example, the small mobile elements called *Tourist* and *Stowaway* are associated with hundreds of flowering plant genes and may have had beneficial effects on the evolution of dicots and monocots (Bureau and Wessler 1994). Among the various kinds of mobile elements, three are best known: transposons, retrotransposons, and mobile introns (see Zeyl and Bell 1996). Transposons tend to be comparatively short DNA sequences that encode for the proteins responsible for their transposition within a genome. The insertion of a transposon results in a duplication of a specific length at the target site and leaves behind a "footprint" of a few base pairs in the site of origin. Transposons are not believed to typically increase in copy number within the genome. In contrast, retrotransposon transposition is replicative, so these mobile elements can increase the size of a cell's genome. Mobile introns are so named because they are similar to introns, the noncoding portions of the genome that interrupt the coding portions of split genes called exons. Normally introns are spliced from exons after DNA transcription (see fig. 8.1), but mobile introns are capable of splicing from nontranscribing portions of chromosomes. A cautious analogy can be drawn between mobile elements and "genetic parasites," because these elements replicate using the genomic machinery of their "host" cells without conferring any immediate or obvious benefit. However, the mutagenic ability of mobile elements and their capacity to spread among chromosomes and establish themselves in populations of outcrossing sexually reproductive species may provide an important source for evolutionary innovation. For this reason it has been suggested that mobile elements can support speciation and, in some circumstances, transspecific evolution (McDonald 1990; Fontdevila 1992; see Zeyl and Bell 1996).

Clearly mutation, even when unopposed by natural selection, is a slow process. Genetic recombination, the general term for the production of offspring that have combined traits of their two parents, can produce variation at a much faster pace. The fusion of sperm and egg from different parents is a pivotal source of genetic variation because of the independent assortment of homologous chromosomes during meiotic cell division giving rise to sperm or egg. Homologous chromosomes have the same gene loci, but they can have different alleles of the same gene. Additional variation is added by inter- and intrachromosomal recombination. Interchromosomal recombination occurs during meiosis when homologous chromosomes pair, nonsister chromatids fragment, and then they rejoin (fig. 1.9). During this process, called crossing over, homologous chromosomes with new allelic combinations may be formed. Subsequent cell division will result in four haploid cells having parental or crossover allelic combinations. If these cells function as sperm or egg cells, new combinations may appear in the progeny. If there are n genes differing in their allelic form, then crossover can produce $n(n - 1)/2$ unique types of gametes in a single generation. As the number of different alleles increases, the number of unique gametes sharply increases.

Genetic variation is also produced by intrachromosomal recombination and exchange between nonhomologous chromosomes (fig. 1.10). Chromosomes can break and segments can reattach to the same chromosome but in the inverse direction (inversion), or they can be permanently lost (deletion). Inversions may alter the appearance of a phenotype when gene expression is influenced by neighboring genes (position effects). Deletions are frequently but not invariably lethal. Segments of linked genes may also be replicated within a single chromosome (duplication). Replicated genes can subsequently diverge in their gene functions, thereby providing the potential for evolving novel phenotypic features. In terms of nonhomologous chromosomes, segments of linked genes may detach from one chromosome and reattach to another (translocation). The transferal of chromosome segments between nonhomologous chromosomes can be equal or unequal (reciprocal or nonreciprocal translocations).

The third source of genetic variation is the migration of genes from a "donor" population to a "recipient" population. Here *migration* means any form of gene transferal. New genes can enter a population when individuals move from one population to another, when individuals from two populations mate and produce viable offspring, when a virus serves as a vector for the transmission of DNA between populations, and so forth. Although land plants normally cannot move from one location to another, their seeds or spores can be carried considerable distances by wind, water, or animals and can introduce genetic variation into widely separated populations. For example, 40% of the matings in a natural stand of Douglas fir, *Pseudotsuga mensiesii*, were the result of pollen transported by wind from sites over 100 m away (Neal 1983), while 20%–40% of the seeds of two insect-pollinated species resulted from

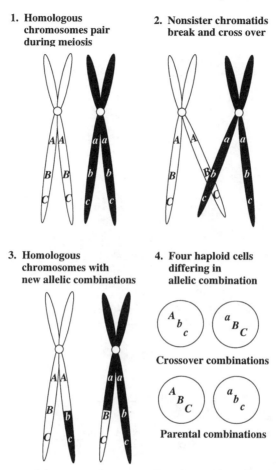

1. **Homologous chromosomes pair during meiosis**

2. **Nonsister chromatids break and cross over**

3. **Homologous chromosomes with new allelic combinations**

4. **Four haploid cells differing in allelic combination**

Crossover combinations

Parental combinations

Figure 1.9. Diagram of the process of crossover. Homologous chromosomes, each consisting of two sister chromatids (denoted by white and black outlines) bear the same sequence of genes, for which each gene has two allelic forms (A, B, C and a, b, c). Homologous chromosomes pair during meiosis (1). Nonsister chromatids cross over, fragment, and then rejoin (2–3). Homologous chromosomes with new allelic combinations are formed (3). Subsequent cell division will result in four haploid cells having parental or crossover allelic combinations (4).

long-distance mating events 0.5–1 km away (Hamrick and Murawski 1990). In general terms, the migration rate of new genes is typically much larger than mutation rates in normal populations of plants.

Nevertheless, mutation, genetic recombination, and migration are limited sources of variation in an evolutionary sense because they generally alter only preexisting biological traits or functions. Significant departures

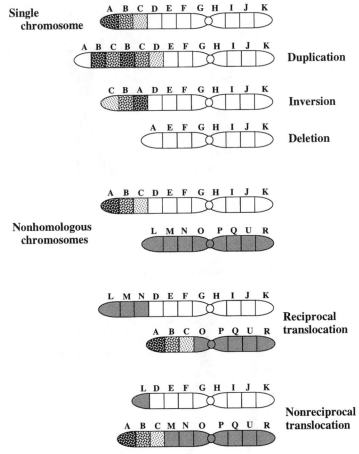

Figure 1.10. Diagrams of intrachromosomal alterations (inversion, deletion, and gene duplication) and interchromosomal alterations between two nonhomologous chromosomes (reciprocal and nonreciprocal translocations).

from existing traits require an expansion of the genetic repertoire. As noted, one way to expand the genome is by gene duplication and subsequent divergence in the roles of the duplicated genes through mutation. (I will discuss a possible example of this in chapter 3 in the context of the evolution of photosynthesis.) A gene may also diverge in function by means of a few base-pair substitutions. Substituting one base pair for another may change a gene product significantly.

Another important avenue for gene duplication in plants is polyploidy. A polyploid individual contains more than one complete set of the normal complement of chromosomes for its species. Polyploidy appears to have occurred frequently in plants, as illustrated by the frequency distribution of haploid chromosome numbers among pteridophyte and dicot species (fig. 1.11). The dominance of even numbers over odd numbers for haploid chromosomes among species within the *same lineage* is expected when polyploidy occurs frequently (Stebbins 1950; Grant 1963). The extremely high haploid chromosome numbers reported for pteridophytes, particularly ferns, suggests that these species either are very ancient or are particularly prone to polyploidization. In terms of their time of origin, polyploids have been classified into "neopolyploids," species with haploid chromosome numbers that are multiples of the basic diploid chromosome numbers of their respective genera, and "paleopolyploids," evolutionarily ancient species that have become repeatedly diploidized and therefore have exceptionally large basic chromosome numbers (Wagner 1954). Polyploid animals also exist, but most of these are insects, flatworms, leeches, or brine shrimp, some of which reproduce parthenocarpically—that is, offspring are produced by the mitotic cell divisions of parent cells. Among higher animal species polyploidy is rare, perhaps because the sex-determination mechanisms in higher animals depend on a delicate balance of chromosome numbers that is easily disrupted by polyploidy (Stebbins 1950; Suzuki et al. 1989).

Of course the inference that polyploidy underpins the dominance of even over odd chromosome numbers must be approached cautiously because of the confounding effects of dysploidy (a change in chromosome number resulting from the breakage and fusion of chromosomes without net gene loss or gain). In a polyploid sequence, lower chromosome numbers are ancestral to higher ones in the sequence. In a dysploid sequence, the chromosome numbers may run in either direction, from low to high (as in polyploidy) or from high to low. In theory the latter is easier. When all the vital genetic material of a chromosome is transferred to one or more other chromosomes, the remnant of the original chromosome is no longer essential and can be lost, and so the original haploid chromosome number is reduced by one. This type of dysploidy appears to have occurred in the flowering plant genus *Echeveria*, which contains 143 species whose haploid chromosome numbers include every number from 12 to 34 (Uhl 1992). Likewise, studies of Australian angiosperm families indicate that species with higher chromosome numbers may be ancestral to those with lower ones. Unlike polyploidy, dysploidy does not lead to gene duplication and subsequent divergence in gene function. Nevertheless, the gene rearrangements resulting from the consolidation of the genetic material of many chromosomes into fewer chromosomes may dramatically affect phenotypic appearance because the influence of a gene

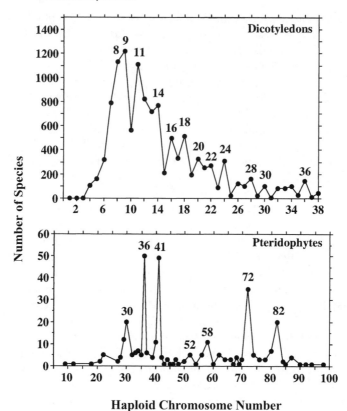

Figure 1.11. Frequency distribution of dicotyledonous and pteridophytic species with different haploid chromosome numbers. Data from Goldblatt 1981.

may be altered by its position in the genome (positional effects). Gene duplication by means of polyploidy followed by gene divergence and position effects caused by dysploidy could have profound evolutionary significance. The role of polyploidy in plant speciation is treated in greater detail in chapter 2.

In summary, the basic message of this section is threefold: (1) evolution, adaptive or otherwise, would essentially come to a standstill unless new variation was introduced into a population by mutation, recombination, or migration; (2) all new variation essentially comes from gene and chromosome mutations; and (3) the rate of mutation, which is the maximum rate of evolution, is held in check by natural selection

Additive Genetic Variance and Heritability

Many factors contribute to the phenotypic variance observed for a trait in a population. In general terms, however, the total phenotypic variance of a trait can be subdivided into two very broad variance components: the total genetic variance and the environmental variance. The total genetic variance results from the presence of different genotypes in the population; the environmental variance results from differences in the physical or biological environment. The extent to which phenotypic variation is heritable is largely governed by the total genetic variance of the trait, although heritability can be substantially reduced when the environmental variance is exceptionally high. In this section we will examine the components of total phenotypic variance. Special attention will be given to the total genetic variance, because this variance can be further subdivided into components in ways that predict the effect of selection on the trait. Specifically, the effect of selection on a trait will be found to be proportional to the genetic variance associated with the average effects of substituting one allelic form of a gene for another allelic form of the same locus. This component of total genetic variance is called additive genetic variance. When the additive genetic variance for a trait is high compared with the total genetic variance and environmental variance, the trait is highly heritable and thus the effect of selection on the trait is high.

Consider the following data for two alleles, A and a, whose three allelic combinations result in phenotypes differing in a quantitative trait, say plant height:

Genotype	AA	Aa	aa
Frequency (f_i)	0.49	0.42	0.09
Phenotype (x_i)	20 cm	14 cm	10 cm

These data show that genetic variance for plant height exists—genotypes with the A allele are taller than those lacking this allele. They also indicate that the A allele is not completely dominant—the mean height of heterozygotic individuals is less than that of those with the AA genotype and more than that of those with the aa genotype. Thus the total genetic variance for plant height, or s_T^2, consists of two components: (1) the variance resulting from substituting one allele for another, which is the additive genetic variance, or s_A^2, and (2) the variance resulting from the partial dominance of the A allele over the a allele, which is called the dominance variance, or s_D^2. Clearly, traits like plant height are subject to environmental and developmental variance. But for illustration we will assume that these components of variance are negligible such that $s_T^2 = s_A^2 + s_D^2$.

To calculate the total genetic variance, we must determine the average effect of each of the two alleles on the phenotype. For each allele, this average effect equals the sum of the number of the allele in each phenotypic class bearing it times the mean height of the individuals with the allele divided by the total number of the allele in the entire population. Because the population consists of diploid organisms, the average effect of $A = \bar{A}$ = [2 (0.49) 20 cm + 1 (0.42) 14 cm] / [2 (0.49) + 1 (0.42)] = 18.2 cm, and the average effect of $a = \bar{a}$ = [2 (0.09) 10 cm + 1 (0.42) 14 cm] / [2 (0.09) + 1 (0.42)] = 12.8 cm. The difference between the average effects of the two alleles on plant height is 5.4 cm. To calculate s_T^2 we also must know the mean height for all the plants in the population. This equals the sum of the average heights of the three phenotypic classes weighted by their frequencies of occurrence. In other words, mean phenotype height = $\bar{x} = \Sigma x_i f_i$ = 20 cm (0.49) + 14 cm (0.42) + 10 cm (0.09) = 16.58 cm.

The total genetic variance of the population equals the sum of the frequency of occurrence of each phenotypic class times the square of the difference between the mean height of each phenotypic class and the mean height of the entire population. Thus $s_T^2 = \Sigma f_i (x_i - \bar{x})^2$ = 0.49 (20 cm − 16.58 cm)^2 + 0.42 (14 cm − 16.58 cm)^2 + 0.09 (10 cm − 16.58 cm)^2 = 5.73 cm + 2.79 cm + 3.89 ≈ 12.42 cm. To calculate the additive genetic variance, we must also compute the frequencies of A and a in the population. Inspection of the data shows that the frequency of $A = f_A$ = [2 (AA) + 1 (Aa)]/2 = [2 (0.49) + 1 (0.42)]/2 = 0.70. Likewise, the frequency of $a = f_a$ = [2 (aa) + 1 (Aa)]/2 = [2 (0.09) + 1 (0.42)]/2 = 0.30. The additive genetic variance equals twice the sum of each of these two frequencies times the square of the average effect of each allele minus the mean height of all phenotypes. Thus s_A^2 = 2 $[f_A (\bar{A} - \bar{x})^2 + f_a (\bar{a} - \bar{x})^2]$ = 2 [(0.7) (18.2 cm − 16.58 cm)^2 + (0.3) (12.8 cm − 16.58 cm)^2] ≈ 12.25 cm. Because $s_T^2 = s_A^2 + s_D^2$, it follows that s_D^2 = 12.42 cm − 12.25 cm = 0.17 cm. In this example the dominance variation, which also includes the variance resulting from the interaction of other gene loci on the gene influencing plant height (i.e., the variance due to epistasis), is small compared with the additive genetic variance.

We are now in a position to appreciate why compartmentalizing the total genetic variance into its various components is important. In the absense of environmental variance or developmental "noise," the heritability of a trait, denoted by h^2, equals the additive genetic variance divided by the total genetic variance observed for the trait. That is, $h^2 = s_A^2/s_T^2 = s_A^2/(s_A^2 + s_D^2)$. Thus h^2 is the difference between a trait measured for selected parents and the mean for the trait measured for the unselected population ("the selection differential") divided by the difference between the offspring of selected parents and the previous generation ("the selection response"). The greater the value of h^2, the greater the difference between the trait measured for selected parents and for the whole population that will be maintained by the offspring of the selected parents. In the preceding example, $h^2 = s_A^2/s_T^2 = 12.25$ cm$/12.42$ cm $= 0.986$, which indicates that selection for Aa individuals will maintain the difference between the height of the selected individuals and the mean height of the unselected population. Thus a plant breeder could cross Aa individuals with Aa individuals and expect a high response to selection. The same cannot be said for the following population:

Genotype	AA	Aa	aa
Frequency	0.49	0.42	0.09
Phenotype	14 cm	20 cm	10 cm

Here, unlike the previous example, Aa individuals are on average taller than those with the AA or aa genotype. Nevertheless, calculations quickly show that $s_T^2 = 11.89$ cm and $s_A^2 = 0.73$ cm, and so $h^2 = s_A^2/s_T^2 = 0.06$. For this population the selection response for crossing Aa individuals will be poor, although, assuming very low environmental variance, hybrids made between previously intensely inbred lines of plants could be used to increase particular quantitative traits (a tactic often used by plant breeders).

Changes in the additive genetic variance for traits can be used to evalute the strength of natural selection. For example, there is little disagreement among biologists that plant herbivores or pathogens can respond adaptively to changes in the defense postures of their host plants, but there is still debate over whether the selection pressure imposed by herbivores or plant pathogens is strong enough to cause plant traits to respond adaptively (see Rausher 1996). This debate, which is about the "co" in plant/enemy coevolution, is not likely to be canonically resolved by one well-crafted study, or even a few. However, if coevolution occurs at all, then the additive genetic variance of plant traits believed to respond adaptively to attack should change in response to the intensity of attack. In this regard Ellen Simms and Mark Rausher (1989) showed that the additive genetic variation for Darwinian fitness (measured as lifetime seed production) was eliminated when herbivores were excluded from natural populations of the morning glory *Ipomoea purpurea*. Studies like this one indicate that herbivory can impose a strong enough selection pressure to cause plant populations to adaptively diverge.

Changes in the additive genetic variance of traits can be used to evaluate a variety of adaptive hypotheses. But it is important to bear in mind that a trait for which high heritability has been reported in one population under one set of experimental conditions can have substantial differences in heritability among naturally populations experiencing different environmental conditions. This follows from the fact that the total phenotypic variance seen in a particular population is the sum of the total genetic variance, the environmental variance, or s_E^2, and the variance resulting from developmental noise, or s_N^2. Because $h^2 = s_A^2/(s_A^2 + s_D^2 + s_E^2 + s_N^2)$, substantial increases in environmental variance will tend to diminish the heritability of a trait. Environmental variance can be much reduced in laboratory-grown populations, but it is often substantially large for populations growing under naturally occurring conditions.

Because there is no guarantee that populations with the same genetic composition will respond in the same way to different environments, there is no reason to believe that evolution is an ineluctably "progressive" process. This erroneous view of evolution is argued to follow directly from R. A. Fisher's (1930) "fundamental theorem of natural selection." This theorem states that the rate of increase in the mean fitness of a population is proportional to the genetic variance in fitness. Because a variance can never take a negative value, Fisher's theorem has been incorrectly interpreted to suggest that fitness will invariably increase so that evolution must be "progressive." However, Fisher's theorem holds true only if the relative fitnesses of genotypes remains constant, which hardly ever happens in real populations. Thus there is no reason to expect natural selection to lead to progressively more fit or more complex organisms, although this may be the case for some lineages during certain episodes in their evolutionary history (Ayala 1988).

The Phenotype

The Hardy-Weinberg equilibrium equation shows, in concrete quantitative terms, how reproductive (Darwinian) fitness is defined and how natural selection can "adapt" a population to a particular environment by removing individuals carrying reproductively unfavorable genotypes. Unfortunately, this equation can give the erroneous impression that natural selection acts directly on the genotypes within a population when in fact it operates on the corporeal manifestation of each genotype, the phenotype, which is the sum of the physical, metabolic, physiological, and behavioral traits that define the individual. For example, in our hypothetical population of white- and red-flowered individuals, selection acted against white-flowered individuals, not against the a allele per se, a point driven home by the fact that the reproductive fitness of all red-flowered individuals was equal, regardless of whether red flowers resulted from the Aa genotype or the AA genotype. Natural selection changes the gene pool of the population because selection pressures identify the phenotypic traits that influence reproductive fitness. Selection against the phenotypes carrying these traits indirectly changes the gene pool only because a correlation exists between the phenotype and its genotype. Thus the true measure of our understanding of how the genetic structure of a population changes owing to selection pressure is our ability to relate particular genotypes to particular phenotypes and to uncover how the environment discriminates among phenotypes.

Even though selection pressures can discriminate among phenotypes with different relative fitness, three features limit our ability to map phenotypes onto their corresponding genotypes. First, many important phenotypic traits are quantitative rather than qualitative. A quantitative trait may be defined as any trait for which the average phenotypic differences among genotypes are much smaller than the variation among individuals within a genotypic class. Traditionally, quantitative traits are assumed to be determined by a large number of genes, each having a small and independent effect on the phenotype. This multiple-gene-factor hypothesis suggests that a quantitative trait is governed by a polygene—a confederation of "small but equal effect" genes. Quantitative traits cannot be studied by simple Mendelian genetics and so pose practical limitations on our assigning a one-to-one correspondence between different phenotypes and genotypes. The second difficulty in mapping each phenotype onto its corresponding genotype is that a single gene may influence more than one phenotypic trait. Such genes are called pleiotropic genes. The influence of a pleiotropic gene on Darwinian fitness can be complex; in general terms, it will depend on whether the positive effects of the pleiotropic gene outweigh its negative effects. Suppose the dominant allele A coding for red flower color also codes for the production of little or no nectar. In these circumstances the overall reproductive fitness of red-flowered plants would depend on whether insect pollinators are attracted more by flower color then by the rewards they receive for inadvertently carrying pollen from one flower to another.

The third difficulty in assigning a one-to-one correspondence between each phenotype and its genotype is that the appearance and structure of an individual organism can vary with local environmental conditions. The color of *Hydrangea* flowers depends on the acidity of the soil in which a particular plant grows and will change if the plant is moved from one soil type to another. The size of the leaves and stems of many plants depends on whether shoots are exposed to sunlight or wind. These and many other examples of "phenotypic plasticity" show that the phenotype is not always rigidly defined but may have a natural range of variation, called the norm of reaction (fig. 1.12). Because the norm of reaction

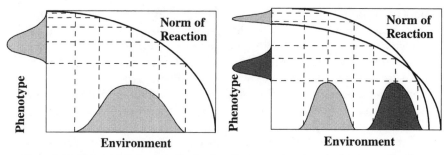

Figure 1.12. Phenotype frequency distribution diagrammed against the frequency distribution of an environmental variable to illustrate the norm of reaction. Phenotypic variance can be altered by environmental changes.

can vary as a consequence of the environment, it is often difficult to determine the relation between genotype and phenotype.

The norms of reaction for vegetative and reproductive traits may differ in ways that make perfect intuitive "adaptive sense." The norm of reaction tends to be comparatively broad for vegetative organs (leaves, stems, and roots), and remarkably different-looking phenotypes can be produced by the same genotype depending on local conditions. This makes adaptive sense because most plants are sedentary organisms with a comparatively long life expectancy. Because most plants cannot move from the location where they began their life and must continue to vegetatively survive in an environment that can change from year to year or decade to decade, it is logical that vegetative organs have broad norms of reactions. Also, the same kind of vegetative organ may continue to function adequately despite what often appear to be significant phenotypic differences. Whether the blade of a leaf has two, three, or four lobes may not significantly alter its ability to intercept sunlight. In contrast, the norm of reaction for reproductive traits tends to be narrow, so that a given genotype results in a very specific phenotype. It may be that reproductive "conservatism" is a hedge against subtle changes in reproductive organs that can have profound negative effects on reproductive success. A change in the hue, shape, or size of a flower may subtract from its ability to attract a pollinator and so reduce the number of seeds it produces. More dramatic departures in the appearance of reproductive organs may make them totally useless. By the same token, a random mutation may dramatically increase reproductive fitness and so inspire bursts in plant evolution.

Although the norm of reaction of vegetative organs may be broad in contrast to that of reproductive organs, the ability of a phenotype to successfully reproduce depends on its ability to vegetatively survive and grow and in turn on the plant's ability to invest metabolites in the formation of reproductive organs and the progeny therein. Clearly, just as the genotype is the sum of all its genes, the phenotype is also the sum of its many vegetative and reproductive traits, each of which can contribute to the reproductive fitness in different, albeit interrelated, ways. Although it may be easier (more convenient) to focus on one or two "obvious" reproductive traits to evaluate how much reproductive fitness varies as a function of variations in these traits, the entire phenotype is the vehicle of evolution.

Modes of Natural Selection

Natural selection can change the prevailing phenotype of a population in one of three modes: stabilizing selection, directional selection, or diversifying (disruptive) selection (which on a large scale is called adaptive radiation). Each mode depends on how the environment changes, as may be shown by plotting the number of individuals sharing the same phenotype against the different phenotypic categories within a population (fig. 1.13) Stabilizing selection results when the interactions between the plant and its environment remain constant and natural selection removes the extreme phenotypes, leaving behind intermediate phenotypes. Stabilizing selection therefore reduces the phenotypic variation in the population. For example, plants with intermediate seed sizes may be selected for when the nutrients stored in very small seeds are insufficient to establish seedlings and when very large seeds become too obvious a source of nutrition to escape herbivory. The extreme phenotypes within populations have traditionally been viewed as homozygous because the reaction of homozygotic individuals to the environment tends, on the average, to be more extreme than that of heterozygotic ones (Haldane 1954; Wallace 1981; see the previous section, however). By causing homozygotic individuals to take on extreme phenotypes, environmental variation tends to expose them to intensive stabilizing selection.

Directional selection results when individuals expressing a particular trait (to either the greatest or the least extent) are preferentially subjected to the greatest selection. This mode of selection, which tends to shift the frequency distribution of phenotypes to one or the other end of the spectrum (and so favors originally rare phenotypes), can continue until the genetic diversity within a population becomes exhausted. Directional selection and the consequences of increasing genetic uniformity are nicely illustrated by considering the yield of crop plants, like oats, subjected to continuous inbreeding by plant breeders. During the early 1900s, the yield of oats (bushels/acre) in the United States was raised by identifying

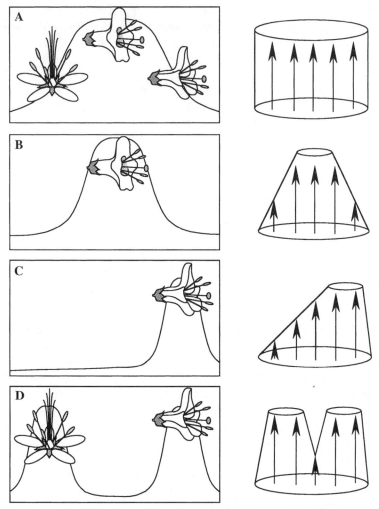

Figure 1.13. Diagram of the change in the frequency distributions of three floral morphologies in a population (A) subsequently experiencing stabilizing selection (B), directional selection (C). and diversifying selection (D).

superior varieties from around the world and importing seed. During the 1920s and 1930s, however, the yield plateaued dramatically when plant breeders sought to improve pure inbred (homozygous) lines by artificial selection. From the mid-1930s through the mid-1960s, the yield of oats was increased once again by programs based on hybridization and the selection of superior individuals among the progeny of successive generations. During these three periods, the rates of improvement in bushels/acre/year were 1.8, 0.1, and 1.5 (Sprague 1967). These data show that the effect of directional selection decreases as genetic materials become uniform.

The final mode of phenotypic selection, called diversifying or disruptive selection, occurs when intermediate phenotypes have a selective disadvantage relative to those of both extremes. Biologists repeatedly emphasize environmental heterogeneity or "graininess" and how well organisms can adapt to different "grains" that subject individuals within the same population to potentially discordant selection pressures. That environmental graininess occurs even in close quarters is amply illustrated by a study of two strains of corn (*Zea mays*), white flint and yellow sweet, growing in the same field (see fig. 2.6). The original strains required virtually the same number of days for flowering (72 days for the tassels and 75 days for the ears). When the varieties were grown together for four generations, however, the flowering times of nonhybrid plants shifted appreciably—the tassels and ears of white flint plants flowered after 67 and 70 days, while those of yellow sweet flowered after 75 and 77 days, respectively (Paterniani 1969). A reexamination of the distributions of flowering times of the original, unselected strains indicates that this response to disruptive selection was to be expected because some plants in the original populations of the two strains flowered either earlier or later that the prevailing phenotypes.

The three modes of natural selection are not mutually exclusive. Successive generations may undergo one mode of selection and then another in response to shifting patterns of environmental change. Phenotypic variation within a population subjected to directional selection will decease and may eventually become stabilized (e.g., the yield of oats in the United States during the first half of the twentieth century; see Sprague 1967). Likewise, diversifying selection favoring phenotypic extremes due to environmental heterogeneity may come to closure with stabilized phenotypes. Likewise, an environment may gradually change in one direction and then gradually change in the reverse direction, as from dry to wet and then wet to dry. If so, then directional selection will shift the mean phenotype, initially favoring one extreme and later the other. Thus the prevailing phenotype that originally conferred the highest fitness may languish and die as its environment changes, while previously less fit phenotypes prosper and increase in number.

In this sense evolution can involve a peculiar kind of hide-and-seek where a randomly changing environment can shift, obscure, or even

eliminate previously prevailing conditions, while to survive and reproduce the organism must simultaneously pursue the niche to which it was
previously adapted. Herein lies adaptive evolution's peculiar danger. By
the time an adaptation to one set of environmental conditions takes
place, the environment may have already changed, making the adaptation irrelevant or, worse, maladaptive. Indeed, this explains why few if
any species are "optimally" adapted. Suppose that under directional selection a population evolves tubular flowers whose length is best suited
to a particular insect pollinator with a commensurately long tongue. The
same population would become poorly adapted if circumstances favored
insects with short tongues and would be at risk should the long-tongued
insect became extinct, perhaps owing to a sudden dramatic change in the
environment. The failure of a population to adaptively evolve at the rate
and in the direction an environment changes poses a danger because it
could lead to the eventual extirpation of the population. That some environments change rapidly or unpredictably and that the adaptive state of
an organism may lag significantly behind a mercurial environment are
possible explanations for the extinction of entire species or higher taxa
(Lynch and Lande 1993; Bürger and Lynch 1995).

One solution to rapidly changing environments is to have a very short
reproductive cycle, in which the genetic composition of a population can
be altered at the same pace as the environment changes (Levins 1968).
Small organisms tend to have shorter reproductive cycles and higher
mutation rates than larger organisms (Levin and Wilson 1976; Prager,
Fowler, and Wilson 1976; Lenski and Travisano 1994; Niklas 1994).
Natural selection could also act directly on the development of an organism to favor individuals with a growth strategy capable of dealing with
rapid environmental change (fig. 1.14). Many kinds of plants have developmental patterns evoking different phenotypes, each of which may be
well suited to a different set of local environmental conditions. This
"phenotypic plasticity" has already been mentioned in relation to the
concept of the norm of reaction. The vegetative organs of plants were
said to typically show a broad norm of reaction, so that very different
phenotypes result depending on the local environmental conditions attending the growth and development of leaves, stems, and roots. This is
particularly advantageous to a sedentary organism like a terrestrial plant
that is incapable of leaving an inhospitable environment for a more
advantageous one.

Another possible solution to highly variable environmental conditions
is to be a "phenotypic generalist," capable of coping almost equally well
with a fairly broad range of environmental conditions. This strategy

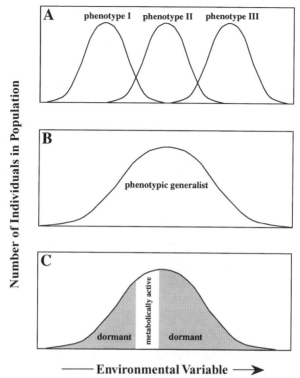

Figure 1.14. Alternative adaptive phenotypic responses to a changing environmental variable illustrated by plotting the frequency distribution of phenotypes against an environmental variable critical to survival or reproductive success: (A) population may consist of individuals with different phenotypes, each adapted to a particular range of the environmental variable; (B) population may consist of a phenotype capable of survival and reproductive success across a broad range of the environmental variable; (C) individuals may suspend metabolic activity (become dormant) when the environmental variable exceeds a critical threshold.

entails the whole phenotype rather than the nature of the constituent organs of the individual. Such "whole organism" adaptations may involve the induction of enzymes, switching from one physiological pathway to another, alterations in phenology, and so forth. Another solution for dealing with a variable environment is dormancy. Many species of unicellular algae suspend their metabolic or sexual reproductive activities when the environment becomes too stringent and form resting cysts or encapsulated dormant spores or zygotes. Changes in patterns of reproduction in response to changes in the environment are not uncommon

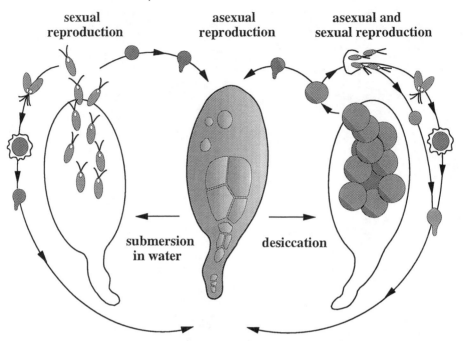

Figure 1.15. The reproductive cycle of the green alga *Protosiphon botryoides* as an example of the influence of environmental factors on sexual and asexual reproduction. When the alga is submerged in water it may form gametes or asexual reproductive cells (left portion of diagram). When it experiences drying conditions it may form spherical multinucleate cells (called coenocysts) that subsequently undergo asexual or sexual reproduction when rehydrated. Adapted from Smith 1955 and Bold and Wynne 1978.

(fig. 1.15). Likewise, the seeds of many terrestrial plants may remain dormant for decades, forming "seed banks," and can reestablish a population when environmental conditions conducive to growth return.

Yet another way of coping with a changing environment is to physically "track" the conditions that best suit the organism. On a small geographic scale, many kinds of plants explore the habitat for sunlight, water, and minerals by means of their growth habit (fig. 1.16). A broad spectrum of plants have a rhizomatous growth habit; that is, they form horizontally oriented stems that can grow either below or above the ground surface. New leaves and stems are produced at the growing tip of the rhizome, while older parts of the rhizome may rot away. By branching ahead and fragmenting behind, a rhizome can grow forward and spread outward, thereby colonizing new areas while leaving parts of itself behind in previously occupied locations. Much the same thing is accomplished by plants like the banyan tree (a species of *Ficus*), whose aerial branches produce prop roots that expand in girth and serve as vertical supports functionally equivalent to vertical stems. When the main trunk of a banyan tree dies and rots away, the aerial spreading branches of the original tree may separate and "walk away" in different directions on their massive, stiltlike prop roots (see fig. 1.16). Another way plants migrate is by means of spores, seeds, and fruits that often can be transported remarkable distances by wind, water, or animals. Plant populations languishing in one location because of deteriorating environmental conditions can colonize more amenable locations by this long-distance dispersal. Indeed, simultaneous asexual and sexual reproduction permits the simultaneous exploration of new habitats and the retention of favorable old ones.

Some of these strategies for dealing with changing environment may have deleterious long-term consequences. It is possible that the phenotypic generalist is a "jack of all trades but master of none"—that it may survive under different conditions but be less successful when pitted against specialist species (Levins 1968). In theory, organisms with broad norms of reaction (or those that go dormant when conditions become unfavorable) can be at risk because they are shielded from the potentially beneficial long-term effects of intense selection pressure. Unconditional deleterious mutations may accumulate in a population experiencing little or no selection pressure, only to surface with catastrophic results when the population is later subjected to intense selection. However, the evidence that phenotypically plastic and thus generalist species are less able to compete with specialized species is weak at best (see Sultan 1992, 1995), suggesting that phenotypically plastic organisms may hold their ground when placed in direct competition even with highly specialized species. Noting that in the absence of sufficient genetic variation even a gradually changing environment may forecast a decline in the overall fitness of a specialist species, leading to its ultimate demise, we see that the ability of a single genotype to produce functionally appropriate phenotypes in response to different environmental conditions may confer a long-term evolutionary benefit.

If, as some believe, phenotypic plasticity is an adaptation sensu stricto, then it must be a heritable property subject to selection pressure. The advantages of phenotypic plasticity are obvious—it provides an opportunity for the same or closely related genotypes to survive and reproduce in very different environments. In contrast, the modus vivendi by which phenotypic plasticity evolves is still a matter of conjecture, because a curious tension exists between evolutionary forces that act to integrate an organism's genome and other forces that act to partially dismantle and adaptively reshape the genome in the face of environmental challenges.

A

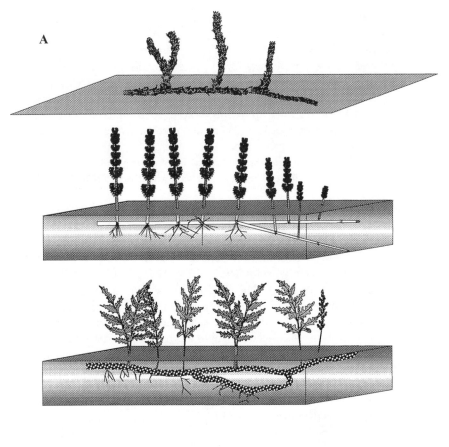

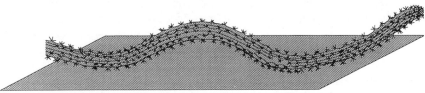

Figure 1.16. Growth habits permitting some plants to explore and asexually reproduce in their immediate habitats. (A) From top to bottom: aboveground rhizome of a lycopod (*Lycopodium lucidulum*); belowground rhizome of a horsetail (*Equisetum arvense*); and rhizome-like growth habit of a cactus (*Harrisia pomanensis*). (B) The aerial prop roots of the Banyan tree (*Ficus*) can be used to much the same effect as a rhizome (see text).

B

The traditional view is that balanced systems of harmoniously interacting (coadapted) genes are required for the survival of the individual.
Each individual is genetically tethered to its ancestors and thus is afforded a well-tested lifeline it would be foolish to abandon. Nevertheless,
even a coadaptive gene network must change to some degree if populations are to evolve and cope with new environmental conditions. Thus a

remarkable dynamic exists between selection pressures that act to build harmonious balanced systems of interacting genes and those that act to modify these systems to cope with new conditions (Waddington 1942; Schmalhausen 1949; Mayr 1963; Bonner 1988). This dynamic is underwritten by *epistasis*—a condition where the phenotypic expression of alleles at one locus depends on the alleles present at one or more other gene loci. Epistasis is profoundly important because it shows us that the expression of genetic and thus phenotypic variation is under genetic control. Epistatically interacting genes can engender phenotypic uniformity despite underlying genetic diversity, but they can also give rise to phenotypic variation whenever large networks of formerly interacting genes are broken up or "modularized" into smaller networks, as when a species is subjected to stabilizing selection for some of its networks of genes and directional selection for other sets of epistatic genes. The tension between evolutionary forces that hold some coadapted gene networks together and dissociate other networks can result in what may be crudely thought of as "genomic mosaic" evolution. That the genomic architecture of most organisms evinces some degree of modularity is shown by the ability of most organisms to adapt to different environments, because an epistatically "megalithic" genome would resist evolutionary change—it would be virtually impossible to move by virtue of its sheer cohesive inertia. Because most organisms are capable of adapting to new environmental conditions, some genomic modularity must exist in most organisms. The way phenotypic plasticity evolves likely has its roots in how epistatic gene networks can become modularized when different selection forces act on different phenotypic traits. Just as some phenotypic traits that were once variable can become fixed, while others may become tightly linked with other characters, still other phenotypic characters can gain variability by compartmentalizing large gene networks into smaller, semiautonomous ones.

The modularity of coadapted gene networks could and probably does permit considerable variation within individuals, especially for plants. Indeed, most plants exhibit a fascinating phenotypic modularity that may well correspond to their genetic modularity. This possibility has only recently been recognized because most models for the evolution of adaptive phenotypic plasticity as well as empirical studies of this effect tend to dwell on the role of coarse-grained environmental variations as the underlying selective agent (Via and Lande 1985; Van Tienderen 1991; Weis and Gorman 1990; Andersson and Shaw 1994). Here coarse-grained environmental variation means that each individual grows and develops essentially under one set of environmental conditions. Much less consideration has been given to fine-grained environmental heterogeneity, in which an individual experiences many different environmental conditions during its lifetime. Although the mature form of many kinds of animals is achieved during a comparatively brief period of development early in the life cycle (coarse-grained environmental heterogeneity), most multicellular plants continue to grow and develop for very long periods and thus typically experience many different physical and biological conditions (fine-grained environmental heterogeneity). During its lifetime a single plant may produce millions of leaves and stem internodes, and the potential for morphological and physiological differences among these organs (within-individual phenotypic plasticity) can be highly adaptive (see Chabot and Chabot 1977; Huner 1985; Winn and Evans 1991; Sultan and Bazzaz 1993). Perhaps for this reason, it is not surprising that some of the most dramatic examples of phenotypic plasticity are drawn from the plant world.

The Fitness Landscape

Perhaps the most elegant way of visualizing most of what has been said up to now was offered by Sewall Wright, who envisioned the consequences of natural selection as an adaptive "walk" over a "fitness landscape" (fig. 1.17). Wright defined the dimensions of the landscape, its breadth and depth, in terms of all possible genotypic combinations within a population. Supposing that each gene has two alleles, the simplest landscape is described by a set of orthogonal axes (one for each allele of each gene). If we draw contour lines around different genotypes conferring the same or nearly the same relative fitness on their phenotypes, the landscape becomes a flat "fitness map." The fitness of all possible genotype combinations may also be plotted along a vertical axis so that the landscape becomes three-dimensional, with fitness "valleys" occupied by the least fit genotypes within the population and "peaks" occupied by the most fit. Because natural selection eliminates the least fit genotypes in valleys and because the genotypes on the slopes of adaptive peaks are more fit, the genetic composition of the population "walks" toward and up fitness peaks. Because the vehicle of this walk is the phenotype rather than the genotype, the fitness landscape can be recast in terms of all the conceivable phenotypes produced by the population.

Each walk on the fitness landscape is a consequence of the combined effects of random forces (e.g., gene and chromosome mutations) and nonrandom forces (natural selection). These two types of forces are not necessarily antagonistic. Random forces are just as likely to push a walk up a peak as to direct it toward regions of lower fitness. Therefore an adaptive walk is the consequence of the simultaneous operation of

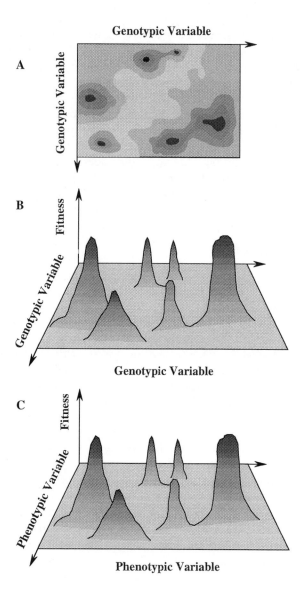

Genotypic Variable

A

Genotypic Variable

B

Fitness

Genotypic Variable

Genotypic Variable

C

Fitness

Phenotypic Variable

Phenotypic Variable

random and selection forces. Five additional points are equally important. First, walks over fitness landscapes are not a conscious process—a
population does not "strive" toward adaptive peaks, nor does it "flee"
lethal valleys. Walks are propelled by random effects and selection pressures, neither of which can divine the location of fitness valleys and
peaks. Second, some adaptive walks may be impossible because natural
selection can act only on the phenotypic variants within a population,
and some conceivable phenotypes may never be developmentally realized
by individuals. Third, although not all phenotypes may be possible,
dissimilar phenotypes may confer similarly high fitness. Consequently a
fitness landscape can have manifold adaptive peaks. Fourth, the three-
dimensional fitness landscape shown in figure 1.17 is a gross oversimplification. Each phenotype is the sum of many vegetative and reproductive
features, each contributing to the total fitness of an individual organism,
and so the phenotypic landscape is more realistically rendered as a multidimensional space. And fifth, in addition to its spatial complexity, the
fitness landscape changes over time as the environment changes biologically and physically. As it evolves, the population changes, and so the
fitness landscape changes because the prevailing phenotype changes. As
the physical environment changes, the relative fitness of an individual
changes. The rendering of adaptive evolution Wright proposed is a
metaphor—an analogy for how adaptive evolution occurs rather than a
faithful picture of an extraordinarily complex process.

Nonetheless, likening adaptive evolution to hill climbing is not as far-
fetched as it may appear. The amount of genetic variation that a population of sexually reproductive organisms can produce is staggering. Consider that a single parent with N number of genes, each with two alleles,
can produce 2^N genetically different sperm or egg cells. Because sexual
reproduction involves two parents, each set can therefore produce an
offspring with one of 4^N different genotypes. Supposing each parent genotype has 150 genes with two alleles each, each parent can give rise to over
10^{45} genetically different sperm or egg cells. Thus a single set of parents
can produce more than 10^{90} genetically different offspring. To be sure,
these are very conservative estimates. They reflect neither the possibility

Figure 1.17. Sewall Wright's metaphor for adaptive evolution depicting a fitness landscape
for all genotypic permutations: (A) two-dimensional display of regions of high relative
fitness (darkly shaded areas) and regions of low relative fitness (lightly shaded areas);
(B) three-dimensional fitness landscape resulting when relative fitness of genotypes is
plotted against a vertical axis; (C) fitness landscape recast in terms of all phenotypic
possibilities. Fitness peaks and valleys for genotypes and phenotypes may not coincide
as shown here.

of chromosomal variations nor the fact that most sexual organisms have orders of magnitudes more genes, many of which may have more than two alleles. Even so, Stephen Hawking (1988) estimates there are only 10^{80} particles in the observable universe. Clearly, not all possible genotypic variants are produced (this is a fundamental reason to reject the notion that adaptive evolution can achieve "perfectly adapted organisms"). But it is clear that some evolutionary process must eliminate a great many variants, because naturally occurring populations typically contain only a small, albeit fascinating, cross section of what is genetically possible. The process by which "all conceivable" variants are winnowed to give rise to what "actually exists" was explained by Charles Darwin. From his vantage point on the evolutionary landscape, Darwin saw that organisms typically produce many more offspring than their local environment can support, that the progeny of a parental pair differ in varying degrees from each other and from their parents, and that the environment extirpates the less fit individuals within the population, leaving the more fit to reproduce. In short, natural selection.

Sewall Wright's metaphor for adaptive evolution can inspire the erroneous notion that evolution inevitably produces perfect organisms. But the occupants of adaptive peaks are never "perfect" because the environment is always changing, because natural selection cannot create new phenotypes, and because the variation that exists in a population is the result of chance genetic mutation, the initial genetic composition of a colonizing population, and other factors. Likewise, the phenotypic variation within the inclusive population of a species reflects an evolutionary history of descent with modification through a long line of ancestors. This legacy cannot be abandoned willy-nilly (Gould 1980; Alberch 1989). To be sure, how far adaptive walks are obstructed by genetic, developmental, or other barriers is relative rather than absolute because it undoubtedly varies among different kinds of organisms and perhaps changes over evolutionary time for each kind. The traditional view is that plants are developmentally more "plastic" than most animals (Scharloo 1991), perhaps because they tend to be sedentary organisms that cannot leave a difficult environment, or because plant metabolism is more closely attuned to the physical environment than animal metabolism, or because plant development relies more heavily on environmental cues than animal development. Another important feature is that plants may not be able to achieve the precise orchestration of development seen in animals. Plants are composed of fewer different cell types than most animals, and cells are not free to migrate within the plant body as happens in animal development. Whatever the reasons, the developmental plasticity of plants is so remarkable that one has the impression that their adaptive walks may be far less impeded by genetic or developmental barriers than those of animals. Nevertheless, it remains that the fitness landscape of plants as well as animals is very much like an obstacle course—some walks are more difficult than others, some may be temporarily blocked, and others may be forever closed.

Yet another limitation on adaptive walks is that each organism must perform many diverse tasks to grow, survive, and reproduce. Undoubtedly some of these tasks have different phenotypic requirements. For example, to grow and reproduce, plants require sunlight. As the surface area projected toward the sunlight increases and the volume bounded by this surface area decreases, the efficiency of light interception increases. Therefore the biological obligation of plants to intercept sunlight is best fulfilled by phenotypes possessing very high ratios of surface area to volume. These phenotypes are easily maintained in an aquatic habitat, where plants are continuously bathed in water and structurally supported by it. On land, water is lost through the external surfaces exposed to the drier atmosphere. Because the amount of water lost through evaporation decreases as the surface area exposed to the atmosphere is reduced, land plants typically have highly reduced leaf areas, particularly in arid habitats. And in general terms the biological obligation to conserve water is best fulfilled on land by plants with extremely low ratios of surface area to volume. An extreme reflection of this response is seen in some species of cacti whose leaves are reduced to nonphotosynthetic spines and whose succulent stems have a low ratio of surface area to volume.

Another example of conflicting phenotypic requirements is seen when light interception and mechanical stability are jointly considered. All plants depend on sunlight for growth, and all terrestrial plants must resist the force of gravity to grow upright. In general terms, light interception is maximized when photosynthetic organs are oriented nearly horizontal to the sun. But this orientation also maximizes the bending forces in stems and leaves. When these forces exceed the strength of leaf and stem tissues, leaves and stems break. Therefore the orientation that maximizes light interception tends to be exactly the opposite of the orientation that minimizes the risk of mechanical (bending) failure. Over the course of their evolution, plants have managed to reconcile these and other conflicting phenotypic requirements by evolving "middle ground designs," that is, phenotypes that work well enough, though not as well as they conceivably could if only one task determined survival and reproductive success.

Although organisms are not machines, the relation between the number of functional tasks an organism must perform to survive and reproduce and the number of phenotypes that confer the same overall fitness is somewhat similar to the relation between the number of equally efficient designs for an engineered artifact and the number of tasks the artifact must perform. Engineering theory and practice show that as the number of tasks an artifact must perform increases, the number of equally efficient designs also increases (Meredith et al. 1973; Gill, Murray, and Wright 1981). At the same time, the efficiency with which any task is performed decreases as the number of tasks increases. If these generalizations hold true for organisms, even in part, then as the number of biological tasks an organism must perform increases, the number of adaptive peaks on a fitness landscape will increase, while the heights (relative fitness) of these peaks will decrease. These expectations are intuitive, at least from an engineering perspective, because, on the one hand, each function likely has a specific design specification that maximizes efficiency in performing the task and, on the other hand, reconciliations among conflicting, highly specific design requirements cause each task to be performed less efficiently. Although there is no a priori reason organisms should subscribe to engineering theory and practice, it does seem reasonable to suppose that they are phenotypic rapprochements in the sense that the phenotype has come at some cost in terms of lower efficiency in performing its many biological tasks. If this supposition is true, and if it is also true that every type of organism is a "multifunctional device," then most suffer the same disadvantage to some degree, and so the evolutionary playing field is level for all.

Curiously, there may be an evolutionary advantage to being more biologically complex. Provided the lessons of engineering carry over into biology, the number of adaptive peaks in a fitness landscape will increase while the topography of the landscape becomes less severe as the number of biological functions an organism must reconcile increases. If so, then adaptive walks may become easier as complexity increases, in the sense that differences in the fitness of phenotypes become less pronounced.

Testing Adaptive Hypotheses

This topic catapults us back to the beginning of this chapter, because the methods used to determine if a feature or set of features is adaptive depend on whether the "state of being" adapted or the "process of becoming" adapted is the principal concern. If the former, then we can take a strictly experimental approach. Such an approach is concerned with how an adaptive feature works in a particular environmental context and what benefits it confers on the organism in terms of survival or reproductive success without reference to the historical changes leading up to its appearance. The virtue of the experimental approach is that it can finely resolve the interactions among biological and physical variables. If the evolution of adaptive features is of interest, however, then a phyletic hypothesis that establishes the chronological appearance of taxa within a lineage is essential.

The most frequently used (and perhaps most accurate) method of constructing a phyletic hypothesis is cladistic analysis. The principal objective of cladistics is to group species or higher taxa sharing a common ancestor. The phyletic hypothesis based on cladistic analysis is depicted as a bifurcating "tree" called a cladogram (Hennig 1966; Wiley 1981). The branch points on a cladogram are called nodes, and taxa linked to the same node are hypothesized to have a common ancestor. The branching topology of a cladogram is determined by the distribution of characters among the taxa included in the analysis. Although systematists have an intuitive and practical grasp of what is meant by "character," numerous attempts have been made to formally define this term. At the risk of oversimplification, a character may be thought of as any heritable trait that serves to identify an organism. Each character can have two or more heritable states. For example, "floral symmetry" has at least two character states, "radial symmetry" and "bilateral symmetry." Which among alternative character states is "ancestral" and which is "derived" is the key issue cladistic analyses attempt to resolve. Ancestral and derived character states may be determined from developmental and historical information, or they may be inferred by designating the character states of each node in the cladogram that minimize the number of character transformations (character reversals) required to explain the observed patterning of character states among descendant taxa. Either method serves to identify the character states of the ancestor shared by closely related taxa.

The fundamental principle governing the assignment of character states to the nodes of a cladogram is *parsimony,* the assumption that the smallest number of character state transitions needed to construct a phyletic hypothesis most likely results in the best (simplest) phylogenetic reconstruction (Wiley 1981). In a sense Occam's razor is used to uncover the evolutionary scheme of relations among taxa that is most likely to be "true." That character reversals can occur is one of the important reasons that many features are used to construct cladograms and that cladistic hypotheses are more robust when they are based on evolutionarily conservative characters. Nonetheless, the principle of parsimony presumes that the rate of overall character change within lineages is slow relative to the rate of speciation. Only in this way can a species retain a phyletic "memory" of its ancestral character states. If the rate of character change is high relative to the number of speciation events, however, then the phyletic memory of a species will progressively "decay" with time, making the retrieval of phyletic relations extremely difficult.

By way of illustration, suppose that the phyletic relations among five species, A–D, are reconstructed based on perianth characters—the number of sepals and petals, how far these perianth parts are fused, and floral symmetry (fig. 1.18). Suppose A–E all have flowers consisting of five sepals and five petals, but that two species, say A and B, have radially symmetrical flowers with unfused petals (actinomorphic and apopetalous flowers), while species C–E have bilaterally symmetrical flowers with laterally fused petals (zygomorphic and sympetalous flowers). In the absence of other characters, the most parsimonious phyletic hypothesis is that A–E share a common ancestor that had flowers with five sepals and five petals; A and B share a common ancestor that had radially symmetrical flowers with unfused petals; and that C, D, and E share a common ancestor that had bilaterally symmetrical flowers with fused petals. But this cladogram is a "closed system" in the sense that there is no a priori way to infer which among the latter two common ancestors reflects the ancestral condition. That is, there is no compelling reason to assume that species with radially symmetrical and apopetalous flowers are more ancient than species with bilaterally symmetrical and sympetalous flowers. The inclusion of additional taxa could "open this system" to further analysis shedding light on whether radial symmetry and fused floral parts are the ancestral or derived character states. Nonetheless, because A and B are linked to the same node of the cladogram and because both have radially symmetrical, apopetalous flowers, the common ancestor to A and B is inferred to have the same character states. Likewise, the common ancestor of C–E is inferred to have had bilaterally symmetrical, sympetalous flowers.

This phyletic hypothesis is extremely crude because it is based on only a few characters (floral symmetry, fusion, and number of parts). The cladistic hypothesis may dramatically change in light of how other characters (and their states) are distributed among species A–E. Suppose that A has numerous stamens and many unfused carpels arranged radially within the perianth, while species B has only five stamens and five fused carpels arranged in a bilaterally symetrical pattern. Further suppose that the character states of the stamens and carpels of species B align perfectly with those of species C–E. Under these conditions, the most parsimonious hypothesis might be that two character state reversals have occurred in the perianth of species B (i.e., bilateral symmetry → radial symmetry and fused petals → unfused petals) rather than that all the other characters had undergone reversals (fig. 1.18). This alternative hypothesis would be further supported by the developmental analyses of species

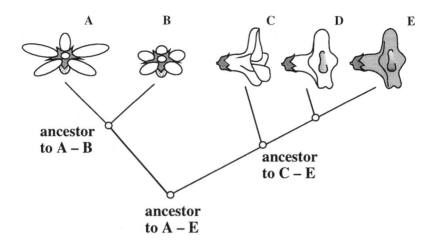

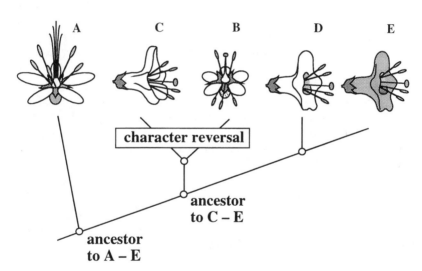

Figure 1.18. Hypothetical cladograms for five extant species of flowering plants (A–E). Nodes of cladograms depicting ancestors shared by species are shown as open circles. Upper cladogram depicts a phylogenetic hypothesis posited by assuming a continuous transformation sequence from radially symmetrical flowers with nonfused floral parts (A) to bilaterally symmetrical flowers with variously fused floral parts (E). Lower cladogram depicts a phylogenetic hypothesis resulting from a presumed character reversal identified by stamen and carpel transformation series.

A–E if data indicated that the character states of perianth parts (sepals and carpels) are ontogenetically less conservative than those of stamens and carpels.

Until recently, cladistic analysis and investigations of adaptive evolutionary changes have focused on very different questions, and so few linkages have developed between these two areas of research. Now, however, the agenda of cladistics has expanded to test hypotheses of adaptive evolution. The general procedure is to map adaptive characters (or associated adaptive characters) on cladograms independently constructed from other characters. The logic is that an independently constructed cladogram that correctly identifies the historical relations among taxa can be used as a test for the hypothesized historical sequence of adaptive characters. To avoid circular reasoning, it is vital that the phyletic hypothesis be based on characters other than those underpinning the hypothesis for adaptive evolutionary change. (Logically, a relation cannot be used to prove its own truth.) Suppose a cladogram is constructed based on the anatomical features of the mature stems of species and that character states resulting in weak stems are inferred to be the ancestral states while character states that result in stiff and strong stems are inferred to be the derived character states. Although the reconstructed phylogeny for these species may be correct, it would be ingenuous to argue that the cladogram supports the view that stem anatomy has undergone adaptive evolutionary changes such that species with stiffer and stronger stems are more recent. As I noted, this logical circularity can be avoided (or at least minimized) by reconstructing the phylogenetic relation among taxa based on characters independent of those implicated in the adaptive hypothesis. Suppose reproductive characters, which tend to be more evolutionarily conservative than vegetative features, were elected for this purpose to test an adaptive hypothesis for stem anatomy. In this case, how far the cladogram based on reproductive characters is concordant with the historical sequence of stem anatomies predicted to be adaptive could be used to evaluate whether evolutionary changes in stems are adaptive.

For example, suppose our object is to test a biomechanical adaptive hypothesis for species A–D regarding the evolution of stem anatomical character states e–h (fig. 1.19). The adaptive hypothesis predicts that selection forces favor mechanically more stable stems, while an experimental protocol indicates that stems with e–h are stiffer and stronger than stems with e–g, stems with e–g are stiffer and stronger than stems with e–f, and so forth. This adaptive hypothesis can be tested by reconstructing the phyletic relations among A–D based on characters independent of e–h, say the reproductive character states a–d. How well the adaptive

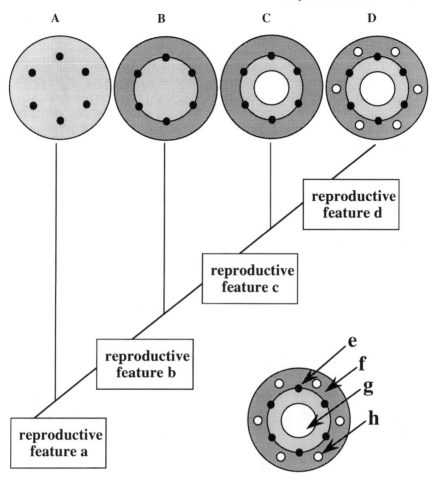

Figure 1.19. Testing an adaptationist hypothesis for mechanical architecture of the stems of four species (A–D). The cladogram depicts a phylogenetic hypothesis based on reproductive features (a–d) that are presumably independent of the mechanical architectural features employed by the adaptationist hypothesis (e–h; see transverse section at lower right).

historical sequence complies with the cladistic historical sequence would provide a reasonable test of the adaptive hypothesis. Further support for the hypothesis would be supplied if additional concordant historical and adaptive patterns could be found in other independent lineages, because this would indicate convergent evolution in unrelated lineages—additional circumstantial evidence that selection forces have led to adaptive evolutionary changes.

Of course the danger exists that the characters selected to construct a phyletic (cladistic) hypothesis are not independent of those used to construct the adaptive hypothesis. The same danger applies to the characters used to construct the cladogram. We have already discussed the effects of pleiotropic genes that can simultaneously influence many phenotypic traits. We have also seen that the phenotype is the physical manifestation of coadapted gene complexes for which small genetic alterations can have potentially manifold or cascading effects. Ontogenetic hitchhiking is yet another danger; that is, some character transformations may be auto-correlated as a consequence of developmental constraints. Thus, even when constructing a cladogram for the sole purpose of inferring phylogenetic relations, the independence of characters and character states must be approached with great caution. In terms of testing adaptive hypotheses, we must be aware of how far adaptive characters are associated with one another in addition to the rate of the branching episodes of speciation in a particular lineage (cladogenesis). Reconstructed phylogenies will shed light on adaptive hypotheses only when the selective associations among characters are weak and the rate of cladogenesis is very high. Notice that the test for an adaptive hypothesis is that the predicted historical sequence of adaptive characters complies with the history of speciation inferred from a cladogram. If the association among adaptive characters is very strong (presumably because of selection forces), then adaptive characters will tend to appear concurrently at the same nodes in a cladogram. This may obscure the "sequence of appearance" of adaptive characters, with the curious result that adaptive hypotheses dealing with characters strongly associated owing to selection forces will tend to be rejected more frequently than hypotheses dealing with characters that are weakly associated.

In this respect, cladistic analyses of plants benefit from the "pas de deux" of the plant life cycle, the alternation of generations between a multicellular diploid organism (the sporophyte) and a multicellular haploid organism (the gametophyte). In many types of plants these two generations have an independent existence and different ecological requirements. Consequently, adaptive hypotheses can be formulated for character states of one generation and tested against the reconstructed phyletic relations among the taxa based on the character states of the other. Of course, when the object of cladistic analyses is simply to reconstruct phylogenetic patterns, the cladograms for the two generations of the taxa can be juxtaposed, and comparisons between the two "independently" constructed cladograms can be used to deduce phyletic relations (e.g., Mishler et al. 1994).

Cladistic analysis has also been proposed to test whether specific character states arose in a lineage through natural selection for their current functions. Here the object is to demonstrate that a particular trait serving a particular adaptive function in a current environment originally spread in the ancestral population of a current species through natural selection because the trait served the same function in the past as it does currently. From prior discussion, recall that a heritable trait (a character in the cladistic sense) can be considered an adaptation for a current function only if the trait was clearly forged by selection forces for its current function in the same environmental context. Thus, to test a hypothesis for the adaptive origin of a trait, we have to map adaptive character states, their current functions, and current environmental contexts onto an independently constructed cladogram (fig. 1.20). That is, the derived character state s_D would have arisen as a consequence of natural selection for its current function f_D only if it could be shown that s_D and f_D map directly above the same node of a cladogram, and if f_D conferred a functional advantage relative to the ancestral function f_A of the ancestral character state s_A. Notice that this test assumes the ancestral and derived character states s_A and s_D have the same current consequences on the fitness of extant taxa as they had on the fitness of the ancestral populations ("phenotypic stasis" in the sense that character states must not change once they make their evolutionary appearance). The test also assumes that the environment in which the ancestral and derived character states function (e_A and e_D) has not changed, because the "functional advantage" conferred by the derived character state s_D over the ancestral character state s_A can be gauged only if both character states function in the same physical and biological environment. If the environment has changed, then an evaluation of whether the derived character state is functionally superior to the ancestral one tells us nothing about whether the derived state arose through natural selection for its current function.

The assumptions underpinning this test for the origin of adaptive traits are likely not to hold true in many cases. Changes in the biological and physical components of the environment can occur over time scales far shorter than those in which taxonomic divergence is measured. In these circumstances, the assumption that the environment has remained constant *relative* to the rate of cladogenesis is less robust than the assumption that the environment changes faster than new species appear. Also, because adaptive evolutionary changes in characters necessarily track environmental changes (at least to some degree), it is very improbable that "phenotypic stasis" is a reasonable expectation, particularly for an adaptive hypothesis. Indeed, if adaptive evolution has occurred, then an

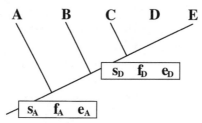

Figure 1.20. Testing an adaptationist hypothesis for five extant species in terms of a phylogenetic hypothesis for ancestral and descendent character states (s_A and s_D, respectively) and their adaptive functions (f_A and f_D) in the context of ancestral and descendent environments (e_A and e_D). The phylogenetic hypothesis depicted by the cladogram must be independently constructed and not incorporate the character(s) used in the adaptationist hypothesis (see fig. 1.19).

adaptive hypothesis *assumes* that character states and the environments in which they function change at least to some degree. Finally, even if the ancestral and derived character states have not changed since the divergence of taxa, we cannot assume that other character states have remained static. Each biological trait functions in the context of the totality of other traits called the organism. It is therefore reasonable to assume that even if some adaptive character states are "phenotypically static," changes in other traits will influence their functional selective advantages.

Yet another caveat is that a phyletic pattern of speciation and extinction may exhibit an "adaptive" evolutionary trend even when many species within a lineage counter the trend (fig. 1.21). Evolutionary trends of this sort may result when selection acts on emergent properties of species rather than on the properties of individuals (i.e., species selection). The species within a lineage that survive the longest may have a higher probability of producing new species than do species with shorter durations, and therefore the longest-lived species within a lineage may determine the direction of major evolutionary trends. (The differential speciation envisaged by this hypothesis is analogous to differential reproductive success in microevolution.) Regardless of whether species selection occurs, attempts to extract a single evolutionary procession of species from an incomplete fossil lineage is very much like trying to describe the shape of a tree by tracing the branching pathways from a few twigs to the base of the trunk (the perception of a tree's shape is governed by both the number and the position of the terminal twigs selected for study). In like manner, in cladistic analyses incorporating fossil species, including or excluding certain species can produce different phylogenetic hypotheses. That the fossil record may not preserve the remains of every species within a lineage and that not all the fossil species within a lineage

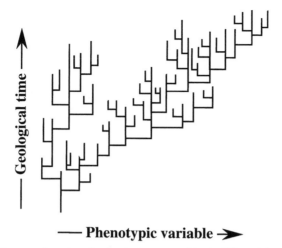

Figure 1.21. Diagram of species selection (a pattern resulting in differential speciation and extinction evincing a directional phenotypic component). As a consequence of differential speciation and extinction, the evolutionary trend in a phenotypic variable may be incorrectly construed as evidence for adaptive evolution.

may have been found owing to limited exploration of the geological column are serious concerns.

A totally pessimistic view regarding the use of fossils to test adaptive hypotheses and the origins and spread of adaptive traits in past populations is not warranted. Although it is naive to assume a priori that phenotypic stasis and environmental stasis hold true, these assumptions can be tested by examining the fossil record, which often supplies sufficient data to reconstruct both past plant morphologies and anatomies and the environments in which plants grew and reproduced. And it is never naive to use paleoenvironmental and stratigraphic data to *test* the assumption that the physical and biological factors influencing the growth and survival of plant species have or have not remained comparatively constant over the course of evolutionary history.

2   Species and Speciation

A large number of genes interact in various ways to produce a smaller number of phenotypic traits which are the direct objects of selection.

Richard Levins

The concept that individuals can be grouped into populations and that populations can be grouped into discrete biological units called species is a central tenet of biology. The idea that species evolve to give rise to new species is the fundamental precept of evolutionary biology. Nevertheless, a precise definition for species uniformly applicable to all kinds of past and present organisms remains elusive, and the absence of a generally agreed on definition lies at the heart of scientific debates over how species arise from preexisting ones.

The issue is not whether species exist or whether new species evolve from old ones. The existence of species as objectively recognizable entities was acknowledged long before Darwin demonstrated beyond doubt that evolution is the conversion of variation among individuals into differences among species. The questions biologists debate center on how species preserve their biologically diagnostic features in spite of heterogeneous and changing environments, while at the same time manifesting the ability to evolve into new species. Other debates concern whether the mechanisms responsible for maintaining variation within a species are the same ones responsible for the origin of new species (speciation) or the origin of higher taxa (transspecific evolution). In this chapter I focus on the data and theories relating to these questions.

One of the difficulties in drawing a portrait of the birth of a species is that speciation involves a succession of events that span such a considerable time that it is rarely possible for a scientist to observe them all directly or to reproduce them experimentally in the laboratory. Another difficulty is that what looks like incipient speciation may not ultimately conclude with the appearance of a new species. In general, speciation is not directly observed but merely adduced after the fact.

Nevertheless, many of the individual events believed to immediately prefigure or follow the origin of a species have been observed in nature or simulated in the laboratory. Some of these events have been observed on island chains like the Hawaiian archipelago. Geographically isolated areas often serve as "living laboratories" for the study of speciation. The geological evidence verifies that the main Hawaiian islands are comparatively young, on the order of only a few million years old. Therefore species found only on these islands must have evolved recently, and their biology gives insight into the mechanisms of speciation. The portrait emerging from the study of islands is that speciation is accelerated by chronic physical or biological disturbance that fragments large populations into smaller, genetically isolated subpopulations. Subsequent genetic divergence of these subpopulations can result in the appearance of new species, often in a remarkably short time. Small organisms that complete their life cycle rapidly appear to be the best candidates for speciation because they tend to have higher mutation rates than larger organisms and because the rate of change in population allele frequencies tends to increase as a function of reproductive rate.

One of the more stunning "laboratory" investigations of speciation is that of Loren Riesenberg and his coworkers, who reproduced the genetic changes leading to the formation of a naturally occurring species of sunflower (*Helianthus anomalus,* an outcrossing diploid restricted to swales and sand dunes in Arizona and Utah). Under laboratory conditions these changes are repeatable across independent experiments. *H. anomalus* appears to have arisen by recombinatoral speciation (a process in which two species hybridize and the mixed genome of the hybrid becomes a third species that is reproductively isolated from both of its ancestral species). The two putative ancestral species of *H. anomalus* are *H. annuus* and *H. petiolaris.* All three species are self-incompatible annuals. The two ancestral species often produce large hybrid swarms whose members are typically semisterile because the species differ by fixed chromosome arrangements that cause meiotic nondisjunction. However, over several generations these arrangements can sort out into a new genome that is fertile but incompatible with its ancestral genomes. Riesenberg et al. (1996) hybridized *H. annuus* and *H. petiolaris* to produce three independent hybrid lines that were subjected to different sib-mating and backcrossing regimes. With the aid of molecular analyses, they found that plants from all three lines converged to nearly identical gene combinations including parallel changes in the nonrearranged portions of chromosomes that must represent gene fitness effects. Although many features of this experiment still need to be understood, it is clear that a complex network of genetic interactions was involved in this laboratory example of microevolution and that despite this complexity, the path of evolutionary change was repeatable in ways suggesting that selection rather than chance governs the genomic composition of hybrids between *H. annuus* and *H. petiolaris*.

Even though the juxtaposition of observations made in the field with experimental manipulations of populations in laboratories provides much of the theory and many of the data for exploring the species concept and the tempos and patterns of speciation, the fossil record is required in order to study the long-term dynamics of speciation. Scientists studying living organisms can observe evolutionary tempos and modes only over comparatively short time scales and only for a small portion of all the species that ever lived. In contrast, the fossil record gives insights into changing species diversity and the volatility or stability of lineages in relation to the long-term challenges offered by the changing physical and biological environment. Just as scientists studying living species must juxtapose experimental data gained from laboratory experiments with observations made in the field, evolutionary biologists must juxtapose the data drawn from the fossil record with the data and theory derived from living organisms.

"Classical" Concepts

The traditional opinion, still widely held, is that a species is a collection of individuals whose members can interbreed and yield viable, fertile offspring but cannot breed successfully with members of other species (Dobzhansky 1937; see also Mayr 1942, 1963, 1969; Stebbins 1950; Grant 1963, 1981). This definition is called the biological species concept. It has two logically independent components: all the individuals within a species actually interbreed (or have the potential to); and all are reproductively isolated from organisms belonging to other species. In general, the biological species concept places greater emphasis on the second component (reproductive isolation) than on the first (the capacity, whether real or potential, for interbreeding).

The biological species concept remains the most generally accepted species definition for organisms with biparental sexual reproduction. It can be applied to many plant species whose members are interfertile and reproductively isolated from other species by strong sterility barriers (Raven 1984). For many plant species, however, the biological accent is more on one of the two components of the biological species concept. Some plant species hybridize freely and so are not reproductively isolated from one another. In other cases interbreeding is limited to immediate neighbors within populations. For these reasons and others, the biological species concept as it applies to plants has been richly criticized (Ehrlich and Raven 1969; Raven 1980).

With its emphasis on reproductive isolation owing to genetic incompatibility, the biological species concept simultaneously affords a hypothesis for the evolutionary phenomenon of speciation. A new species evolves from a preexisting one as a consequence of genetic processes that reduce or completely block gene flow between subpopulations originally sharing the same ancestral gene pool. The concept posits gene flow resulting from random mating as the glue that binds all the members of a species into a discrete biological entity and suggests that species evolve from a preexisting ones when this glue is eroded by any mechanism resulting in reproductive isolation (fig. 2.1).

This mode of speciation is easily visualized in terms of a fitness landscape with a cluster of fitness promontories that identify genetically distinct subpopulations belonging to the same species. Because genetic dissimilarities must occupy different promontories, the average member of each subpopulation differs somewhat from the average member of other subpopulations. As long as gene flow is maintained, all the subpopulations will share much of the same ancestral gene pool. If the subpopulations diverge far enough genetically and phenotypically owing to different selection pressures acting on each of them, however, barriers to gene flow between two or more subpopulations may become established. If low gene migration rates persist, reproductive isolation will intensify and foster further genotypic and phenotypic divergence. If this persists, the subpopulations may eventually diverge until they warrant separate species designations.

The subpopulations of a species are expected to diverge over time because of the combined effects of the stochastic (random) processes of mutation and random changes in gene frequencies (genetic drift) and the deterministic (nonrandom) process of natural selection. In theory, the influence of random processes will be more pronounced in small subpopulations than in large groups of individuals. Nevertheless, natural selection will countermand random changes whenever the mean fitness of a subpopulation is reduced for long. Because subpopulations occupy different fitness promontories, which by definition can be reached only by genetic and phenotypic differentiation, the selection pressures acting on different subpopulations must differ in kind or degree. These different pressures have the potential to pull the entire population apart. In a randomly mating (panmictic) population, the fracturing of subpopulations is reduced because all members of the species have reproductive access to one another. The traditional view is that gene flow among subpopulations living in close proximity (sympatry) reduces the extent to which subpopulations genetically diverge (see fig. 2.1). When subpopulations are geographically isolated from one another (allopatry), however, gene

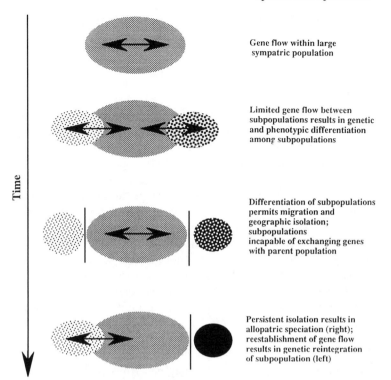

Gene flow within large
sympatric population

Limited gene flow between
subpopulations results in genetic
and phenotypic differentiation
among subpopulations

Differentiation of subpopulations
permits migration and
geographic isolation;
subpopulations
incapable of exchanging genes
with parent population

Persistent isolation results in
allopatric speciation (right);
reestablishment of gene flow
results in genetic reintegration
of subpopulation (left)

Time

Figure 2.1. The traditional model for speciation. Two subpopulations diverge genetically and phenotypically from the parental sympatric population as a consequence of barriers to gene flow (vertical lines). The barrier to gene flow between one subpopulation and the parent population disappears, and the subpopulation becomes genetically reintegrated (left). The barrier to gene flow between one subpopulation and the parent population persists, and allopatric speciation may occur (right). See figure 2.12 for an alternative speciation model grounded on the proposition of a genetic revolution.

flow is reduced and different mutations or allele frequencies produced by chance may be fixed in populations. The paucity or absence of gene flow among allopatric subpopulations can result in speciation.

The portrait of speciation owing to geographic isolation is called allopatric speciation. It involves three sequential steps: first, subpopulations become spatially isolated from the main range of their parent population; second, the subpopulation(s) genetically diverge(s) from the main population; and third, selection pressures lead to a large number of genetic or physical differences that preclude subpopulations' interbreeding with one another and with their parent populations (Dobzhansky 1937; Mayr 1963). In sum, a physical barrier sequesters subpopulations so that they cannot interbreed, allowing them to diverge genetically. At some point the divergence may be so great that a new species comes into existence.

Distractions

The biological species concept states that gene flow in panmictic populations holds the members of a sympatric population together genetically, that species are maintained as discrete biological entities by reproductive isolation from other species, and that isolated subpopulations undergo genetic and phenotypic differentiation as a prelude to speciation. Several factors cast doubt on the general applicability of the biological species concept and its corollary model for speciation. Indeed, they suggest that the broader issue of the relations among genetic differentiation, phenotypic divergence, and the mechanisms that establish genetic incompatibilities leading to reproductive isolation has not been satisfactorily resolved. For example, early studies of plant populations showed that gene flow may be spatially limited even among the members of a sympatric population (Ehrlich and Raven 1969). In these cases gene flow may genetically bind together the individual plants composing subpopulations, but it cannot account for the features characterizing the entire sympatric population (Raven 1980). Early studies also showed that plant species that are radically different morphologically may nevertheless produce viable hybrid swarms of such complexity and size that parental gene complexes become nearly or totally unrecognizable (Turesson 1922; Clausen, Keck, and Hiesey 1940; Tucker 1953). Consider the goldenrod species *Solidago rugosa* and *S. sempervirens* (Goodwin 1937a, b). *S. rugosa* is a tall, strongly hairy plant with thin, dentate leaves, widely distributed in most of the eastern United States and Canada. *S. sempervirens* is a short, glabrous plant with fleshy and entire leaves that grows along the Atlantic seaboard, penetrating inland only along tidal rivers. Yet when populations of these two species make contact, they freely interbreed to produce fertile hybrids. In turn these hybrids produce fertile later-generation and backcross hybrids that differ so much from the two parent species that they are accorded species status as *Solidago* × *asperula*. As a pair of entities, *S. rugosa* and *S. sempervirens* are so phenotypically and ecologically different that few taxonomists would attribute them to the same species. Nevertheless, the two species freely interbreed in nature and give rise to an apparent third species.

Cryptic or sibling species also detract from the proposition that phenotypic divergence is a necessary consequence of reproductive isolation. Sibling species are genetically separated by strong barriers to sexual reproduction yet are barely or not at all distinguishable on morphological or anatomical grounds (Keck 1935; Tobgy 1943; Clausen 1951). Here reproductive barriers may have set in suddenly or progressively, but their establishment has not (or has not yet) been supplemented by phenotypic characters sufficient to distinguish easily among sibling species. Accordingly, the mechanisms underpinning phenotypic and genotypic differentiation are neither invariably identical nor correlated with one another. If one admits that the essence of speciation is the formation of genic differentiation and reproductive isolation, then sibling species show that phenotypic differentiation is neither an attribute of species status nor invariably the end result of speciation, even though it is undeniable that in most cases, in both plants and animals, speciation has been accompanied by or has followed phenotypic differentiation.

Another difficulty with the biological species concept is that to be called a species organisms must pass the test of reproductive isolation. As we have seen, some sexually reproductive species fail this test because they can hybridize with other species. By the same token, asexual life forms, many of which are formally recognized as bona fide species, also fail the test of reproductive isolation. The bacteria are perhaps the most familiar unicellular asexual life forms. Bacteria reproduce asexually by means of binary fission (see chapter 3). Although genetic information can be transferred among different strains of bacteria (e.g., bacterial conjugation), genetically isolated lineages of bacteria have arisen through chance mutation amplified by asexual multiplication. Genetically different clones have clearly diverged through the combined effects of mutation and natural selection to evolve into different species. Even if one argues that species definitions for prokaryotes must be based on criteria entirely distinct from those for eukaryotes, a substantial number of plants, including flowering plants, reproduce exclusively or predominantly by asexual means. For example, many flowering plants reproduce by apomixis; they produce seeds without the fusion of sperm and egg. The apomictic seed develops within the ovary of the flower, as in sexually reproducing species, but the embryo within the seed originates by mitotic division of a somatic cell of the ovule. Except in special cases, the embryo is genetically identical to the parent plant, and so populations of apomicts are genetic clones. Darlington (1940) believed that "apomixis is an escape from sterility, but an escape into a blind alley of evolution." The implication is that all apomicts are doomed to extinction. Nevertheless, apomixis has been reported in at least thirty families of flowering plants including more than three hundred species, some of which, like the dandelion, *Taraxicum offiniale,* are highly invasive and ecologically successful organisms. Those who subscribe to the biological species concept may summarily dismiss uniparental organisms as legitimate species. Ironically, one may also argue that obligate asexual reproduction provides the ultimate reproductive barrier between species. But the dismissal of asexual life forms as species is impotent in the face of parthenocarpic animals and apomictic plants that are in every other respect bona fide species, while the notion that asexual reproduction is a "barrier" to gene flow between species in the sense of the biological species concept is logical legerdemain.

Alternatives and Differences

As valuable as the biological species concept has been to the study of tax-onomy and evolutionary biology, its central tenets continue to be chal-lenged by various biological phenomena (e.g., limited gene flow in sym-patric populations, the formation of complex and vast hybrid swarms, sibling species, and asexual life forms). It should not come as a surprise, therefore, that alternative species definitions have been advanced. Here we will review the strengths and weaknesses of some of them.

The mate recognition concept defines a species as the most inclusive population of a sexually reproductive organism that has a common fertilization system. This species definition emphasizes reproductive adap-tations that have maximized successful mating among members of an inclusive population rather than reproductive isolation barriers (Paterson and Macnamara 1984; Paterson 1985). Here the view is that the sexual isolation the biological species concept rests on is a derivative feature of reproductive adaptations to recognizing mates. In this respect it asserts the opposite side of the coin. Recall that the biological species concept defines a species as "groups of actually or potentially interbreeding pop-ulations, which are reproductively isolated from other groups" (see Mayr 1963, 19). The mate recognition definition focuses on the reproductive mechanisms cloistering a species into a separate and internally coherent biological unit. By doing so, it draws attention to traits presumably directly acted on by natural selection during the process of speciation. This emphasis successfully deals with the problem of hybrid species. Provided organisms share the same fertilization mechanism, they belong to the same species regardless of the pathology or success of their hybrids. However, this assumes that species are biparental organisms, just as does the biological species concept, and excludes a number of kinds of organ-isms that are asexual or uniparental life forms (e.g., apomictics or self-fertilizing plants).

An alternative definition is the morphospecies concept, according to which a species is any kind of organism identifiable by one or more unique and heritable phenotypic traits (Cain 1963). This concept successfully copes with asexual life forms as well as species hybrids, but it is conceptually challenged by the biological reality of sibling species and may fail to discern whether a collection of morphologically introgressive organisms is a set of subtly different species or a single species with extreme phenotypic variation resulting from a broad norm of reaction.

Yet another definition of species is a single lineage of ancestral-descendant populations of organisms that maintains its identity from other such lineages and has its own evolutionary tendencies and historical fate (Simpson 1961; Wiley 1978). This definition, called the evolutionary species concept, emphasizes evolutionary modification through descent. It benefits and suffers from many of the same theoretical and practical reasons as the morphospecies concept.

The advent of cladistic analyses has led to the phylogenetic species concept, according to which a species is the smallest aggregation of populations (of a sexual organism) or lineages (of an asexual organism) diagnosable based on a unique combination of character states found in all members of the aggregate (Eldredge and Cracraft 1980; Cracraft 1983; Nixon and Wheeler 1990; see Baum 1992 for a variation on the theme). This species concept relies on genetically fixed molecular or phenotypic characters that identify each terminal branch of a cladogram as the smallest discernibly different aggregate. Because each terminal branch is a group divergent from all others in the cladogram by virtue of its combination of diagnostic characters, it constitutes a unique phylogenetic entity or "species." Much like the biological species concept, the phylogenetic species definition emphasizes genetically fixed character states reflecting limited gene flow among divergent population systems. But because it focuses on how every discernible difference, even those resulting from a single molecular or morphological marker, distinguishes a phylogenetic species, this concept amplifies the subtleties in the pattern of divergence and so tends to identify many more entities as species (see Davis and Manos 1991). In this sense the phylogenetic species concept is a more modern version of the morphospecies concept. The strength of this approach is its ability to bring to the forefront the intricate genetic diversification patterns that may preface the origin of biological species. Its drawback is its potential to partition diverging population systems into evolutionarily meaningless biological entities. Applying the phylogenetic species concept would separate broccoli, brussels sprouts, cabbage, and kohlrabi within the context of *Brassica oleracea*, but it is dubious that they warrant individual species recognition (fig. 2.2). Put differently, the phylogenetic species definition renders the smallest discernible taxonomic units (aggregates of populations or lineages), but well-reasoned

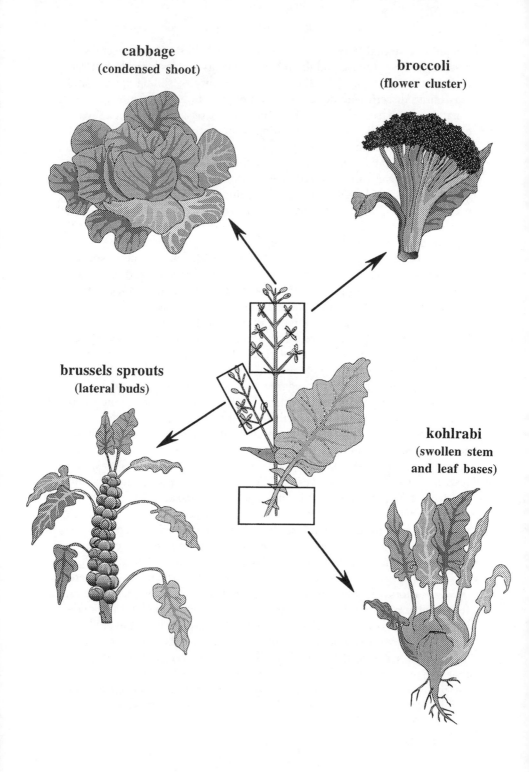

cabbage
(condensed shoot)

broccoli
(flower cluster)

brussels sprouts
(lateral buds)

kohlrabi
(swollen stem
and leaf bases)

debate may fail to reach a consensus on whether these smallest "units" are in fact "species."

Space does not permit a full discussion of all the species definitions so far advanced. The quest for a universally applicable species definition, one that will fit all manner of different fossil and living plants and animals, continues to remain something of a Holy Grail in systematic biology. Virtually every definition has its merits, but the principal difficulty with each is the attempt to reduce the tremendous diversity of life and its multifarious biological entities to a single, uniformly applicable verbal formula. Ultimately each species can be diagnosed by a basic set of genes that all its members share. In many cases but not all, these shared genes have translated into shared morphological features as a consequence of reproductive barriers. Thus genetic and phenotypic markers and reproductive barriers are common elements in many species definitions. But reproductive barriers and morphological divergence are lacking or weakly developed in what many consider to be good species. For this reason reproductive isolation and phenotypic differentiation are best seen as secondary rather than principal attributes of species. Each may or may not accompany or follow the process of speciation. Seen in this light, it may be unrealistic to expect all kinds of organisms to fit neatly into a single species definition and much more practical to employ the concept best suited to understanding the biology of the particular group of organisms of interest (see Cronquist 1978; Ereshefsky 1992).

Most attempts to define what is meant by species are based on animal biology, and many have neglected the conspicuous differences between plants and animals. Five of these differences suggest that species definitions (and the models for speciation they evoke) suffer whenever plant biology is neglected. First, plants tend to be developmentally simpler than animals. The integration and delicate balance between the various organs and organ systems required for the motility, sense perception, and coordinated behavior in animals are far greater than anything existing in even the most morphologically and anatomically complex plants. Thus it is possible that different developmental patterns may be more easily

Figure 2.2. Morphological differences among cabbage, broccoli, brussels sprouts, and kohlrabi depend on few gene differences. These cultivated types, each strongly self-incompatable, represent botanical monstrosities that would have had little chance of survival in nature. Each is derived from a comparatively small genetic difference of the putative ancestor, colewort (*Brassica oleracea*), shown at the center of the diagram. Developmental amplification of different parts of the ancestral plant resulting from small genetic differences obtains each cultivated type (e.g., condensation of the entire shoot gives rise to cabbage).

integrated in plant species hybrids than in animal hybrids (Grant 1971). Second, the motility, sense perception, and coordinated behavior of animals provide opportunities to quickly establish barriers to species interbreeding. In contrast, many groups of plants promiscuously cross-pollinate, and so reproductive isolation may be more difficult to achieve or maintain (Baker 1959). Third, many plants have an open system of growth based on the retention of perpetually embryonic tissues (meristems) that, in addition to amplifying the number of body parts and overall size, makes possible great individual longevity and asexual (vegetative) propagation. Given the capacity for continued growth in size and vegetative propagation, a single mutant or hybrid plant has the potential to establish a large clonal population without sexual reproduction. In contrast to most animal species, low sexual fecundity or sexual sterility does not preclude the establishment of a plant population, nor does it afford an intrinsic obstacle to speciation. The open system of growth by means of apical meristems also provides the opportunity for genetic mosaics through random mutation of meristematic cells. An individual plant could generate a population of genetically different descendants. Fourth, sterile species hybrids may become fertile when the chromosome number in their germinal cells spontaneously doubles (polyploidy). By this mechanism, a sterile hybrid can give rise to fertile progeny that may serve as the founding members of a new species. And fifth, many plant species are capable of self-fertilization. For these hermaphroditic species, uniparental reproduction is the normal method of sexual reproduction.

Other differences exist between plants and animals, but these five highlight the possibility that reticulate evolution may be far more common among plants. Reticulate evolution refers to the anastomosing from time to time of formerly separate branches on the evolutionary tree. Separate branches can graft together when species successfully hybridize and the resulting hybrids evolve into new species. Speciation may occur immediately by virtue of polyploidy or later when populations of sterile organisms persist for long periods (e.g., apomictic populations) and subsequently acquire the capacity for sexual reproduction (Babcock and Stebbins 1938). In either case the evolutionary tree will bear a number of interweaving and merging branches in addition to conventional dichotomous branches.

Reticulate Evolution (Hybridization)

Numerous studies of plant populations verify that hybridization occurs naturally and suggest that hybrid zones may evolve into stabilized, genetically distinctive populations and even species (Stebbins 1959; Riesenberg and Ellstrand 1993; Riesenberg et al. 1996; Arnold and Hedges 1995). Hybridization may also give rise to new adaptations or combinations of adaptations by transferring and combining phenotypic features among species (Riesenberg 1995; see Harrison 1993 and references therein). In this section I will review some examples of species hybrids, emphasizing the genetic distance among the progressively higher ranks in the taxonomic hierarchy (species, genera, and so forth).

Although all the members of a species are closely related genetically and presumably are reproductively compatible, no two subpopulations of the same outcrossing species are identical in every respect. Taxonomic categories below the species rank, called infraspecific units, are frequently established to deal with recognized differences between subpopulations in one or more heritable traits. The most widely used infraspecific unit is the subspecies. Subspecies inhabit different geographical areas or habitats within the range of their species. Nevertheless, they can interbreed and so may genetically and phenotypically intergrade, particularly where their ranges normally make contact or when they are artificially brought together (fig. 2.3). Experiments verify that the ability of subspecies to interbreed reflects differences in divergence times between subspecies. In this respect subspecies may be viewed as biological entities that have taken the first steps toward the genetic and phenotypic differentiation traditionally posited to preface the evolution of a new species, even though gene flow typically is sufficient to hold subspecies together genetically and phenotypically so that they are taxonomically recognizable as belonging to the same species.

Infrageneric taxonomic groups are also recognizable. Infrageneric taxa are groups of closely related species assigned to the same genus. The most commonly recognized is the species group called the section. The species assigned to one section are presumably more closely related to one another than to species assigned to another section within the same genus. Thus the extent to which species within and across sections of the same genus can interbreed illustrates simultaneously the genetic and phenotypic divergence attending the evolution of species and higher taxa. Numerous studies verify that the potential for gene flow between species in the same section varies over a continuum but is generally greater than the potential for gene flow between species belonging to different sections within the same genus. A good example is seen in the North American genus *Ceanothus*, which ranges from southern Canada to Guatemala and is principally found along the Pacific coast. The genus consists of two sections, *Cerastrus* and *Ceanothus* (which was formerly called the section *Euceanothus*). Species within each section can interbreed and produce vigorous hybrids that may extend into the second generation

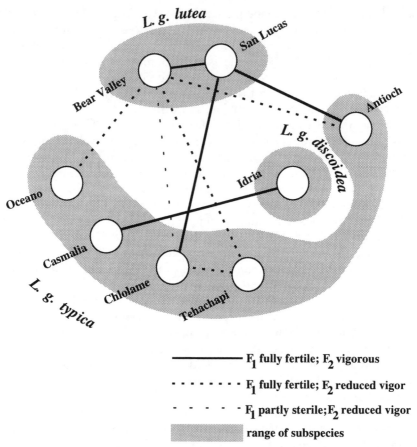

Figure 2.3. Different levels of fertility and sterility in artificial hybrids between various subspecies of the flowering plant *Layia glandulosa* (family Asteraceae), native to California. Geographic locations of subspecies populations are shown (e.g., Bear Valley). F_1 denotes the first filial generation obtained by crossing parent generations P_1; F_2 denotes the second filial generation produced by crossing the F_1 generation. Adopted from Clausen 1951.

(fig. 2.4). However, interbreeding between species belonging to different sections typically yields either sublethal hybrids or hybrids that fail to set seeds even under otherwise favorable garden conditions (Nobs 1963). Presumably, species in the two sections have genetically and phenotypically diverged so far that hybrids between them are inviable, whereas vigorous hybrids are possible between more closely related species in the same section.

Section *Cerastrus*

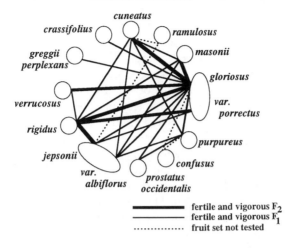

crassifolius
cuneatus
ramulosus
greggii perplexans
masonii
gloriosus
var. porrectus
verrucosus
rigidus
jepsonii
purpureus
var. albiflorus
prostatus occidentalis
confusus

━━━━━ fertile and vigorous F_2
───── fertile and vigorous F_1
·········· fruit set not tested

Section *Cerastrus*

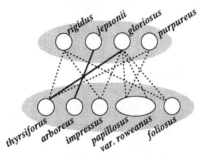

rigidus   jepsonii   gloriosus   purpureus

thyrsiforus   arboreus   impressus   papillosus var. roweanus   foliosus

Section *Ceanothus*
(formerly *Euceanothus*)

───── sublethal hybrids
·········· no seed set
▓▓▓▓▓ sectional affiliation

Figure 2.4. Different levels of fertility and sterility in artificial hybrids between various species of the flowering plant genus *Ceanothus*. The genus contains two species groupings formally recognized as sections (section *Cerastrus* and section *Ceanothus*; the latter was formerly called section *Euceanothus*) The diagram at the top shows how far species in the section *Cerastrus* can hybridize. The diagram at the bottom shows how far species in the section *Cerastrus* can hybridize with the species in the section *Euceanothus*. F_1 denotes the first filial generation obtained by crossing parent generations P_1; F_2 denotes the second filial generation produced by crossing the F_1 generation. Adapted from Nobs 1963 and Grant 1971.

The situation in the genus *Ceanothus* is far from unique. Many intrageneric hybrids are known, and they are sometimes so genetically complex and phenotypically distinctive that the boundaries between species nested in the same genus may be blurred out of existence and hybrids may properly be accorded species status. This has happened for the two species *Clarkia nitens* and *C. speciosa* (Lewis and Bloom 1972). Naturally occurring intrageneric hybrids of oak are also well known (Tucker 1953; Grant 1971). Intrageneric hybrids may result when species that formerly were geographically isolated are brought into artificial contact. For example, the sycamore trees of the eastern United States (*Platanus occidentalis*) and the eastern Mediterranean (*P. orientalis*) are geographically separated and so do not normally interbreed. But when planted together they can produce fertile progeny that have been designated as a separate species (*P. acerifolia*). Habitat disturbance also can result in hybridization. Perhaps the classic example of the "hybridization of habitat" is seen in species of *Iris* inhabiting the Mississippi Delta. Before the human disturbance of their native habitats, two distinct sympatric species, *Iris fulva* and *I. hexagona* var. *giganticaerulea* existed, respectively, in well-drained forests along the edges of rivers and streams and in poorly drained swamps. The advent of deforestation and the draining of swamps produced habitats combining the features of both parental habitats, resulting in an enormous diversity of hybrids and hybrid derivatives between the two species (Anderson 1949).

Hybrid populations are not accorded species status unless they maintain their unique biological identity in successive generations. The evolution of a new species from a hybrid has been called hybrid speciation. As defined by Verne Grant (1971), hybrid speciation occurs when a small proportion of the descendants of a partly fertile species hybrid become stabilized and give rise to a population within which free gene exchange is possible, though it is not possible between either parental type. Among flowering plants, one way this can happen is when the pollination syndrome of the hybrid diverges from that of either of its parents. This has occurred for *Delphinium gypsophilium,* a species resulting from the hybridization of *D. hesperium* and *D. recurvatum,* and for *Penstemon spectabilis,* which evolved from a hybrid between *P. centranthifolius* and *P. grinnellii.*

The ability of species to hybridize has received close attention because interspecific crosses are one way to introduce new traits into commercially important crops. Plant breeding programs therefore shed considerable light on the genetic barriers among related species and genera. The genus *Lycopersicon* consists of two major species complexes, the "*esculentum*" complex (to which the commercial tomato *L. esculentum* belongs) and the "*peruvianum*" complex (fig. 2.5). A major barrier to hybridization separates these two complexes, although hybrids between "*peruvianun*" and "*esculentum*" species have been reported on very rare occasions. This barrier is more easily circumvented by obtaining intermediate hybrids with two sibling species in the "*minutum*" species complex. Highly involved programs of hybridization and embryo culturing techniques have introduced numerous genes conferring disease resistance and stress tolerance isolated from the "*peruvianum*" complex into commercial "*esculentum*" tomato lines (see Stevens and Rick 1986; Taylor 1986). Intergeneric hybrids have also been achieved between the tomato and *Solanum lycopersicoides,* a species related to the potato *S. tuberosum.* These hybrids show that the genetic barriers between genera are not absolute, although they frequently result in extreme hybrid sterility.

There is little doubt that species in different genera are genetically and phenotypically more divergent than species within the same genus. Thus viable intergeneric hybrids tend to be rare. Nevertheless, intergeneric hybrids do occur, and there is good evidence that some can evolve into new species as a consequence of introgression, or backcrossing with subsequent fixation of backcross types. One example of a stabilized introgressant occurs between the two shrub species *Purshia tridentata* and *Cowania stansburiana* (Strutz and Thomas 1964). Hybrids between these genera apparently backcrossed with their parent species, and the backcrossed derivatives became genetically and phenotypically fixed in the form of the new species *P. glandulosa,* which has a wide geographic distribution.

It was formerly believed that plant hybrids typically are morphologically intermediate and less fit than either of their parent species. It was also believed that the parental characteristics of hybrids were likely to remain associated or correlated in segregating hybrid progenies. Recently, however, these beliefs have been challenged by mounting evidence that hybridization can act as a catalyst for speciation (Ford and Gottlieb 1989, 1990; Wilson 1992; Riesenberg 1995). Currently, few doubt that some species have evolved from hybrids. What remains unanswered, however, is the relative frequency of speciation by this means (Endler 1977; Riesenberg 1991, 1995). By providing an expanding arsenal of analytical techniques that permit high-resolution descriptions of the distributions of molecular markers in hybrid populations from which the genetic dynamics of populations may be inferred (Harrison 1990; Arnold 1993; Riesenberg et al. 1996), molecular techniques may eventually resolve this issue. These new techniques may also shed light on the genetic mechanisms and ecological circumstances responsible for transspecific evolution, the origin of higher taxa. The differences between species drawn

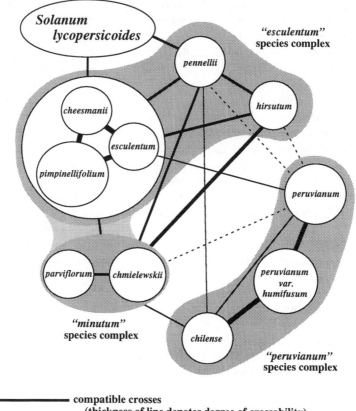

Solanum
lycopersicoides

"esculentum"
species complex

pennellii

hirsutum

cheesmanii

esculentum

pimpinellifolium

peruvianum

parviflorum — chmielewskii

peruvianum
var.
humifusum

"minutum"
species complex

chilense

"peruvianum"
species complex

————— compatible crosses
(thickness of line denotes degree of crossability)

- - - - - - - - - - - F_1 via embryo rescue and culture

Figure 2.5. Summary of interspecific and intergeneric crossing relations among *Lycopersi-con* species and *Solanum lycopersicoides* (a relative of the domestic potato). The genus *Lycopersicon* consists of two large species complexes, the "*esculentum*" complex and the "*peruvianum*" complex. Another complex, the "*minutum*," consists of two sibling species and is recognized by some authorities as distinct from but related to the "*esculentum*" complex. Hybrids between the two large species complexes are difficult to achieve: viable seeds are sometimes produced on plants (e.g., one viable seed was produced from among hundreds of seeds set by a hybrid between *L. esculentum* and *L. peruvianum*), but most can be achieved only through rescuing hybrid embryos by growing them in culture. Hybrids between the two large species complexes are more easily achieved by making inter-mediate hybrids with species in the "*minutum*" complex. Intergeneric hybrids between *S. lycopersicoides* and *L. esculentum* and *L. pennellii* indicate that the genetic barriers between these two genera are not absolute. (*L. pennellii* was formerly considered a species of *Solanum*.) Adapted from Stevens and Rick 1986, fig. 2.2.

from very different taxonomic groups are so impressive that some investigators have suggested that higher taxa arise through mechanisms distinct from the microevolutionary processes traditionally viewed as responsible for speciation. The argument made is that the mechanisms responsible for variation within a species are not the same as those leading to the appearance of higher taxa. This proposition has led to some of the major controversies in contemporary evolutionary theory. Nevertheless, the examples of species hybrids reviewed here cast doubt on the need to invoke unknown (and unspecified) genetic phenomena for transspecific evolution.

Reproductive Barriers

Reproductive isolation can occur in several ways. Traditionally, the barriers to gene flow are grouped into two broad categories: those that prevent cross-mating or cross-fertilization and those that preclude reproductively viable hybrids (table 2.1). Because the zygote is the cell resulting from the fusion of egg and sperm, its formation (a discrete biological event in the sexual life cycle of all biparental species) conveniently allows us to distinguish mechanisms that preclude the formation of a zygote from sperm and egg, each drawn from a different species (prezygotic barriers), from those that exclude the development of a zygote into a hybrid organism (postzygotic barriers). Here I shall review the evidence for prezygotic barriers and explain how the failure of these mechanisms can inspire speciation. The topic of postzygotic barriers is treated later.

Geographic isolation, as emphasized in the allopatric speciation model, is the most readily understood of all the barriers to sexual reproduction. Island chains are showcases of allopatric speciation because they are physically isolated from the mainlands that supply colonizing plants and animals, because they tend to be colonized by only a few individuals (a small gene pool), and because ecological niches are initially unoccupied (Carlquist 1980). These features are well illustrated by the Hawaiian archipelago, which is home to the largest number of endemic plant species concentrated in any one geographic area (Wagner, Herbst, and Sohmer 1990). An endemic species is one found nowhere else in the world. The amount of endemism is particularly impressive in the Hawaiian flora: 89% of the 956 native flowering plants are endemic species, while 15% of the 32 native genera are endemic (table 2.2). In terms of the number of colonizing species, it is estimated that the native flora is derived from as few as 272 successful species introductions, migrating predominantly from the southwest of the Hawaiian archipelago (most of

Table 2.1   Barriers to Successful Reproduction between Species

---

A. Barriers to mating or fertilization of eggs (prezygotic barriers)
  1. Geographical isolation—physically separated populations
  2. Temporal isolation—reproduction occurs at different times
  3. Mechanical isolation—structural differences in reproductive organs
  4. Gametic incompatibility—sperm and egg fail to fuse
B. Barriers to the formation of viable hybrids (postzygotic barriers)
  1. Hybrid inviability—embryos fail to develop; hybrids fail to reach sexual maturity
  2. Hybrid sterility—hybrids fail to produce functional reproductive organs
  3. Hybrid breakdown—progeny of hybrids revert to parental types; progeny are inviable or sterile

---

the native flowering species have a Malaysian taxonomic affinity). The rate of allopatric speciation has been inferred based on potassium-argon dating of Hawaiian volcanic rocks, which indicates that the oldest currently habitable island (Kauai) is between 4 and 7 million years old.

It is tempting to believe that the taxa endemic to the Hawaiian islands could not be older than 6 million years. However, this is incorrect because the Hawaiian island chain extends far to the west to include formerly large and much older volcanic islands that have been worn down by erosion and are now reduced to low atolls and reefs barely breaking the ocean's surface (e.g., Kure Atoll and Bensaleaux Reef; see fig. 7.3). Species evolving on these much older western islands may have colonized younger islands forming in the east, hopscotching from older to younger islands. Thus precise estimates of allopatric speciation rates for the Hawaiian island chain are problematic at best. Nevertheless, some species are endemic exclusively to the younger Hawaiian islands and so could not possibly have evolved in a time less than the age of these comparatively newly formed islands (see Sakai et al. 1995). For example, 79 "single-island" endemic species are known only from Hawaii, which is at most only about 1 million years old. Likewise, 158 species are endemic only to Maui, which formed about 2 million years ago. These and other data indicate that the rates of allopatric speciation on geographically isolated islands can be very high indeed.

Although islands provide clear opportunities for allopatric speciation, a single species may also consist of a graded spectrum, or cline, of subpopulations that differ genetically and phenotypically along a continuous geographic range. Neighboring subpopulations will be interconnected by gene flow, but interbreeding diminishes as the distance between subpopulations increases. The failure of some subpopulations to interbreed even when given the opportunity in laboratory settings suggests that subspecies may evolve along clines. Differences in soil conditions or in the availability of water may provide selection pressures favoring different

Table 2.2   Summary of Endemic and Native Hawaiian Flowering Plant Species and Genera

	Dicots	Monocots	Total
Total species	822 (86%)	134 (14%)	956
Endemic species	759 (92%)	91 (68%)	850 (89%)
Total genera	165 (76%)	51 (24%)	216
Endemic genera	31 (19%)	1 (2%)	32 (15%)
Families	73 (84%)	14 (16%)	87
Total taxa	947 (87%)	147 (13%)	1,094
Total endemism	888 (94%)	107 (73%)	995 (91%)

Source: Wagner, Herbst, and Sohmer 1990.

genotypes in subpopulations along a cline or in patchy microgeographic settings. For example, soil moisture has been correlated with different allele frequencies for a total of six loci coding for enzymes and seed color in subpopulations of the wild oat (*Avena barbata*; see Hamrick and Allard 1972); races of various grasses (*Agrostis, Festuca,* and *Anthoxanthum*) tolerant and intolerant to heavy metal soil toxins are reported to grow in very close proximity (Bradshaw 1972); and different species of *Ceanothus* occupy neighboring sites differing in soil acidity and moisture (Nobs 1963). Differences in the physical environment may also evoke temporal reproductive isolation. Soil moisture and temperature affect the time when reproductive organs develop, mature, and shed pollen. A good example of temporal isolation is seen in the pine species *Pinus radiata* and *P. muricata* (Stebbins 1950; Mirov 1967). Staggering the time of year flowers are receptive to pollen can prevent or greatly reduce gene flow among closely related sympatric species (see Gentry 1974; Levin 1978).

Experiments verify that reproductive isolation owing to temporal displacement can occur rapidly even among sympatric populations when selection pressures are high. Paterniani (1969) planted a mixed population of two varieties of corn (*Zea mays* var. white flint and yellow sweet) and removed all hybrid seeds from successive populations over a period spanning six generations. Even though the original mixed population had a comparatively high rate of intercrossing (35.8% for white flint and 46.7% for yellow sweet), the rate of intercrossing was dramatically reduced by the sixth generation (4.9% and 3.4%). The result of artificial disruptive selection for plants exclusively representing intravarietal matings was a substantial alteration in the flowering times of the two corn varieties after only a few generations (fig. 2.6).

The displacement of morphological characters can provide an effective mechanical prezygotic barrier to reproduction (Brown and Wilson 1956). Character displacement is sometimes observed among closely related sympatric species that flower at approximately the same time of

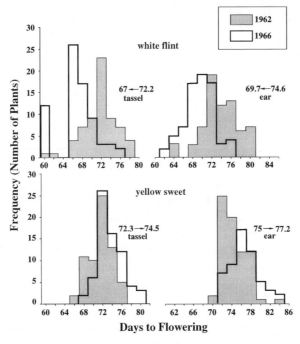

Figure 2.6. Change in the flowering times of the tassel (pollen-producing) and ear (ovule-producing) flowers of two varieties of corn (*Zea mays* var. white flint and *Z. mays* var. yellow sweet) after four generations of artificial selection againt hybrid seeds. The original population was planted in 1962 and consisted of an equal mixture of the seeds from both varieties. The mean flowering time for the tassel flowers was 72.2 for white flint and 72.3 days for yellow sweet; the mean flowering time for the ear flowers was 74.6 days for white flint and 75 days for yellow sweet. After intense artificial (disruptive) selection against hybrid seeds, the mean flowering time of the 1966 white flint population was reduced by approximately one week compared with the 1962 population (from 72.2 to 67 days for tassel flowers and from 74.6 to 69.7 days for ear flowers). The mean flowering times of yellow sweet was increased by approximately two days (from 72.3 to 74.5 days for tassel flowers and from 75.0 to 77.2 days for ear flowers). Reproductive isolation between the two varieties resulted from selection against the genomes of plants flowering at approximately the same time, thereby favoring genomes (represented in the original population) of white flint plants that flowered later and genomes of yellow sweet that flowered later. Data from Paterniani 1969, fig. 5.

year. For example, the lengths of styles and anthers in the flowers of two nightshade species, *Solanum grayi* and *S. lumholzianum,* are similar for both species in allopatric populations in Arizona. However, when the two species coexist in close quarters, as in the Sonora Desert, the anthers and styles of *S. grayi* are significantly shorter (63% and 73%) than those of *S. lumholzianum* (Levin 1978). The difference in style and anther length reduces the probability that the sympatric species will ultilize the same pollinator and so reduces the likelihood that their reproductive processes will interfere.

Gametic incompatibility involving genetically controlled pollen-stigma interactions is well known among flowering plant species (see fig. 8.11 and attending discussion). Space precludes a detailed discussion of these complex interactions here; interested readers are referred to the excellent treatments of this topic provided by Frankel and Galun (1977), Newbigin, Anderson, and Clarke (1993), and Nasrallah and Nasrallah (1993).

Postzygotic barriers to sexual reproduction also help to maintain species as discrete biological entities even when prezygotic barriers fail. Postzygotic barriers preclude the formation of viable hybrid zygotes or the survival of viable hybrids (see table 2.1). When these barriers break down, viable hybrids capable of perpetuating their own kind can provide plants with a mechanism for the evolution of a new species.

A hybrid can be aborted at any stage in its life or, if it survives to maturity, may fail to perpetuate its own kind. The hybrid zygote may die immediately after sperm and egg fuse, or the developing plant may die before reaching maturity (hybrid inviability). In flowering plants, the survival of the zygote and its successful development into an embryo typically depend on the formation of endosperm in the seed. The endosperm is a product of "double fertilization," so called because one sperm fuses with the egg and the second sperm enters another cell that will normally develop into the endosperm. The abortion of hybrid zygotes or embryos is frequently the result of anomolous endosperm growth (e.g., *Datura, Iris,* and *Gossypium*). Even when hybrids develop into plants, they may be vegetatively weak and easily killed by competition with their neighboring parental forms for nutrients and space. Conversely, some hybrids exhibit "hybrid vigor," that is, they grow more luxuriantly and are more competitive than either parent. But the hybrid may fail to develop functional reproductive organs (hybrid sterility) or may produce progeny that either are sterile or, if reproductively competent, genetically segregate back to the parental types (hybrid breakdown). In either case the hybrid fails to perpetuate its own kind and so cannot establish a population of individuals.

The postzygotic barriers to hybridization embrace a wide variety of effects involving disharmony in gene and chromosome interaction that may be expressed in vegetative growth (somatic effects leading to poor growth) or during the formation of reproductive organs (gametic effects that result in hybrid sterility or breakdown). One classic example involving chromosomal disharmony must suffice. In 1927 G. Karpechenko attempted to hybridize the common radish (*Raphanus sativus*) with cab-

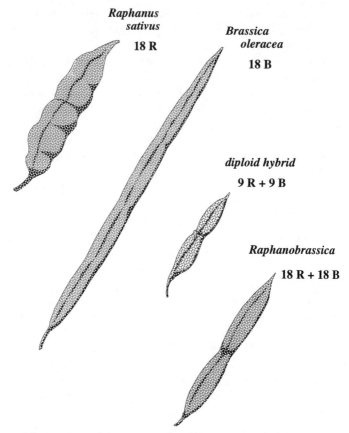

Figure 2.7. The genealogy of the genus *Raphanobrassica,* an amphidiploid resulting from a diploid hybrid between the radish *Raphanus sativus* and the cabbage *Brassica oleracea.* Diploid cells of the radish parent have eighteen chromosomes (18R). Diploid cells of the cabbage parent also have eighteen chromosomes (18B). The diploid hybrid made between the radish and cabbage has eighteen chromosomes, nine from each parent (9R + 9B). Most diploid hybrids are sterile because the 9R + 9B chromosomes fail to pair during meiosis. However, some viable seeds develop on diploid hybrid plants and grow into *Raphanobrassica* plants. These seeds are a consequence of allopolyploidy (see fig. 2.9), the spontaneous doubling of chromosome numbers in the germinal cells of the diploid hybrid (i.e., *Raphanobrassica* plants have 18R chromosomes + 18B chromosomes).

bage (*Brassica oleracea*). Both genera belong to the mustard family (Brassicaceae), and both species have the same number of chromosomes in their diploid cells (2n = 18). Karpechenko's attempt to produce a hybrid was successful (fig. 2.7). The hybrid plants had nine pairs of chromosomes in their diploid cells. However, the nine chromosomes from one parent failed to recognize their homologues from the second parent and so did not pair during meiosis. As a consequence, the division products of meiosis had varying numbers of chromosomes, mostly ranging between six and twelve per cell. Because normal sperm or eggs could not be formed in flowers, the hybrid was sterile as a consequence of chromosomal disharmony (the failure of homologous chromosomes to pair).

The experiments of Karpechenko also show how a new species can evolve from a sterile hybrid whose cells spontaneously double their chromosome numbers. Karpechenko found that some of his "sterile" hybrid plants produced a few viable seeds that grew into fertile plants whose diploid cells had thirty-six chromosomes (see fig. 2.7). Apparently these viable seeds were produced when the chromosome number spontaneously doubled in the germinal tissue, giving rise to cells with two chromosomes of each kind from both parents. This doubling permitted the pairing of partner chromosomes and the formation of gametes with eighteen chromosomes. Fusion of these gametes gave rise to fertile plants whose somatic cells contained thirty-six chromosomes. Karpechenko had "synthesized" a new species because it has no possibility of gene exchange with either of its two parent species. He called this organism *Raphanobrassica;* unfortunately, it had the roots of a cabbage and the leaves of a radish. Nonetheless, Karpechenko had witnessed the birth of a new species in the passage of three plant generations.

Polyploidy and Sympatric Speciation

Karpechenko's experiments showed that a new species can evolve through polyploidy. The cells of a polyploid individual contain extra sets of chromosomes. There are two ways a polyploid species can evolve. The first and simpler to understand is autopolyploidy (fig. 2.8), which results from the spontaneous doubling of chromosome number in a germline cell to produce tetraploid cells. A species could arise through autopolyploidy when the sperm and eggs of members of a population with accidentally doubled chromosome numbers interbreed and produce tetraploid offspring. When these offspring mate with one another, they can produce fertile tetraploid progeny that are genetically isolated from the rest of the population. Thus, in a single generation autopolyploidy can establish a barrier to gene flow between a fledgling species and its parent species. Autopolyploidy can result from either meiotic or mitotic nondisjunction. Meiotic nondisjunction is the failure of homologous chromosomes to move apart during meiotic cell division. Mitotic nondisjunction is the failure of sister chromatids within chromosomes to move apart during mitotic cell division. In the case of meiotic nondisjunction, an autopolyploid species

Meiotic nondisjunction

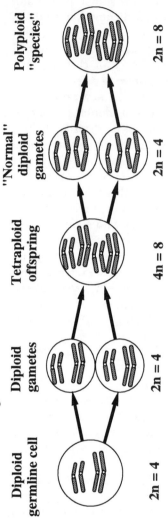

Diploid germline cell · Diploid gametes · Tetraploid offspring · "Normal" diploid gametes · Polyploid "species"

2n = 4 · 2n = 4 · 4n = 8 · 2n = 4 · 2n = 8

Mitotic nondisjunction

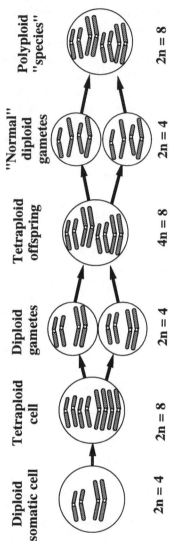

Diploid somatic cell · Tetraploid cell · Diploid gametes · Tetraploid offspring · "Normal" diploid gametes · Polyploid "species"

2n = 4 · 2n = 8 · 2n = 4 · 4n = 8 · 2n = 4 · 2n = 8

Figure 2.8. Two ways autopolyploidy can occur. Autopolyploidy is the spontaneous doubling of the chromosomes in a cell to produce a tetraploid cell. Autopolyploidy can occur as a result of meiotic nondisjunction, which produces gametes from a diploid germline cell (upper diagram), or mitotic nondisjunction, which produces a tetraploid cell that can undergo normal meiosis (lower diagram). In both cases the offspring have an unreduced chromosome number that is twice that of the parent species.

may evolve when normal diploid germline cells begin to divide meiotically but homologous chromosomes fail to separate. The resulting diploid gametes could fuse to produce a tetraploid zygote that would develop into a self-fertile tetraploid plant. Likewise, speciation through autopolyploidy could result when somatic cells within a diploid species become tetraploid because sister chromatids fail to separate during mitosis (mitotic nondisjunction). After meiosis, the resulting diploid gametes could fuse to produce a tetraploid zygote that develops into a self-fertile tetraploid. Speciation by means of autopolyploidy was first discovered by Hugo De Vries while he was examining the genetics of the evening primrose. The parent species, *Oenothera lamarckiana,* has fourteen chromosomes; the autopolyploid, *O. gigas,* which cannot backcross successfully with *O. lamarckiana,* has twenty-eight. Many commercially important crop plants are autotetraploids (e.g., tobacco and horticultural strains of snapdragon).

The second way a sympatric species can evolve through polyploidy is by allopolyploidy, which has occurred frequently among ferns and the "fern allies" as well as flowering plants (e.g., the lycopod *Selaginella* [see Webster 1990] and the fern genus *Polystichum* [see Barrington 1990]). An allopolyploid results when two species successfully mate to produce tetraploid offspring containing the chromosomes of both parents (fig. 2.9). These interspecific tetraploid offspring would be sterile whenever the haploid set of chromosomes from one species excludes the possibility of meiosis to form functional gametes, which can happen whenever the tetraploid has an odd number of chromosomes (i.e., not all chromosomes have homologues) or when homologues do not exist among even numbers of chromosomes. Two mechanisms permit the evolution of a viable species. The first possibility is that mitotic nondisjunction occurs in the germline tissues of the sterile diploid hybrid to produce unreduced polyploid gametes. This mechanism was the route to the new species observed by Karpechenko. Another example is seen in *Triticale,* a polyploid organism resulting from a cross between wheat (*Triticum,* 2n = 42) and rye (*Secale,* 2n = 14) combining the high yields of wheat with the hardiness of rye. Commercially important species have been produced in this way from three *Brassica* species. Known as the *Brassica* "species triangle," the allopolyploids and their parent species are as follows: *B. oleraceae* (2n = 18) and *B. nigra* (2n = 16), which give rise to *B. carinata* (2n = 34, the Abyssinian mustard); *B. oleraceae* (2n = 18) and *B. campestris* (2n = 20), which give rise to *B. napas* (2n = 38, the oil rape and the rutabaga), and *B. campestris* (2n = 20) and *B. nigra* (2n = 16), which give rise to *B. juncea* (2n = 36, the leaf mustard).

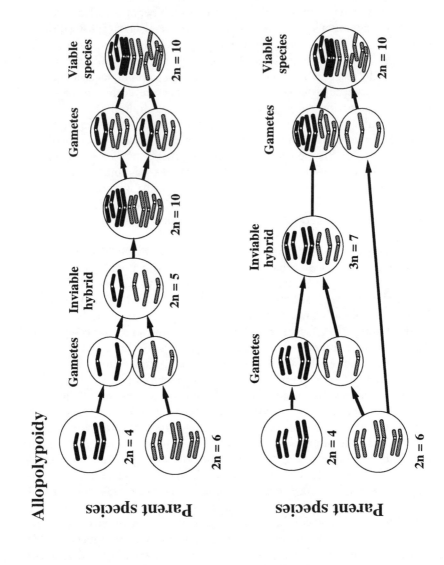

Allopolypoidy

The second way allopolyploidy can occur, which may be more common than the first, is when meiotic nondisjunction in one of two parental diploid species produces an unreduced "diploid" gamete that fuses with a normal gamete from the second parent species (see fig. 2.9). Even though the resulting hybrid is sterile, meiotic nondisjunction can produce unreduced triploid gametes that, when backcrossed with one of the parent species, produce viable offspring. The allopolyploid species resulting from these offspring has a chromosome number equal to the sum of the chromosome numbers of the two parent species. It is estimated that between 25% and 50% of all plant species have evolved from allopolyploidy. Many of these polyploid species are commercially important crops (e.g., oats, cotton, and potatoes).

The birth of a new species through polyploidy does not require a geographic barrier, so it can occur within the home range of the parent species. Thus speciation by means of polyploidy provides an example of how sympatric speciation may occur. "Sympatric" literally means "home range." That polyploidy is comparatively rare among animals yet comparatively common among plants suggests why some zoologists believe allopatric speciation is the most prevalent mode, and why many botanists believe the reverse is true. Although both modes of speciation theoretically are equally likely, from a practical point of view the issue revolves around the "relative frequency" of the two modes. This can be judged only within the context of the type of organism being studied. That plants differ from animals in many important respects is an undeniable fact of life. That plants likely have also evolved and speciated differently from most animals may also be true. But in terms of "relative frequency," plants constitute over 90% of the world's present (and past) biomass, so it is possible that sympatric speciation, on the average, is more common than allopatric speciation when plants and animals are viewed together.

Regardless of whether new species predominantly evolve sympatrically or allopatrically, the role played by random forces in organic evolution is readily apparent in both models. Geographic barriers that splinter populations and that preface allopatric speciation are created randomly. Likewise, polyploids or single-point mutations (see below) arise not as a consequence of natural selection but rather from genetic accident. But the

Figure 2.9. Two ways allopolyploidy can give rise to a new species. In both cases, two species hybridize. The hybrid may be reproductively sterile when the parent species do not share the same chromosome number. However, the chromosome number in the hybrid cells may spontaneously double due to meiotic nondisjunction (see fig. 2.8) to produce a viable species (upper diagram), or an unreduced gamete of an inviable hybrid may backcross with one of the parent species to produce viable offspring (lower diagram).

role played by natural selection is also readily apparent. Novel individuals can increase to form populations of incipient species only if selection forces do not extirpate them. Random forces can engender unique genotypes and phenotypes, but these novelties are fixed in populations only if they are at least as fit as their competitors. Equally important is that random forces—geographic isolation, mutations, and so forth—alter the fitness landscape envisaged by Sewall Wright. These forces disrupt the original frequencies of alleles in populations, thereby changing the frequency distributions of phenotypes. The appearance of any genetic novelty automatically alters the biological environment and therefore the fitness landscape. Random forces can shift the location and relative height of adaptive peaks and change their number. Once these alterations to the fitness landscape occur, selection forces come into play, tending to push the allele frequencies and therefore the prevailing phenotypes of populations upward onto adaptive peaks.

Portraits of Speciation

The evolutionary process of speciation has been portrayed in two contrasting ways. The first is by the gradual accumulation of selectively favorable or neutral genetic variation in spatially isolated populations, leading to genetically incompatible groups of organisms that, although derived from the same original gene pool, are distinguishable as separate species. The second portrait is by a rapid evolution inspired more by the random processes of mutation and genetic drift than by natural selection. The first portrait of the tempo and mode of speciation, which posits the gradual divergence between ancestor and descendant species owing to selection pressures, is called *phyletic gradualism*. This is the traditional Darwinian view. The alternative portrait envisions speciation as a rapid transition between ancestor and descendant owing to the random processes of mutation and genetic drift. When cast in the context of a genetic model, this view of speciation has been called *genetic revolution*. In the context of paleontological patterns of speciation, it may result in what is called *punctuated equilibrium* (see below).

I discuss the two contrasting views of speciation again in chapter 8, where I deal with the topic of extinction. Here speciation will be visualized with the aid of Sewall Wright's fitness landscape. The genetic model for phyletic gradualism depicts a slow transition of an ancestral phenotype to a new form as a series of steps propelled by the combined effects of random and nonrandom evolutionary forces operating in a gradually changing environment (fig. 2.10). Nonetheless, the force of selection is balanced by mutation and other processes of random genetic change.

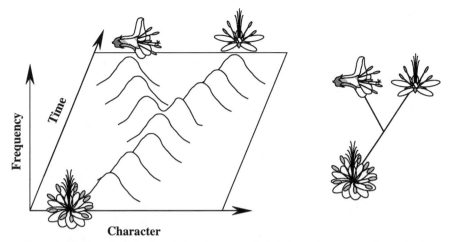

Figure 2.10. Visualization of speciation by means of phyletic gradualism on a fitness landscape. Over time, environmental heterogeneity and natural selection result in genetic and phenotypic divergence between populations sharing the same ancestral gene pool. Progressive divergence gradually results in new species (left diagram). Compare with figure 2.15.

Natural selection will tend to move the evolving population toward fitness peaks—that is, to regions of optimum fitness at a given time and place. Random genetic changes may by chance also increase the mean fitness of the evolving population, or they may by chance reduce it, causing the population to descend to a region of lower mean fitness. The model of phyletic gradualism sees natural selection as the principal evolutionary force directing the walk over the fitness landscape. As the gene pool of the population adaptively changes, the population may diverge into new genotypic and phenotypic entities, each associated with its own fitness peak. Isolated populations would continue to differentiate as a result of the selection pressures acting on each population in its particular environment. Even if the populations could interbreed at their borders, the progeny would likely be less fit than either parent population and thus would fail to become established, because these borders would occur in regions of lower mean fitness. Thus the evolving populations would continue to differentiate genetically and phenotypically owing to sexual or geographic isolation and might assume the status of separate species.

Sewall Wright showed that as a consequence of genetic drift a small population occupying an adaptive peak may diverge into two or more populations occupying different fitness peaks in the *same* fitness landscape. Because the movement down an adaptive peak was considered an arduous process for a very large population (owing to the "struggle" against natural selection and the negligible effects of genetic drift), genetic revolutions were though to be the only reasonable vehicle for rapid speciation in very large populations. But mathematical models derived from conventional population genetics show that even a moderately large population may move quickly from one fitness peak to another. These models show that the decrease in the mean fitness attending the downhill movement may be so rapid that natural selection has no opportunity to act on the population (Lande 1985; Newman, Cohen, and Kipnis 1985). And as the population reaches the foothills of a fitness peak, selection pressures can quickly carry the population uphill, where it increases in size (fig. 2.11). Because large populations are expected to remain phenotypically stable for long periods owing to the inability of random genetic changes to overcome the pull of natural selection, the rapid transit of a moderate-sized population from one fitness peak to another is expected to be followed by little or no phenotypic change unless the environment changes.

The portrait of speciation by means of a genetic revolution posits that genic or chromosomal variations can be rapidly fixed in very small populations by the combined effects of natural selection and random fluctuations in gene frequencies, and that the low fitness of heterozygotes prevents significant gene flow among such small populations. In small populations, random genetic changes may become rapidly fixed provided by chance they confer higher fitness. The result is that a small population may abruptly shift from one adaptive peak to another. The isolation of subpopulations occupying different fitness peaks would favor continued phenotypic divergence that in turn could produce one or more new species. One of the first renditions of speciation by genetic revolution was the peripatric speciation model of Ernst Mayr (1942). According to this model, small isolated subpopulations may be founded by a few individuals at the periphery of the main population of a species. These small subpopulations undergo rapid peak shifts on the fitness landscape owing to random changes in gene frequencies. Because each small subpopulation occupies a geographic area subject to selection pressures that differ from those acting on other subpopulations or the main population, extensive and diverse modifications of the subpopulation gene pools may occur. These modifications may lead to genetic incompatibilities between subpopulations and the main population that could preface speciation.

Ernst Mayr's peripatric speciation model is sometimes called the "founder effect" because it emphasizes the role played by a small number of individuals, which can lay the genetic foundation for a new species. First, genetic drift destabilizes adaptive gene complexes in small, geographically peripheral subpopulations, and selection pressures subsequently

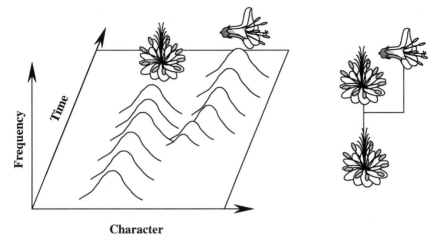

Character

Figure 2.11. Visualization of speciation by means of an abrupt speciation event on a fitness landscape. Even a moderately large population residing on an adaptive peak can suddenly shift to another adaptive peak created when the environment changes suddenly. The transition from one peak to another is so swift that natural selection has no time to act on the temporary drop in the relative fitness of the population. Compare with figure 2.10.

stabilize randomly inspired genetic differences to accord with different environmental conditions. As originally proposed, the peripatric speciation model presupposes a dramatic loss of genetic variance in founders. Because isolated individuals would contain only a small fraction of the total genetic variance of their main population, small populations of founders were assumed to be at risk. However, Russell Lande (1985) showed that the loss of variance in a founder population is equal to one-half the effective population size, that is, $1/2N_e$. Thus, with as few as four pairs of founders ($N_e = 16$), as much as 94% of the original variance would be retained. Nevertheless, population genetics theory suggests that small founder populations may rapidly accumulate slightly deleterious mutations that may eventually cause extinction (Kondrashov 1995).

The peripatric speciation model also assumes that the initial changes in the gene frequencies of a founder population would lead to simultaneous changes in many loci, with cascading effects on the phenotype. More recent speciation models based on genetic revolutions have abandoned this assumption. For example, Alan Templeton (1980) suggests that changes in the allele frequencies of a few founders resulting from random genetic drift may affect only one or at most a few loci that have major phenotypic effects. Once these gene frequencies become stabilized in small subpopulations, natural selection is expected to act on other alleles of interactive loci, modifing phenotypic traits and adapting them to local environmental conditions. Thus the mutation of one or two genes may produce reproductive barriers between ancestral and descendant populations provided the phenotypic effects of the mutation are stabilized by modifier alleles. The speciation model proposed by Hampton Carson (1983, 1987) involves cycles of severe reduction and rapid increase in population size, that is, population "bottlenecks and flushes" (fig. 2.12). According to this model, some blocks of strongly interacting genes will resist change even if other loci change readily as a result of natural selection. Because genetic recombinants of the alleles of these "selection resistant" loci will have lower fitness, the ongoing population is predicted to retain its adaptive gene frequencies. During a bottleneck and flush cycle, however, the selection resistant blocks of loci may become destabilized under conditions of relaxed selection pressure, and even recombinants with low fitness may significantly increase in number. With the onset of another bottleneck, intense selection would act on these variants, picking out those best adapted to the new environmental conditions. The cycles of population flushes and bottlenecks leading to the genetic and phenotypic differentiation of successive subpopulations engendered by the ancestral gene pool has the potential to evoke new species.

The requisite conditions for genetic revolution are frequently found on active volcanic islands, where repeated lava flows can fragment large populations into many smaller ones, some of them essentially biotic islands in a sea of cooling lava. The surviving individuals on each of these biotic islands may subsequently differentiate genetically as their population size increases and as they colonize unoccupied territories until new lava flows reiterate the cycle of population size "bottleneck and flush." On the island of Hawaii, major lava flows occur roughly every 250 years—an extremely short interval in geological or evolutionary terms. Clearly, another factor contributing to genetic revolutions is the time it takes an organism to complete its life cycle, which in turn is related to body size. As noted, the rate at which allelic frequencies change in a population is influenced by the generation time, which tends to decrease with decreasing body size. The Hawaiian islands are home to many small plant and animal species that complete their life cycles comparatively quickly. The suggestion that speciation occurs rapidly for isolated populations of small, rapidly growing organisms living in chronically disturbed or very stressful environments is supported by the fact that the Hawaiian islands have the highest number of endemic plant and animal species known for any geographic area on Earth.

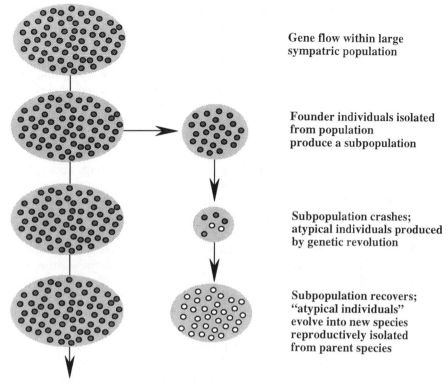

Gene flow within large
sympatric population

Founder individuals isolated
from population
produce a subpopulation

Subpopulation crashes;
atypical individuals produced
by genetic revolution

Subpopulation recovers;
"atypical individuals"
evolve into new species
reproductively isolated
from parent species

Figure 2.12. Speciation by means of a genetic revolution. A few "founder" individuals establish a subpopulation isolated from the parent sympatric population. The subpopulation "crashes," and a few atypical individuals serve to reestablish a subpopulation that is reproductively isolated from the parent species. Compare with figure 2.1.

Monogenic Speciation

The genetic speciation models of Mayr, Templeton, and Carson are all grounded on the founder effect. In contrast to phyletic gradualism, each model envisions speciation as a rapid process initially inspired by random genetic events that offer an opportunity to shift rapidly from one fitness peak to others in the same fitness landscape. Although each model accords natural selection an initially unimportant role, selection pressures are nevertheless required to stabilize the phenotypes of populations occupying different peaks. In this sense the provocateur of speciation is the sudden and dramatic departure from the prior genetic state of affairs—a genetic revolution of some sort. Clearly, however, genetic revolutions are not all equal. Some may result from many concurrent genetic

alterations; others may come about from as little as a single genic mutation. I have reviewed examples of the former kind of genetic revolution.
Here I shall give attention to single-gene or "monogenic" mutations that
have the potential to inspire speciation.

The evidence for simple one-step mutations that engender significant
mechanical or temporal differences in reproductive organs is overwhelm-
ing, particularly for flowering plants (Bachmann 1983; Garcia-Bellido
1983; Hilu 1983; Gottlieb 1984). Among the outstanding examples are
simple one-step mutations that alter the number, position, symmetry, and
fusion of floral parts and even the sexuality of flowers. Because systematists often use such features to characterize taxonomic groups above the
level of the genus, these mutations could conceivably account for the sudden appearance of floral structures so novel that taxonomists would be
compelled to assign them to new genera or even higher taxa (families,
orders, and so forth).

For example, flowers lacking petals (apetalous flowers) are typically
wind pollinated or self-pollinated, while flowers with numerous petals
and other floral parts (polypetalous) are generally pollinated by animals.
Single-gene mutations resulting in apetalous, fertile flowers have been reported for mountain laurel (*Kalmia latifolia*), evening primrose (*Oenothera parodiana*), tobacco (*Nicotiana tabacum*), and a variety of annual
chrysanthemum species, all of which normally have large-petaled flowers
that are insect pollinated. Conversely, monogenic mutations resulting in
flowers with supernumerary petals occur in the mountain laurel, gera-
nium (*Pelargonium hortorum*), soybean (*Glycine max*), gloxinia (*Sin-
ningia speciosa*), garden nasturtium (*Tropaeolum majus*), and petunia
(*Petunia hybrida*). In the case of the soybean mutant, the total number of
floral parts in mutant flowers is the same as that of the normal flower. An
X ray–induced mutation of soybean studied by Singh and Jha (1978) resulted in flowers with two or more carpels instead of the single carpel
typical of the bean family (Leguminosae). Flowers with supernumerary
carpels, which developed normal fruits, were the result of a single recessive gene. Although atypical for the bean family, the flowers of *Swartzia
ignifolia* and *S. littlei* typically have two or three ovaries. In theory, the
single-gene mutation observed for soybean could explain the multi-
carpellate condition of these two "anomalous" species.

The number of flowers in an inflorescence, which can distinguish related species and genera, can also be influenced by single-gene mutations.
The corn inflorescence (*Zea mays* subsp. *mays*) differs from that of its
wild ancestor teosinte (*Z. mays* subsp. *mexicana*) in a number of ways.
But one of the most important differences is that the female spikelets of
corn are paired, whereas those of teosinte are single. This phenotypic

trait is controlled by a single gene whose mutation could account for the difference between these two subspecies.

Monogenic mutations may establish mechanical isolation among subpopulations by changing floral symmetry, thereby potentially altering the mode of pollination. A simple mutation alters the bilaterally symmetrical flower of the insect-pollinated snapdragon (*Antirrhinum*) into an almost radially symmetrical flower similar in appearance to that of the mullein (*Verbascum*), which is in the same family, while another mutation in the snapdragon confers a spurred petal similar to that on the flowers of the butter-and-egg plant (*Linaria vulgaris*), also in the same family (see Hilu 1983). Species within the family Asteraceae are distinguished in part by whether their inflorescences contain radially symmetrical "disk" flowers, bilaterally symmetrical "ray" flowers, or both. Yet by performing artificial crosses between two composite species, *Haplopappus aureus* and *H. venetus* subsp. *venetus*, which have rayed and rayless florets, respectively, Jackson and Dimas (1981) found that the presence and absence of ray florets is controlled by a single gene. Thus a single mutation may determine the difference between these two species.

Evidence also suggests that single-gene mutations can establish or remove temporal or mechanical barriers to sexual reproduction. For some flowering plant species, a temporal barrier to hybridization can be created by a change in day length or temperature. Examples of monogenically inherited effects on the time of flowering are the difference between day-neutral and day-sensitive forms of *Salvia splendens,* the formation of flowering versus vegetative branches in the peanut (*Arachis hypogaea*), and early flowering in the sweet pea (*Lathyrus odoratus*). In terms of mechanical barriers to hybridization, Honma and Bukovac (1966) showed that gibberellin-induced heterostyly in tomatoes is governed by a single gene. (Heterostyly is the formation of flowers differing in the lengths of their styles so that flowers with one style length are predominantly pollinated with pollen from a flower with a different style length.) Another mechanical barrier to hybridization is the formation of flowers that never open and therefore are self-pollinated (cleistogamy). In the cucumber, a single recessive gene mutation results in cleistogamy, as it does in a mutant of the soybean. Finally, genetic analyses of the sexual expression of flowers and the sex ratios of populations indicate that these phenomena are often controlled by one or only a few genes, although their effects are modulated by other modifier genes and environmental factors (e.g., light intensity, day length, temperature, soil conditions, and plant health).

Mutations affecting floral development have received a great deal of attention because alterations in the pattern formation of reproductive organs provide insights into the evolution of new species from old ones. A bewildering array of mutations is now known to alter the pattern of flower formation. Nevertheless, relatively few genes appear to specify the type of floral organ that will normally develop. Mutations in most of these genes cause homeotic transformations. Homeotic loci are believed to contain the genetic information required to shunt developing cells along a particular pathway. Thus homeotic mutations switch the development of cells from one fate to another and may replace one type of organ with another that is perfectly normal in every respect except its location in the overall body plan. Some of the best understood floral homeotic mutations occur in the mouse-ear cress, *Arabidopsis*, and the snapdragon, *Antirrhinum*. Like many other angiosperms, these two plants have four concentric whorls of floral organs, of which the outermost develops into sepals and the innermost develops into carpels. Homeotic mutations result in dramatic departures from the normal pattern of flower development. Most of these mutations alter the type rather than the number of organs formed, suggesting that the pattern is regulated at several levels, with some genes regulating organ identity and others organ number. For example, mutations of the *AP3* or *PI* genes of *Arabidopsis* or the *DEF* gene of snapdragon cause petals to be replaced by sepals, and stamens by carpels (Koornneef et al. 1983; Bowman, Smyth, and Meyerowitz 1989; Sommer et al. 1990). Mutations in the *AG* gene of *Arabidopsis* and the *PLENI* gene of the snapdragon convert stamens into petals and carpels into sepals (Carpenter and Coen 1990). Because gene flow within populations of flowering plant species may be maintained by the ability of flowers to attract specific pollinators, changes in the appearance of flowers may alter the kinds of pollinator species visiting them. Thus, because they have the potential to establish reproductive barriers, homeotic mutations may serve as a mechanism for character displacement, genetic divergence, and subsequent speciation.

Naturally, a balanced view must be maintained regarding the role of monogenic mutations. Although these mutations have the *potential* to engender novelties that could evolve into new species, there is little evidence that the monogenic mutations previously reviewed *did*, whereas there is ample evidence that floral characters can change, sometimes very rapidly, in response to environmental changes. For example, Candace Galen (1996) showed that plant populations of the alpine wildflower *Polemonium viscosum* rapidly adapt to abrupt changes in pollinator assemblages. Her data indicate that the broadly flared flowers of the bumblebee-pollinated *P. viscosum* could have evolved from narrower ones in a single generation because corolla flare increased by 12% from populations pollinated by a wide assemblage of insect pollinators to those pollinated only by bumblebees.

Regulatory Genes and Heterochrony

The development of plants as well as animals depends not only on structural genes encoding for proteins, but also on regulatory genes that coordinate and direct the activities of hundreds of structural genes. Regulatory genes encode for proteins that either activate or repress the ability of other genes to express themselves. Regulatory genes are usually constitutive; that is, they produce repressor proteins continuously but at comparatively slow rates. Even though these repressor proteins are always present, they do not always block the transcription of their targeted genes, because the protein's ability to function depends on the chemical environment within the cell, which in turn is indirectly self-regulated by the activity of the genome as a whole as well as subject to changes in the external environment. Regulatory and structural genes therefore operate in vast, complex feedback systems that receive and respond to physiological cues generated from within and without the organism.

Networks of regulatory genes help to directly coordinate the activities of the genome, so they directly or indirectly influence cell metabolism, growth, and development. Their role may be likened to the fingers of a violinist. Virtually every piece of music ever written can be played on a violin, even though this instrument can play only a limited number of musical notes. The versatility of the violin comes from the musician's ability to vary the sequence, duration, and volume of the notes played. In like manner, the genome of any organism consists of a finite number of genes, each encoding for a very particular gene product. By varying the sequence, duration, and activity of individual genes, regulatory genes control the combinations of gene products simultaneously existing in a cell. The same gene product thus can play many different roles depending on its physiological or developmental context. Naturally, regulatory genes are themselves influenced by this context, and so the music of the cell is a fugue that turns in upon itself in complex and strangely beautiful ways.

Advances in molecular genetics have provided some insight into how some regulatory gene networks influence growth and development. One network has recently been postulated to control the time of transition from vegetative to reproductive growth and the development of floral organs in the mouse-ear cress *Arabidopsis thaliana* (Yang, Chen, and Sung 1995). Two genes, called the *EMBRYONIC FLOWERING* or *EMF* loci, figure prominently in this network (fig. 2.13). In wild type plants, the activity of *EMF* genes gradually declines in the normal course of vegetative development. Once the activity level drops below a critical threshold, the shoot apical meristem undergoes a transition from vegetative to reproductive growth. Other genes, such as the *EARLY FLOWERING* and *CONSTANS* (*ELF*

Developmental stages of wild type *Arabidopsis*

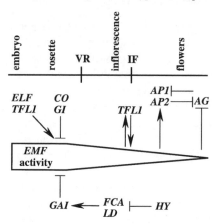

Fig. 2.13. A model proposed for a gene network influencing the transition time from vegetative to reproductive growth (VR) and the time of initiation of flower development (IF) in *Arabadopsis thaliana*. The developmental stages for the wild type are shown from the time the embryo germinates to form a rosette to the formation of mature flowers. The decrease in the activity level of *EMF* gene loci is indicated by the thickness of the wedge-shaped trapezoid. Gene loci repressing other genes are indicated by ⊣; gene loci promoting other genes are indicated by →. External environmental factors known to influence gene activity are not indicated. Gene loci names are as follows: *EMF* = EMBRYONIC FLOWERING, *ELF* = EARLY FLOWERING, *TFL* = TERMINAL FLOWERING, *CO* = CONSTANS, *GI* = GIGANTEA, *AP* = APETALLA; *AG* = AGAMOUS, *GAI* = GIBBERELLIC ACID INSENSITIVE, *FCA* = FLOWERING, *LD* = LUMINIDEPENDENS; *HY* = LONG HYPOCOTYL. Adapted from Yang, Chen, and Sung 1995 and Haughn, Schultz, and Martinez-Zapater 1995.

and *CO*) genes, are believed to regulate the rate at which the *EMF* products decline, either extending or reducing the transition time from vegetative to reproductive growth. These genes, together with *EMF* genes, provide a feedback loop (see Haughn, Schultz, and Martinez-Zapater 1995).

Rendering this gene network as shown in figure 2.13 is misleading. The linear sequence of "cause and effect" with focus on the *EMF* genes obscures the notion of a network of interacting genes wherein the activity of no particular gene takes precedence because the role and influence of each gene can be defined only in the context of the whole system. Likewise, the number of genes involved in the transition from vegetative to reproductive growth and the formation of flowers in *Arabidopsis* or any other flowering plant is undoubtedly much greater. Indeed, the genes involved in the transition from vegetative to reproductive growth and the ultimate development of flowers most likely encompasses the entire genome. Thus figure 2.13 illustrates our current lack of understanding as much as our comprehension of an intrinsically complex facet of plant growth and development.

Someday our understanding of regulatory genes and the consequences of their mutation, such as those briefly described for *Arabidopsis thaliana,* may help us uncover the genetic mechanisms resulting in the phenomenon called heterochrony (De Beer 1958; Gould 1977; Raff and Wray 1989). Heterochrony is broadly defined as any evolutionary change resulting from any alteration in the sequence, duration, or timing of developmental events. Unfortunately this cavalier definition tends to invite unbridled speculation about the ability of heterochrony to engender phenotypic novelties that may preface transspecific evolution. Strictly speaking, heterochrony refers only to the morphological expression of adult form with respect to the presumed ancestral adult condition. Therefore it is a purely descriptive term lacking precise reference to the genetic or developmental mechanisms responsible for a phenotypic novelty. Consider that there are two general classes of heterochrony: (1) paedomorphosis (from the Greek *pais, paidos,* "child," and *morphe,* "form, shape"), which is the retention of juvenile (ancestral) features by adults of the species, and (2) peramorphosis (from the Greek *pera,* "beyond, across," or *peraiteros,* "farther"), which occurs when adult size or shape is developmentally extended beyond that of the ancestral condition. However, each of these two classes of morphological expression can be achieved by a number of categories of developmental modifications (table 2.3). For example, paedomorphosis can result from neoteny or progenesis. Neoteny occurs when somatic growth is delayed or retarded so that the reproductively mature adult retains juvenile (ancestral) features. Progenesis achieves the same result but by accelerating the rate of sexual development with respect to somatic (vegetative) development.

The *emf* mutants of *Arabidopsis thaliana* may show how progenesis produces paedomorphosis because they result in the formation of flower-like structures on embryos (Sung et al. 1992; Yang, Chen, and Sung 1995). Before seed germination, *emf* mutant embryos either produce a shoot apical meristem whose features are similar to those of the apex that gives rise to the inflorescence in mature wild type plants, or they produce a short-lived vegetative shoot apex that quickly changes to a reproductive shoot apex. Experiments verify that *emf* mutations can suppress the expression of a number of other regulatory genes such as the *AG* homeotic gene that converts stamens into petals in *Arabidopsis thaliana.* Because *emf* mutants are sexually mature when they are somatically (vegetatively) embryonic, they result in paedomorphosis via the developmental route of progenesis.

Table 2.3   Two Classes of the Morphological Expression of Heterochrony and Their Developmental Categories (Each Category Is Described in Terms Referring to the Presumed Ancestral Condition)

Class of Morphological Expression	Developmental Category	Alteration		Growth
		Sexual	Somatic	Rate Onset
Paedomorphosis				
	Neoteny	Same	Delayed	
	Progenesis	Accelerated	Same	
	Postdisplacement			Same Later
Peramorphosis				
	Acceleration	Same	Accelerated	
	Hypermorphosis	Delayed	Same	
	Predisplacement			Same Earlier

It is very tempting to speculate that gene mutations like *emf* may produce developmental "monsters," like flower-bearing embryos, that are nevertheless "hopeful" in the sense that they may have some selective advantage over their normal counterparts and so may ultimately evolve into a new form of life. Indeed, neoteny and progenesis have been implicated in the evolution of various plant and animal lineages. In some cases the evidence for the roles of these and other heterochronic phenomena is marginally compelling (Takhtajan 1958, 1959; Gould 1977; Hébant 1977). As a general rule, however, "developmental monsters" are anything but "hopeful"—most are incapable of surviving, let alone reproducing (fig. 2.14). Although *Arabidopsis emf* mutants have sexual characteristics when in the embryonic condition, they do not survive because they lack functional foliage leaves and their flowers are completely sterile.

Punctuated Equilibrium

The view that macroevolutionary phenomena cannot be explained by microevolutionary processes has inspired continued controversy, one expression being the theory of punctuated equilibrium, which emphasizes the pattern of speciation observed in the fossil record (Eldredge and Gould 1972). Its salient ingredients are few, even though the theory as originally cast was subsequently modified in a series of papers (Gould and Eldredge 1977; Gould 1981; Eldredge 1985).

The theory begins with the observation that typically species appear rapidly in the fossil record and do not subsequently change appreciably in appearance. These observations are taken at face value and are argued to be inconsistent with the leisurely and steady tempo of speciation and

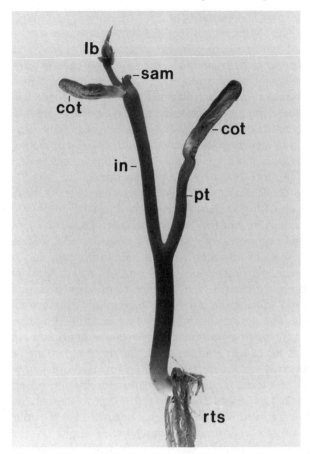

Figure 2.14. Anomalous seedling of common kidney bean (*Phaseolus vulgaris*) resulting from spontaneous genetic mutation. The two embryonic leaves, or cotyledons (cot), which are normally sessile to the stem and attached to the same node, are separated by an intervening stem internode (in). One cotyledon is supported by a petiole (pt) extending from the shoot axes. In the axil of the other, more apical cotyledon, an axillary bud has developed into a short lateral branch (lb). The shoot apical meristem (sam) has aborted, as has the apex of the embryonic root (which is obscured from view by numerous small roots, rts). This seedling "hopeful monster" never produced normal foliage and died after fifteen days of cultivation.

the gradual but persistent change in phenotypes predicted by phyletic gradualism. To explain the pattern of speciation seen in the fossil record, the theory of punctuated equilibrium argues that small local subpopulations split from the the range of their parent species, rapidly diverge genetically and phenotypically, become reproductively isolated as a consequence, and subsequently reenter the main range of the ancestral species, where they may successfully compete for similar resources and space (fig. 2.15). The similarity between the theory of punctuated equilibrium and Ernst Mayr's model of peripatric speciation is readily apparent.

Some aspects of the theory of punctuated equilibrium are consistent with experimental observation, as for example the experimental results of Richard Lenski and his colleagues. Working with the bacterium *Escherichia coli,* Lenski and Travisano (1994) grew populations founded by a single cell taken from a previous population subjected to nutrient starvation just before serial transfer to a fresh growth medium. Because each new population grew from a single cell, the only genetic variation that could be introduced between sequentially cultivated populations was due to chance mutation. One of the advantages of working with a bacterial population is that descendants could be grown alongside ancestors retrieved at will from frozen storage. Thus the extent to which descendants could compete with their ancestors could be studied. Experiments of this sort verify that bacterial cell morphology and fitness (measured in terms of competition with members of ancestral populations) evolve rapidly for the first 2,000 generations, then plateau and remain nearly "static" during the last 5,000 generations from a total of 10,000 generations. The time scale and environments studied by Lenski and his colleagues are microevolutionary in scope, and no new species were evolved. But the salient features of the theory of punctuated equilibrium were supported—the bacterial phenotype evolved quickly, was more fit than those of ancestors, and subsequently leveled off, becoming nearly static later on. Lenski and Travisano (1994) speculate that "the most reasonable interpretation for the eventual stasis . . . is that the organisms have 'run out of ways' to become much better adapted to their environment." In other words, the bacterial populations initially capitalized on and subsequently exhausted their genetic variation.

The experiments of Lenski and Travisano illustrate an obvious criticism of the theory of punctuated equilibrium: "sudden events" measured in the time scale of geological history are gradual events to the geneticist studying populations of living organisms. For example, the average duration of fossil flowering plant species in the geological record is about 3.5 million years, and new species appear about every 0.38 million years (Niklas, Tiffney, and Knoll 1983). Taking these numbers at face value, the average speciation event that can be adduced from the fossil record is roughly 10% of the average lifetime of a flowering plant species. This may appear rapid to a paleontologist, but the average generation time of a flowering plant is about five years. Thus, if speciation occurs every 0.38 million years, it involves roughly 76,000 *generations,* during which gradual changes in allele frequencies and corresponding changes in the

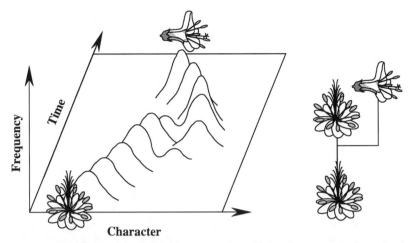

Character

Figure 2.15. Depiction of the theory of punctuated equilibrium by means of a fitness land-scape. A subpopulation splinters from the parent population, genotypically and phenotyp-ically diverges, and then reenters the home range of the parent species, where it success-fully competes for resources. Compare with figures 2.1 and 2.11.

prevailing phenotype of a population can occur (see fig. 1.8). This criti-cism is reinforced when we turn to estimated rates of speciation for iso-lated island chains such as the Hawaiian archipelago. Recall that many of the species endemic to these chains are single-island species; that is, they are found on no other island and thus could not have migrated from older to younger islands, thereby casting doubt on their point (and time) of ori-gin. Ann Sakai and her colleagues report the number of such single-island endemic species as well as the total number of species and the age of the islands on which these species are found (Sakai et al. 1995). Taking these data at face value, speciation on the Big Island, Hawaii, occurs every 0.42 million years, or roughly every 84,000 flowering plant generations.

Leaving aside the issue of geological versus genetic time scales, pheno-typic stasis and the absence of intermediate forms bridging descendant and ancestral species are easily explained in terms of the modern synthe-sis. Once a population reaches an adaptive peak on a fitness landscape it will be kept there by the force of natural selection, which discourages the downward walk to a fitness valley. The rarity of a series of transitional forms is expected because they will have lower fitness than either their ancestral or descendant forms occupying fitness peaks. Thus transitional forms will exist in relatively small populations for brief periods. This kind of organism will not be found routinely in the fossil record, which is biased in favor of preserving long-lived species existing in large popu-lations. Conversely, phenotypic stasis in the fossil record may in some instances be illusory. Because fossil species are typically distinguished by morphology and anatomy, there is something tautological about the claim that most morphological changes accompany speciation events because fossil species can be distinguished only by such changes. That morphological changes sometimes attend speciation is not in doubt. In fact, lists of isolating mechanisms (see table 2.1) invariably include "structural" or "mechanical" (morphological) barriers to sexual reproduction that may preface speciation (e.g., Dobzhansky 1951; Dobzhansky et al. 1977; Mayr 1963). But it is also true that physiological or behavioral traits can evolve undetected in the fossil record yet be of sufficient stature to create reproductive barriers among related organisms. Among the types of reproductive barriers that exist, the paleontologist can generally recognize the existence of a temporal reproductive barrier only when morphospecies occur in rock strata differing significantly in age, or the existence of a geographic reproductive barrier only when temporally contemporaneous morphospecies occur in rocks differing widely in location. In any circumstances, the coexistence in time and space of fossils sharing very much the same morphology and anatomy cannot always be taken as prima facie evidence that the fossils are all members of the same biological species.

The tempo of speciation in the fossil record unquestionably agrees with that predicted by the theory of punctuated equilibrium. The theory also complies with genetic models for speciation evoked by genetic revolutions, which show that chance genetic changes may evoke rapid speciation, at least as measured by paleontological time scales. Nevertheless, the highly charged debate over punctuated equilibrium versus phyletic gradualism unnecessarily polarizes the relative importance of random versus nonrandom evolutionary forces. Randomly generated variants would disappear from populations without the intervention of additional genetic changes (e.g., stabilizing modifier alleles) required to fix phenotypic traits in populations, and stabilized variants would neither proliferate nor eventually assume species status without the beneficial effects of natural selection. The critical question is the intensity of selection: Is it always strong, as the traditional Darwinian view holds, or is selection sometimes very weak, as the theory of punctuated equilibrium suggests? If the former, then every attribute of every species is an adaptation. If the latter, then new species are, on average, neither more nor less fit than their parent species—they are merely different and survive because their differences go undiscovered by environmental pressures. As of now there is no canonical answer to which of these evolutionary paradigms is the more general, and for this reason the theory of punctuated equilibrium remains a useful intellectual gadfly.

Part 2 Life's Chronicles: The Fossil Record

Plate 2. The genus *Cooksonia* constitutes the fossil remains of what may be the earliest bona fide vascular plant. Considering its comparatively simple shape, small size, and typically fragmentary state of presentation, however, some of the fossils that have been assigned to this genus may represent different kinds of plants, only some of which may have had true vascular tissues. This particular specimen, which is from the late Silurian Bertie Formation of Herkimer County, New York, was described in 1973 by Harlan P. Banks. Specimens of *Cooksonia* are also known from early Devonian sediments. The genus is characterized by dichotomizing, leafless axes bearing terminal ovoid sporangia containing spores with Y-shaped (trilete) markings indicative of meiotic cellular division. Based on sporangia containing these meiospores, *Cooksonia* is the fossil remains of the sporophyte generation.

3 Origins and Early Events

*The central problem of evolution . . . is that of a
mechanism by which the species may continually
find its way from lower to higher peaks.*

Sewall Wright

In contrast to the previous chapters, which dealt with
evolution at the level of population and species, this
chapter begins with an outline of plant history in terms of broad physiological or morphological innovations. This outline is difficult to sketch
because the complex history of plant life begins with the evolution of the
first living cells roughly 3.6 billion years ago. Nevertheless, eight events
in the history of plant life appear monumental because each has profoundly shaped the subsequent course of organic history. These events
are the evolution of the first forms of life, the origin and development of
photosynthetic cells, the appearance of cells with organelles, the acquisition of a multicellular construction, the ecological assault of plants on
land, the evolution of conducting tissues, the evolution of the seed habit,
and the rise of the flowering plants. These eight events are treated in this
chapter and the following one. The division of effort is somewhat arbitrary, since history is a continuous process. However, this chapter treats
the origin of life to the appearance of multicellular plant life; chapter 4
deals with the invasion of land by plants and concludes with the evolution of flowering plants.

Compared with the origins of life, subsequent historical events seem
pale. But each of these is of comparable importance to life's beginnings
because without them the world as we know it would not exist. Perhaps
the most obvious of these events is the evolution of photosynthesis, the
chemically intricate process that converts light energy into chemical energy. In one way or another, nearly all forms of life depend on the consequences of photosynthesis, either for food or for the recycling of atmospheric gases. Additionally, the evolution of photosynthetic organisms,
particularly those that release oxygen as a by-product, dramatically
changed the physical and biological environment of the ancient Earth.
Equally important to the history of all life was the evolution of cells

containing the microscopic membrane-bound objects called organelles, whose internal structure and chemical composition are highly specialized to participate in life's various biochemical and genetic processes. The absence or presence of organelles in cells divides all living things into two great organic camps: the prokaryotes, organisms whose cells lack organelles, and the eukaryotes, whose cells contain them. The prokaryotes comprise the bacteria; the eukaryotes encompass the forms of life most familiar to us—plants and animals. The bacteria are the more ancient and undoubtedly the more physiologically resilient and ecologically successful forms of life. Current evolutionary theory states that the first eukaryotes evolved from ancient loose symbiotic confederations of bacterialike forms of life. Thus the "great organismic wedge" between prokaryotes and eukaryotes was not as sharply defined in the distant past as it is now.

Organisms capable of sexual reproduction likely appeared collaterally with or shortly after the first eukaryotes evolved, although some believe a significant time gap exists between these two events. What can be said with certainty is that sexual reproduction required the evolution of meiosis, the "two-stage" type of cell division producing cells with half the chromosome number of the cells they arise from. Meiosis underwrites genetic recombination, whose evolutionary consequences were treated in chapter 1. Unlike the bacteria, which principally rely on chance mutations to introduce genetic variation into their populations, the eukaryotes can mix and blend the traits of different parents through sexual reproduction and meiosis. Although it is somewhat surprising, biologists continue to debate the importance of sexual reproduction to evolution, and there is no general agreement about the benefits conferred on organisms by genetic recombination. The bacteria continue to be the most successful forms of life without benefit of sexual reproduction, but genetic recombination continues to be an effective way of adapting organisms to new habitats and adopting new and different forms of biological organization. Sex may have evolved to deal with the decreasing mutation rates associated with increasing organismic size. Larger organisms reproduce at a slower rate than smaller organisms. Arguably, sexual reproduction may have been required to engender sufficient genetic variation as rates of mutation declined in proportion to the increasing body size of eukaryotes. What can be said without doubt is that the appearance of unicellular eukaryotes presaged multicellular organisms and the evolution of the most visibly conspicuous forms of aquatic and terrestrial plants.

The following is an all too brief exploration of the early history of plants. Here the goal is to paint the history of life with broad brushstrokes against the backdrop of geological time (fig. 3.1).

Life's Beginnings

Precisely how life began is still a mystery, particularly since the prevailing conditions of the early Earth were inimical to life as we know it today. The embryonic Earth was a protoplanet devoid of atmosphere, cracked by volcanic upheaval and lightning, flooded by lethal ultraviolet light, and rained on by the rocky debris of a juvenile solar system still actively in the process of birth—a world whose original surface was molten rock and whose first atmosphere, largely composed of hot hydrogen gas, was vented into space almost as quickly as it emerged from the Earth's radio-active core. Although hard to conceive of, these were the conditions when Earth formed roughly 4.6 billion years ago (fig. 3.1). Geologists call Earth's first age the "Hadean," reflecting the netherworld status of our juvenile planet. Nonetheless, with the passage of about 600 million years, Earth's version of Hades cooled enough to retain the first oceans formed from torrential rains when water condensed in the atmosphere. Although possibly diminished, sources of energy like lightning, volcanoes, and ultraviolet light continued to inspire molecular upheaval, breaking and forming molecular bonds in the atmosphere and the shallow oceans. Based on the gases currently vented from volcanoes, Earth's first atmosphere likely consisted of a mixture of water vapor, carbon dioxide, carbon monoxide, nitrogen, methane, and ammonia, but its precise composition is still debated (Cairns-Smith 1982; Schopf 1983). Most textbooks categorically state that Earth's first atmosphere was a strongly reducing one, in large part because the first successful laboratory experiments to abiotically synthesize amino acids used a strongly reducing atmosphere (see below) and because the standard free energies required to abiotically synthesize simple organic molecules like amino acids from raw ingredients such as carbon dioxide and ammonia are much lower in a strongly reducing atmosphere (e.g., methane, ammonia, water vapor, and hydrogen) than in a strongly oxidizing one (e.g., carbon dioxide, water vapor, nitrogen, and oxygen). However, recent geochemical and astronomical evidence suggests that the primitive atmosphere was not as reducing as formerly believed but may have been very much like that of today without, naturally, the effects of life (an atmosphere dominated by N_2 and CO_2, with a small ration of noble gases along with some H_2 and CO).

For convenience, the continuous evolution of early life has been sorted into four stages: (1) the accumulation of abiotically (nonliving) synthesized small organic molecules (monomers) such as amino acids and nucleotides; (2) the abiotic joining of these monomeric chemicals into polymers such as proteins, lipids, and nucleic acids; (3) the origin of the heredity molecules ribonucleic acid (RNA) and deoxyribonucleic acid

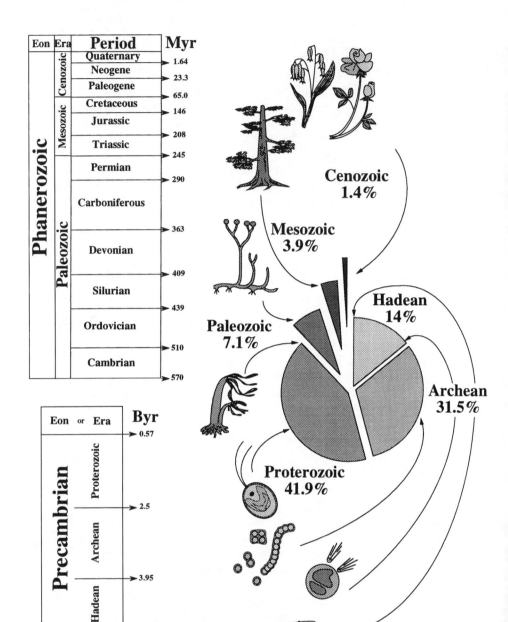

Eon	Era	Period	Myr
Phanerozoic	Cenozoic	Quaternary	1.64
		Neogene	23.3
		Paleogene	65.0
	Mesozoic	Cretaceous	146
		Jurassic	208
		Triassic	245
	Paleozoic	Permian	290
		Carboniferous	363
		Devonian	409
		Silurian	439
		Ordovician	510
		Cambrian	570

Eon	or	Era	Byr
Precambrian		Proterozoic	0.57
			2.5
		Archean	3.95
		Hadean	4.57

Cenozoic 1.4%

Mesozoic 3.9%

Paleozoic 7.1%

Hadean 14%

Archean 31.5%

Proterozoic 41.9%

(DNA); and (4) the aggregation of abiotically synthesized molecules into cell-like membrane-bound droplets (protobionts) that had an internal chemical environment differing from the external chemical environment.

Of these four stages, the first is the most easily simulated in the laboratory. Organic monomers can be abiotically synthesized with comparative ease using a variety of energy sources and a range of atmospheric compositions believed to mimic early Earth conditions. This simulation was first widely publicized with the landmark experiments of Stanley Miller and Harold Urey (1953, 1959), who created a strongly reducing "mock" atmosphere of water vapor, hydrogen gas, methane, and ammonia (probably too strongly reducing, according to recent thinking) suspended over a mock ocean of heated water (fig. 3.2). After continuously subjecting the atmosphere to an electrical discharge (simulating lightning), Miller and Urey chemically analyzed their miniature ocean and found a variety of organic compounds. The most abundant of these was glycine, a common amino acid. Additionally, many of the other standard twenty amino acids were identified, although in lesser amounts than glycine. Subsequent experiments using different recipes for Earth's ancient atmosphere have abiotically synthesized several sugars and lipids as well as the compounds generated in the original Miller-Urey experiment. Equally (if not more) important was the discovery that the purine and pyrimidine bases found in RNA and DNA could also be synthesized, often under comparatively mild chemical conditions. Also, when phosphate was added to a mock ocean, adenosine triphosphate (ATP) could form spontaneously. ATP is the major energy-transferring molecule in living cells. The demonstration of the abiotic synthesis of ATP under conditions mimicking those of early Earth is important to any theory of the origin of life.

Regarding the second stage, Sidney Fox has been able to abiotically synthesize polypeptides, called proteinoids, by heating amino acid mixtures on hot sand, clay, or rocks and subsequently flooding them with water. Polypeptides and other organic polymers may have been synthesized during the early history of the Earth when mixtures of monomers carried by ocean waves were repeatedly dehydrated and rehydrated on the surfaces of a cooling volcanic shoreline. The abiotic synthesis of complex polymers may also have occurred on the surface of cool clay, which

Figure 3.1. Chronology of some major adaptive events in the history of plant life. The geological column (shown to the left) consists of two eons, the Precambrian and the Phanerozoic (or botanically speaking, the Phanerophytic), each further subdivided into eras, which in turn are more finely subdivided into geological periods. The geological column is rendered as a pie diagram showing the percentage of total time occupied by the eras.

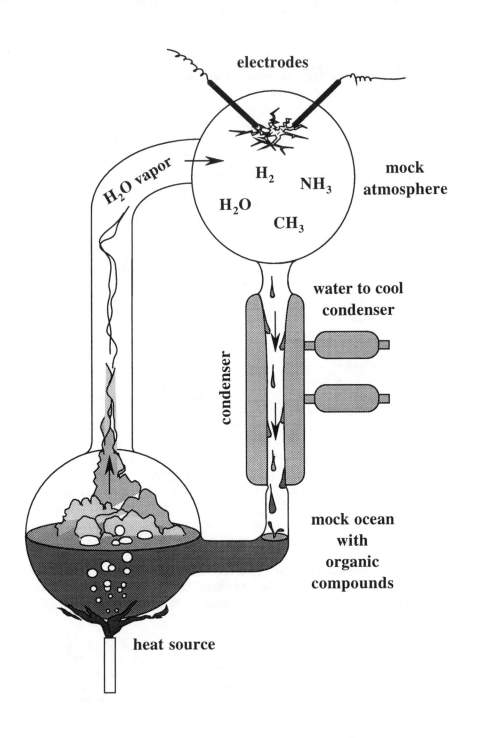

has the capacity to concentrate amino acids and other monomers from dilute aqueous solutions. Metal atoms, such as iron and zinc, found in a variety of clays are known to expedite dehydration reactions that precede the polymerization of monomers into larger, more complex chemicals (Cairns-Smith 1982, 1985). More recent experiments show that clays can also mediate the abiotic synthesis of RNA from its constituent monomers.

Clearly, from a genetic perspective the most complex and important biopolymers are the two nucleic acids, RNA and DNA. The evolution of the genetic code was obviously a pivotal event, and because all living organisms carry their genetic codes in the form of RNA and DNA, there is every reason to believe that these heredity molecules are extremely ancient and coevolved with the last common ancestor to all living things (Woese, Kandler, and Wheelis 1990). Among current organisms, DNA is the "master heredity molecule" of the two nucleic acids. It transcribes its genetic information into RNA, which subsequently translates this information during the biosynthesis of specific enzymes and other proteins that regulate and participate in the chemical reactions within cells (see fig. 1.7). Despite the deferential role RNA plays in current cells, DNA and RNA essentially use the same genetic code—sequences of nucleotides arranged in strands. Each nucleotide in RNA and DNA consists of a sugar, a phosphate group, and one of four nitrogen-containing bases. In DNA the sugar is deoxyribose and the four bases are adenine (A), guanine (G), cytosine (C), and thymine (T). In RNA the sugar is ribose and the base uracil (U) substitutes for thymine (fig. 3.3). The four bases pair very specifically in the double-stranded molecular architecture of DNA. Specifically, adenine pairs with thymine (A-T), while guanine pairs with cytosine (G-C). Adenine on the DNA molecule is transcribed as uracil on the RNA molecule (A-U). The bases in DNA constitute the genetic "alphabet," triplets of bases forming the genetic "words" that encode the information for protein structure. For example, the triplet CTT in DNA is transcribed as CUU in RNA, where it designates the information to add the amino acid leucine to the growing strand of a polypeptide. The "universality" of these DNA and RNA features suggests that the last common ancestor of all living things stored its genetic information in some kind of nucleic acid that could specify all of life's intricate biochemical machinery by encoding the information for protein structure.

Figure 3.2. Diagram of the apparatus Miller and Urey (1959) used to simulate the environmental conditions of early Earth and to synthesize organic compounds under these conditions.

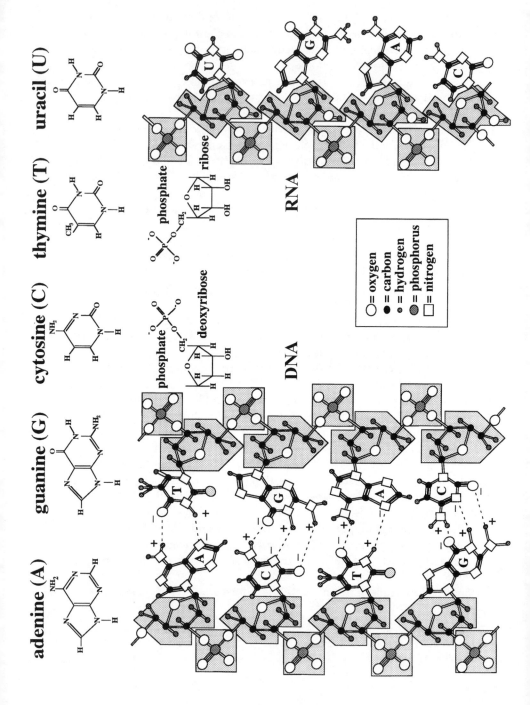

adenine (A)   guanine (G)   cytosine (C)   thymine (T)   uracil (U)

phosphate   deoxyribose   phosphate   ribose

DNA   RNA

= oxygen
= carbon
= hydrogen
= phosphorus
= nitrogen

Given that there are two nucleic acids, DNA and RNA, one of the most intriguing questions is whether DNA was the first heredity molecule or whether RNA served this function only to become displaced by DNA later in the history of life. Many scientists favor the latter hypothesis and speak of an ancient "RNA world" (see Gesteland and Atkins 1993; Orgel 1994). Indeed, as time goes by the concept of the RNA world gains increasing credibility. For example, although the Urey-Miller type of experiments showed that the purine nucleic acid bases (adenine and guanine) form readily under laboratory-simulated early Earth conditions, an efficient, plausible chemical pathway for the abiotic synthesis of the pyrimidine bases remained elusive. Recently, however, Michael Roberston and Stanley Miller (1995) have shown that as its concentration increases in water, urea can react with cyanoacetaldehyde to form cytosine, which in turn forms uracil. These authors speculate that urea and cyanoacetaldehyde, which are easily synthesized under Earth's early conditions, may have reacted in pools of water slowly evaporating under a hot sun to spawn the chemical reactions requisite to abiotically synthesize cytosine and uracil. The Urey-Robertson-Miller experiments show that Earth's ancient conditions were likely very conducive to the abiotic synthesis of all the bases required to make RNA.

The RNA-world hypothesis received further support with the discovery of "ribozymes," enzymes composed of RNA. Until the discovery of ribozymes, the term "enzyme" was typically reserved for proteins that serve as catalysts. Now the concept must be expanded to include ribozymes, which join together short strands of RNA and can synthesize RNA strands up to forty nucleotides long when monomers of RNA are mixed with zinc in a test tube. It is now also known that the RNA in ribosomes acts as an enzyme, rather than the ribosomal protein as formerly believed. Ribosomes consist of protein and ribosomal RNA (denoted

Figure 3.3. The structure of DNA and RNA. These two nucleic acids contain five bases. The five bases fall into one of two chemical categories: the purines (adenine and guanine) and the pyrimidines (cytosine, thymine, and uracil). DNA contains the two purine bases and the pyrimidine bases cytosine and thymine. RNA also contains the two purine bases and the pyrimidine base cytosine, but uracil replaces thymine. Each base is linked to a sugar along a backbone composed of sugar phosphate. Each base and its companion sugar are called a nucleotide; each sugar phosphate is called a nucleoside. The sugars in the backbones of DNA and RNA differ. The sugar in DNA is deoxyribose; the sugar in RNA is ribose. In addition to their chemical differences, DNA and RNA also differ structurally. DNA has a double-stranded molecule; each strand has a helical shape. The two DNA helices run in opposite (antiparallel directions). The companion nucleic acids (called base pairs) of the two antiparallel strands always consist of one purine and one pyrimidine and are held together by "lock and key" hydrogen bonds (shown as dashed lines ending in − and +). RNA is a single-stranded molecule.

rRNA). Ribosomes move along strands of messenger RNA (mRNA) and link amino acids (specified by the mRNA trinucleotides) to assemble complex proteins. It is now clear that rRNA can by itself catalyze some reactions, giving greater credibility to the hypothesis that RNA was one of the first autocatalytic molecules to appear in Earth's history. To be sure, short-chain polymers of RNA, called oligonucleotides, can be abiotically created in the test tube with comparative ease, and recently the in vitro evolution of the first RNA known to catalyze a bond-forming reaction on a substrate not containing nucleotides has been reported (Wilson and Szostak 1995). If, in the ancient oceans of the Earth, some RNA molecules acquired the ability to link other oligonucleotides together, as ribozymes do today, and if some of these "protoribozymes" replicated autocatalytically, then they could have served as an extremely primitive genetic apparatus (fig. 3.4). The discovery of the catalytic properties of RNA demonstrates that a single molecular species can function simultaneously as a genome and an enzyme (doing much to resolve the "chicken and egg" paradox of which came first, functional proteins or information-carrying nucleic acids). In turn, this discovery strongly suggests that the first genes were not on DNA molecules but on short strands of RNA capable of self-replication.

How the RNA world evolved into the DNA world with cells like those of today is not known, but researchers have provided a very plausible history based on chemical and mathematical laws. This history assumes that early Earth's oceans contained a jumble of kinds of autocatalytic RNA molecules with different reaction rates. Over time, one molecular species of RNA eventually came to govern the reaction rates of its autocatalytic compatriots. In doing so, this species assumed the role of the "master" autocatalytic molecule (Eigen 1971, 1987). Nevertheless, all the RNA molecules were essentially "naked genes"—they replicated for no purpose but their own replication. Experiments and theory also show that very short strands of RNA cannot self-replicate with high fidelity without protein assistants (enzymes in the traditional sense). Consequently, for life to evolve, self-replicating molecules must have evolved specifying proteins assisting in the very process of self-replication.

The solution to this seeming paradox is called the "hypercycle" (Eigen and Schuster 1977; Eigen 1992). A hypercycle consists of a complex network of mutually reinforcing chemical processes—crudely akin to "chemical A supports the synthesis of chemical B, chemical B supports the synthesis of chemical C, and chemical C supports the synthesis of chemical A"—until the point is reached where the whole process creates a molecular environment that expands its synthetic capacity to produce more complex molecules (see Eigen 1983, 1992). In order for it to expand,

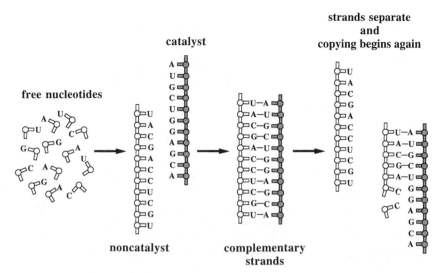

Figure 3.4. Hypothetical sequence showing how an autocatalytic, self-replicating molecule may evolve based on the chemistry of RNA.

however, some kind of information-storage molecule must emerge from a hypercycle. In this respect DNA is thermodynamically far more stable than RNA, and so DNA is a superior molecular repository for the genetic code. Theoretical treatments of hypercycles show that, once evolved from RNA, DNA will take over as the master heredity molecule by virtue of its remarkable storage capacity. The chemical differences between RNA and DNA are comparatively small and can be achieved by relatively small chemical mutations. Once DNA emerged as the master heredity molecule, the length of its nucleotide strands could be increased, spelling out the alphabet of base pairs containing all the information required for the machinery of life.

The fourth stage in the early evolution of life was the aggregation of this machinery into a cell-like protobiont, a structure whose interior chemically differs from its external surroundings. Protobionts have an essential property of life called "boundary": the segregation of chemical reactions from the external environment by some sort of organic "container." Protobionts form spontaneously in the laboratory under early Earth conditions. They also can become coated with phospholipids, which are natural membrane builders. Likewise, cooling solutions of proteinoids, like those synthesized by Sidney Fox, can self-assemble into small spheres that, when coated with a lipid membrane, swell or shrink in solutions of different salt concentrations, store energy in the form of a membrane potential, and catalyze chemical reactions when enzymes are added to the solution (e.g., maltose can be synthesized from glucose-phosphate in the presence of amylase and phosphorylase). Protobionts and microspheres grow in size by absorbing proteinoids and lipids from their local environment, subsequently splitting into companion cell-like structures after reaching a critical, apparently unstable size.

A precise definition of "life" has eluded biologists. But it is clear that the protobionts and microspheres thus far created in the laboratory are not living things. They lack the sophisticated ensemble of biosynthetic pathways coordinated by enzymes synthesized from the genetic instructions encoded on DNA molecules by means of intermediary RNA molecules. Nevertheless, at some point in history the intricate machinery of life was assembled within a cell-like entity. This assembly need not have occurred all at once. The metabolism of the first living entities on Earth could have relied to some degree on externally supplied, abiotically synthesized monomers and simple polymers. Over time, however, the external supply of these abiotically synthesized chemicals must have gradually diminished, perhaps as they were used up by expanding populations of the first forms of life or as Earth's environment changed and the rates of abiotic synthesis in the proto-oceans declined. In any circumstances, mutation and natural selection would have inspired and favored biosynthetic variants of cells that were metabolically more self-sufficient (Horowitz 1945; Horowitz and Miller 1962; Gest and Favinger 1983). In this way, over countless years, living matter progressively distanced itself from its ancestral physiological dependency on abiotic synthesis. This distancing of life required an amplification of the genetic library to encode for chemical pathways originally occurring outside the cell. In a very crude sense, the expansion of the genome allowed life to turn the chemical machinery of the abiotic world outside in. By doing so, early life acquired four prominent features shared by all contemporary living cells: the capacity to grow, self-replicate, inherit traits from ancestors, and mutate. With these four properties life emerged, crossing the thin, albeit profoundly important, line separating a purely chemical system from a strictly living and evolving one.

One plausible consequence of the passage from the inanimate to the animate world may have been viroids, virusoids, and viruses. Modern viroids and virusoids are the smallest known autonomously replicating molecules, and some of them have exactly the same enzymatic capabilities envisioned for the first self-replicating RNA molecules. Indeed, some workers believe viroids are "living fossils" that trace their ancestry back to the ancient RNA world. The viruses are extremely diverse in size, structure, and genomic organization. Nevertheless, every known virus is an obligate intracellular parasite whose genome consists of DNA or RNA or both and encodes for an encapsulating protein coat (see Strauss, Strauss, and Levine 1996). Three theories have been proposed for the origins of these life forms: (1) the regressive theory, which holds that viruses are degenerate forms of intracellular parasites; (2) the host-cell RNA theory, which argues that viruses arose from normal cellular components that gained the ability to autonomously replicate and thus evolve independently; and (3) the theory that viruses originated and evolved along with the most ancient molecules that had the ability to self-replicate. Whether one or more of these three contending theories is correct may never be known with confidence. Nevertheless a reasonable case has been made that some RNA viruses are the evolutionary descendants of extremely ancient life forms that coevolved with the first self-replicating RNA molecules. This hypothesis argues that the antecedents of the RNA viruses were originally integrated parts of the normal metabolic machinery of the first prokaryotic cells but became obligate intracellular parasites later in their evolutionary course (see Joyce 1991).

That life arrived at its present level of biochemical complexity through initial sequences of abiotic chemical reactions is a remarkable hypothesis. Even more remarkable is life's precociousness. Although geologists do not yet know the exact timing, the first atmosphere and oceans formed during the closure of the Hadean, roughly 3.9 billion years ago (fig. 3.1). Paleontologists also tell us that the most ancient fossil cells found occur in Archean rocks, which are at least 3.6 billion years old. It is worth remembering that "most ancient" means "minimum age." Exploration of the fossil record conceivably could push the first occurrence of cells still further back in time. At most, therefore, only 300 million years separate nature's "trial and error" experiments with abiotic chemical synthesis and the first rudimentary forms of cellular life. True, by human standards this is a long time. But when dealing with evolution, human standards of time must be abandoned. We measure our lifetimes in years or decades and recorded human history in thousands of years. But the paleontologist typically measures time in hundreds of millions of years. Given the intrinsic complexity of biological metabolism and self-replication, the evolution of the inaugural forms of life occurred in a "brief instant."

Martian Chronicles

The geological precociousness of early life on Earth and the fact that complex organic molecules can easily be abiotically synthesized with different mixtures of chemicals under laboratory conditions continue to fuel scientific speculation about the probability and appearance of extraterrestrial life. Many scientists agree that life on other planets is not only possible but highly probable. This belief has been dramatically reinforced by biochemical and mineralogical studies of a meteorite from Mars called the Allan Hills 84001 meteorite, or simply ALH84001 (McKay et al. 1996). This rock, found in Antarctica in 1984, is about 4.5 billion years old. It is one of only a dozen meteorites presumably ejected from Mars when a large asteroid collided with our neighboring planet roughly 15 million years ago. These rocks traveled through space in the inner solar system before landing on Earth about 13,000 years ago when Stone Age humans were beginning to develop agriculture. ALH84001 is the oldest of these meteorites and the only one containing significant amounts of carbonate minerals that NASA scientists believe formed on the Martian surface about 3.6 billion years ago.

The meteorite, which is about the size of a large Idaho potato, is riven with tiny fractures believed to have formed from impacts with other rocks before it was ejected from the Martian surface. Thin sample sections that included some of these preexisting fractures contain polycyclic aromatic hydrocarbons, strings of small ovoid and elongated structures, and globules of iron sulfide and magnetite (McKay et al. 1996). A variety of careful contamination checks and control experiments indicate that the ALH84001 carbonate material is not indigenous to Earth and thus not the result of terrestrial contamination. The Martian origin of ALH84001 is based on the chemical composition of gases trapped in very small pockets within the meteorite. The mixture and proportions of these gases closely match the chemical composition of the current Martian atmosphere, which the unmanned *Viking* spacecraft sampled in 1976.

At issue is the interpretation of the data reported by NASA scientists. Polycyclic aromatic hydrocarbons can form nonbiologically during early star formation, or biologically through the activity or subsequent fossilization of living organisms. On Earth these organic compounds are abundant in sedimentary rocks containing fossils, but they also form spontaneously during the partial combustion of organic materials, as when a candle burns or food is grilled. Perhaps the most controversial line of evidence for life on ancient Mars consists of the strings of sluglike microstructures seen on fracture surfaces under the transmission electron microscope. These collections of ovoid and elongated objects bear a vague similarity to very small bacteria, called nanobacteria, that are found on Earth. Moreover, fine-grained magnetite and iron-sulfide particles are found in association with these ovoid objects. These mineral particles are chemically, structurally, and morphologically similar to magnetosome particles produced by some bacteria on Earth. Magnetite and iron sulfide are formed by the oxidation and reduction of iron, respectively, and are often (but not necessarily) produced by living organisms.

Clearly, each of the chemical, mineralogical, and morphological lines of evidence is inconclusive when considered separately. The ovoid and elongated particles found in ALH84001 have not been sectioned to determine whether they have an internal cellular structure. Many structures seen in very ancient terrestrial rocks have been interpreted as fossils, only to be found later not to be organic in origin. Thus the sluglike structures in ALH84001 may be "pseudofossils." Nevertheless, based on the whole body of evidence, NASA scientists have concluded that these structures are the remains of Martian bacteria-like life forms that lived about 3.6 billion years ago

The age of these putative life forms correlates remarkably well with the oldest currently known terrestrial fossils. It also tallies well with the facts that Earth and Mars consolidated as planets at the same time, likely had similar ancient atmospheres, and undoubtedly had large expanses of free-flowing water early in their geological history. Indeed, Mars was once a warm and wet planet. Photographs sent back to Earth by the *Mariner* 9 spacecraft reveal branching canyons and braided channels that can only be the result of extensive water erosion. Thus it is not unreasonable to believe that life on Earth and Mars evolved concurrently and may have initially proceeded along very similar biochemical and morphological lines.

However, if life did evolve on Mars as many believe it did, evolution had scant opportunity for the organic elaboration and diversification witnessed on Earth. Mars has a mass less than 11 percent that of Earth, and the force of gravity at its surface is only 38 percent of Earth's. For these reasons Mars lost most of its atmosphere and all of its oceans to deep space. The current Martian atmosphere is remarkably thin, consisting mainly of carbon dioxide and a mere trace of nitrogen with an atmospheric pressure only about 1 percent of Earth's. These remnants of a once denser atmosphere provide little if any thermal insulation—the average daily temperature at the Martian surface varies over a range of about 150°F.

Bacteria-like life forms may still exist on Mars, perhaps under the soil surface or in the crevices of rocks in the Martian "tropics," where the noontime summer temperatures reach 50°F. Certainly many bacteria on Earth survive and grow under remarkably hostile conditions. Some cyanobacteria grow at 160°F water temperature; others cope with many times greater levels of X-ray and gamma radiation that cause the death of eukaryotic cells. A nostocacean cyanobacterium was revived after more than a century of dry storage on a herbarium sheet; another blue-green bacterium can be maintained in its growth phase at pH 13. Bacteria grow in the Antarctic ice sheet, in hot springs, in deserts, and buried deep in the ocean floor. Given this remarkable ecological range, some sort of bacterial life on Mars would come as little surprise to most biologists, although the existence of morphologically complex and large Martian life forms is ruled out based on information sent back to Earth by the *Mariner* and *Viking* spacecraft in the 1960s and 1970s.

Although the contention that ALH84001 contains Martian fossils remain problematic, it is clear that abiotically synthesized organic compounds are widespread in our solar system, so much so that some scientists believe that organic materials contained in meteors and comets falling through Earth's early atmosphere may have participated in life's beginnings (Oró 1961). Two lines of evidence foster this belief: the chemical analyses of stony meteorites containing carbon compounds, called carbonaceous chondrites, and studies of the cratering of our celestial neighbors, which indicate that large numbers of meteors and asteroids collided with Earth in the ancient past. Perhaps the best-known carbonaceous chondrite is called the Murchison meteorite. In 1969, fragments of this meteorite were collected on the day it landed in Australia. Chemical analyses of these fragments revealed numerous organic compounds including amino, hydroxy, monocarboxylic, and dicarboxylic acids, urea, amides, ketones, aldehydes, alcohols, and amines (Kvenvolden et al. 1970). Some of the compounds, especially the amino acids, had more or less equal proportions of left- and right-handed configurations, which is characteristic of abiotic synthesis (amino acids from living organisms have a left-handed configuration). Thus the amino acids found in the Murchison meteorite were not terrestrial contaminants and must have formed abiotically in space.

Large numbers of carbonaceous chondrites like the Murchison meteorite may have "rained" down on Earth during Hadean and early Archean times (see fig. 3.1). The evidence for this comes largely from studies of the chronology and intensity of cratering on the lunar surface that, by inference, shed light on the rate of deliver of meteors and asteroids to Earth. Most of the moon's large craters are quite old, ranging from slightly less than 4 billion to about 4.6 billion years old. The frequency of cratering is estimated to have been more than a thousand times the present rate. However, lava flows on the lunar surface, which are between 3.2 and 3.9 billion years old, are modestly pockmarked, indicating that the intensity of meteorite showering decreased (to near present-day levels) shortly after 4 billion years ago. The date for the diminution of cratering coincides remarkably well with the prebiotic phase of organic evolution on Earth.

The amount of extraterrestrial rock colliding with Earth's surface some 4.5 billion years ago may have been truly impressive—estimates range between 1 and 10 million kilograms of material per year (Chyba et al. 1990). Researchers calculate that roughly two tons of Martian material currently rains down on Earth every year and that roughly the same amount of terrestrial rock smashes into Mars. The real questions, however, are the amount of organic materials contained in these rocks and whether these organic materials survive atmospheric entry. Ancient Earth's atmospheric pressure may have been ten times higher than today's. If so, then much of the organic materials in meteors and asteroids falling through the atmosphere would have simply burned up. Thus the notion of panspermia—that "biological seeds" once peppered the ancient solar system and took root on ancient Mars and Earth, which is advocated by the distinguished British astronomer Sir Fred Hoyle—is not seriously entertained by most scientists.

We may never know if the organic compounds contained in carbonaceous chondrites and other extraterrestrial rocks survived their fall through the atmosphere and "seeded" Earth's proto-oceans. Nor do we know the extent to which those that survived the fall contributed to the prebiotic phase of terrestrial life. What can be said with some certainty is that studies of carbonaceous chondrites, the moons of Jupiter and Saturn, and far more distant interstellar gases reveal abundant amounts of abiotically synthesized organic compounds in outer space. Indeed, the chemical diversity and quantity of these compounds suggest that the chemistry of the prebiotic phase prefiguring life on Earth is a natural, possibly ubiquitous consequence of the evolution of inorganic matter.

Photosynthesis

Life gained tremendous physiological independence when it evolved the ability to convert light energy into chemical energy. This ability, called photosynthesis, is found in current representatives of both the eukaryotes and the prokaryotes. For the purposes of this book, the photosynthetic eukaryotes are considered plants and encompass the algae and the embryophytes. With few exceptions, all photosynthetic prokaryotes belong to the eubacteria, a kingdom containing ten clades, five of which have one or more photosynthetic representatives. These five clades are the purple bacteria (which include the sulfur and nonsulfur bacteria), the green sulfur bacteria, the filamentous green nonsulfur bacteria, the gram positive bacteria, and the cyanobacteria (Woese 1987). Of these, only the cyanobacteria release oxygen as a by-product of photosynthesis. In contrast to the oxygenic cyanobacteria, all other photosynthetic bacteria are non-oxygen-evolving (anoxygenic) photosynthetic organisms.

Although the first living entities may have been chemoautotrophs that evolved in hot, chemically active microenvironments (Bengtson 1994), the evidence available suggests that the photosynthetic prokaryotes evolved early in Earth's history (Schopf 1983; Woese 1987; Schidlowski 1988; de Duve 1991; Knoll 1992). Fossils discovered in Archean rocks roughly 3.6 billion years old are morphologically very similar to present-day photosynthetic eubacteria (see Walter 1983; Awramik 1992), and an isotopic "signature" characteristic of autotrophic carbon fixation as a consequence of photosynthesis is reported for rocks 3.8 billion years old (Schidlowski 1988). Whether these early Archean organisms were oxygenic photosynthetic prokaryotes, like the cyanobacteria, or anoxygenic photosynthetic prokaryotes, like most of the eubacteria, is uncertain because of the morphological similarities among otherwise physiologically very diverse bacteria (Oyzizu et al. 1987; Ward et al. 1989). Because the atmosphere of the early Earth was more reducing than the present atmosphere, and because the vast majority of present-day photosynthetic bacteria neither produce nor consume molecular oxygen—in fact, most are strict anaerobes that are poisoned by free oxygen—it is generally believed that anoxygenic photosynthetic prokaryotes evolved first. Furthermore, these organisms very likely were obligate anaerobes or extremely inefficient aerobic respirers. What can be said with assurance is that by the Archean prokaryotes had physiologically diversified, some achieving a sophisticated level of metabolic organization capable of using the energy of light to manufacture organic compounds.

Attempts to reconstruct the evolution of photosynthesis necessarily focus on comparisons among the eubacteria, because these organisms possess a number of features believed to be ancient rather than derived and because the fossil record does not preserve the chemical history of photosynthesis in sufficient detail. Among the eubacteria, three features are almost universally shared among photosynthetic representatives: an "antenna/reaction center" organization of the photosynthetic unit; the use of chlorophyll-based pigments as the principal light-gathering molecule; and the presence of a heterodimeric protein core within the reaction center. These shared, presumably ancient features may inspire the belief that all photosynthetic prokaryotes evolved directly from a single common ancestor. However, the evolutionary origin and development of photosynthesis appears to have involved the lateral transfer of genes between different lineages of bacteria; that is, different parts of the photosynthetic apparatus at present found in the cyanobacteria and plants trace their ancestry to at least two prokaryotic lineages, each of which has an analogue in very different eubacteria. This in turn lends compelling support to the hypothesis that the chloroplasts of eukaryotes evolved when a cyanobacteria-like organism gained entry and took up permanent residence within a heterotrophic prokaryotic "host" cell. This "endosymbiotic hypothesis" is treated in greater detail later (see figs. 3.10–11). For now the important point is that the origin and "fine-tuning" of photosynthesis did not proceed along a linear evolutionary pathway.

The nonlinearity underlying the evolutionary history of photosynthesis is best appreciated by returning to the three "universalities" mentioned earlier. Photosynthesis involves the participation of pigment molecules that absorb light energy and give up electrons as a consequence of their elevated energy states. Photosynthesis also involves protein molecules that accept and transfer the electrons stripped from pigment molecules excited by light energy. These proteins and pigments are organized into a "photosynthetic unit" consisting of a light-gathering antenna system of pigment molecules attached to the electron transfer proteins arranged in what is called a reaction center. The antenna/reaction center organization of the photosynthetic unit is an almost universal feature of all photosynthetic organisms, both prokaryotes and eukaryotes. The sole exceptions to this generality are the halobacteria (Stoeckenius and Bogomolni 1982; Blankenship 1992). These photosynthetic prokaryotes live in extremely saline habitats, some species growing in water ten times as salty as seawater. The halobacteria are unique among the prokaryotic photoautotrophs because they use a retinal-containing protein, bacteriorhodopsin, to trap the energy of light. Bacteriorhodopsin is directly bound to membranes and is not organized in an antennae/reaction center configuration as in other photosynthetic organisms. Because there is no direct comparison between the photosynthetic apparatus of the halobacteria and the other photosynthetic apparati of other photosynthetic prokaryotes, it is safe to say that photosynthesis, in the literal sense of the word, has independently evolved at least twice.

The second virtually universal feature of photosynthetic organisms is that they contain one or more types of the class of pigments known as the chlorophylls (Beale and Weinstein 1990; Blankenship 1992). Additional pigments, the carotenoids and bilin pigments, are also used to trap and channel light energy to the reaction center at the heart of the photosynthetic unit. These "accessory" light-harvesting pigments account for differences in the color of photosynthetic cells (e.g., photosynthetic bacteria and eukaryotes may be brown, red, or purple). But the role of the accessory photosynthetic pigments does not color the fact that the principal underlying pigment in all photosynthetic organisms (other than the

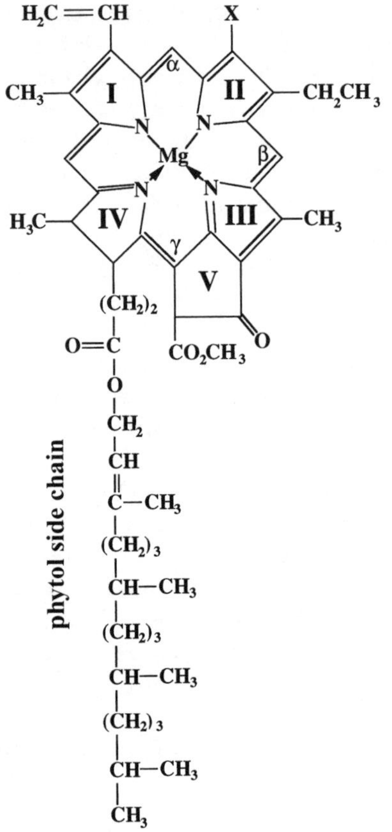

Figure 3.5. Basic molecular structure of the chlorophyll-based pigments. The pigments differ in terms of the functional group attached at the site marked X (e.g., X = $-CH_3$ in chlorophyll *a* and X = $-CHO$ in chlorophyll *b*).

halobacteria) is chlorophyll based (fig. 3.5): the chlorophylls and the bacteriochlorophylls. The type of chlorophyll-based pigment found in photosynthetic prokaryotes correlates with whether the organism is an anoxygenic or oxygen-evolving photoautotroph. All oxygen-evolving prokaryotes contain chlorophyll *a*; all anoxygenic prokaryotes lack chlorophyll *a* and instead use bacteriochlorophyll to trap the energy of light.

The presence of bacteriochlorophylls in four anoxygenic prokaryotic phyla, which differ in many other important respects, suggests that these pigments are more ancient than chlorophyll. Curiously, however, chlorophyll *a* is an intermediary rather than a final step in the biosynthesis of the bacteriochlorophylls and chlorophyll *b* (fig. 3.6). Thus the biosynthetic pathways of the chlorophyll-based pigments present something of

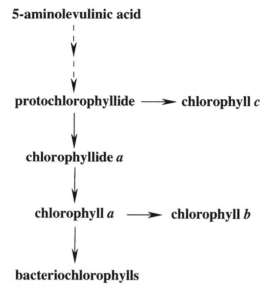

5-aminolevulinic acid

protochlorophyllide ⟶ chlorophyll *c*

chlorophyllide *a*

chlorophyll *a* ⟶ chlorophyll *b*

bacteriochlorophylls

Figure 3.6. Redaction of the biosynthetic pathway to the chlorophyll-based pigments. Bacteriochlorophylls occupy terminal steps of the pathway.

a phylogenetic dilemma. If "biosynthesis recapitulates phylogeny"—that is, if intermediates in a biosynthetic pathway are the final products of biosynthetic pathways of earlier evolutionary forms (see Granick 1965)—then prokaryotes employing chlorophyll *a* evolutionarily preceded those using the bacteriochlorophylls as the principal photosynthetic pigment. Alternatively, the metabolic pathway for the chlorophyll-based pigments may have evolved backward. That is, if organisms typically evolve ways to synthesize essential chemical compounds from progressively simpler compounds as the available stocks of essential molecules are depleted in the environment (see Gest and Schopf 1983), then photoautotrophs with bacteriochlorophylls are truly more ancient than those using the chlorophylls as their photosynthetic pigments.

Upon reflection, both the biosynthetic "bootstrapping" and the "deconstructionist" hypotheses seem wanting. On the one hand, it is difficult to imagine that a molecule like chlorophyll (or its precursor, protochlorophyllide) could have been abiotically synthesized; on the other, the simplest precursors of the chlorophyll-based pigments are amino acids and other colorless compounds that could not have served as pigments (e.g., 5-aminolevulinic acid; see Beale and Weinstein 1990). One plausible solution is that the biosynthetic pathway for the chlorophyll-based pigments evolved from the middle outward (Mercer-Smith and Mauzerall 1984; Mauzerall 1990; see also Blankenship 1992). The recently discovered heliobacteria, which contain the previously unknown pigment bacteriochlorophyll *g* (Gest and Favinger 1983; Madigan 1992), adds credibility to this "middle outward" biosynthetic hypothesis. Bacteriochlorophyll *g* has the chemical properties of chlorophyll *a* and bacteriochlorophyll *b*. Therefore it could have served as a possible intermediate pigment derived from an ancient biosynthetic pathway that was subsequently elaborated upon and fine-tuned to give rise to the chlorophylls on one hand and the bacteriochlorophylls on the other. The heliobacteria are anoxygenic photosynthetic organisms and therefore fall in line with the generally held belief that the first prokaryotes were obligate anaerobic bacteria. Unfortunately, little is known about the biosynthesis of bacteriochlorophyll *g*, and therefore the validity of the "middle outward" hypothesis is uncertain.

An intriguing hypothesis has recently been advanced suggesting that bacteriochlorophyll evolved in ancient bacteria living near hydrothermal systems as an adaptation to finding and staying near their source of food (Nisbet, Cann, and Van Dover 1995). Molecular evidence supports the belief that the earliest eubacteria were thermophilic organisms, well suited to living near hot, turbulent hydrothermal vents, which were widespread on Earth during the early Archean. Similar vents existing today provide a rich assortment of nutrients to chemotrophic bacteria, which risk starvation whenever they lose touch with their nutrient supply. Thus organisms like the purple bacterium *Rhodospirillum centenum* are thermotactic and move toward the infrared light emitted by vents. Euan Nisbet and colleagues (1995) point out that the absorption spectra of bacteriochlorophylls match the thermal emission spectra of a hot body submerged in water. They speculate that the chemical precursors to the bacteriochlorophylls may have evolved in thermophilic, chemotrophic bacteria as an adaptation to locating and staying near hydrothermal vents. Ancient thermotactic bacteria could have evolved into phototactic and photosynthetic organisms using the infrared part of the spectrum of sunlight in shallow bodies of water and H_2 and H_2S provided by hydrothermal vents. For an organism that is already surviving as a chemotroph, a comparatively small evolutionary change to allow infrared phototaxis is an adaptive advantage. In turn, infrared phototaxis could "preadapt" an organism for optional infrared photosynthesis based on bacteriochlorophyll and, still later, the development of chlorophyll to use visible light.

Regardless of the evolutionary routes that give rise to them, all the chlorophyll-based pigments absorb visible light efficiently because of their many conjugated bonds. Furthermore, these pigments can delocalize the energy absorbed from photons (i.e., the energy can be spread throughout their electronic structures). When seen in this light, the bacteriochlorophylls and chlorophylls appear to have a "selective advantage" over other candidate pigments for photosynthesis. Yet the chlorophyll-based pigments are far from being "ideal" light-absorbing compounds. Some of the energy they absorb is lost as either heat or fluorescence. That the chlorophyll-based pigments are "adequate" pigments is witnessed by the fact that they underpin Earth's food chain.

In terms of the evolution of the reaction center, Robert Blankenship (1992) points out that the reaction centers of all existing photosynthetic organisms fall into two general categories—those that contain pheophytin and a pair of quinones as the early electron acceptors, and those that use iron-sulfur clusters as early electron acceptors. For example, the green gliding bacteria and the purple nonsulfur bacteria have the pheophytin-quinone type of reaction center, and the green sulfur bacteria and the heliobacteria have the iron-sulfur type of reaction center. Despite their differences, both types of reaction centers share some important features. Notable among these is that their protein cores are composed of two related yet distinct proteins. That is, the cores of both types of reaction centers contain a "heterodimeric protein complex." It is interesting that the two types are found in very different kinds of eubacteria, suggesting that the ancestor of modern-day photosynthetic prokaryotes may have had a much simpler protein core, perhaps a monomeric protein core. Indeed, it has been suggested that gene duplication in this hypothetical ancestor may have led to a homodimeric protein core whose two protein components subsequently diverged to become heterodimeric (Blankenship 1992).

The notion that the otherwise diverse phyla comprising the eubacteria are all derived from a common photosynthetic ancestor is also supported by another deep similarity between the pheophytin-quinone and the iron-sulfur types of reaction centers. In both types, a quinone serves as an early electron acceptor and is preceded by a chlorophyll-type pigment molecule in the electron transfer chain. Therefore, in addition to the heterodimeric protein core, the early steps in electron transfer within both types of reaction centers are similar among all known photosynthetic prokaryotes. This similarity suggests that all photosynthetic prokaryotes shared a last common ancestor with a reaction center having a chlorophyll-type of pigment molecule preceding a quinone electron acceptor.

The pheophytin-quinone and iron-sulfur types of reaction centers are associated with two very different photosynthetic systems that coexist today only in the cyanobacteria and the photosynthetic eukaryotes. The simplest evolutionary scenario for the occurrence of the two types of reaction centers in the cyanobacteria, which are the simplest oxygen-releasing organisms (Bryant 1987; see also Blankenship 1992), is that two kinds of bacteria, one possessing the pheophytin-quinone type of reaction center and another kind of bacteria possessing the iron-sulfur type, fused genetically to obtain a photosynthetically chimeric prokaryote. Initially the two photosystems may have operated independently, only later becoming functionally linked as they are in the cyanobacteria.

This hypothesis helps explain how the "light reactions" found in the cyanobacteria and plants may have evolved. The light reactions involve the participation of two photosystems, called photosystem I and photosystem II, each associated with its own chain of electron acceptors and donors (fig. 3.7). Photosystem I has an iron-sulfur type of reaction center, while photosystem II has a pheophytin-quinone type. With the exception of the cyanobacteria, prokaryotes possess photosynthetic apparati that are analogous to either photosystem I or photosystem II but not both. The most parsimonious hypothesis is that the cyanobacteria evolved from some kind of genetically chimeric prokaryote resulting from the fusion of two kinds of bacteria. Recent evidence indicates that photosystem I may optimize the efficiency of photosystem II under aerobic conditions. If so, then the fusion of two kinds of bacteria, each containing only one of the two photosystems, may have been particularly adaptive in a photosynthetic world with an atmosphere initially lacking much oxygen.

It is not known whether the chimeric ancestor to the cyanobacteria released oxygen as do present-day cyanobacteria or subsequently evolved the ability to use water as the electron donor molecule to reduce carbon dioxide and so release oxygen as a photosynthetic by-product. Among oxygenic photoautotrophs, water molecules are "split" to yield electrons, hydrogen ions (protons), and molecular oxygen. In chemical notation, $2H_2O \rightarrow 4e^- + 4H^+ + O_2$. The final electron-acceptor is carbon dioxide, which is "fixed" in the Calvin cycle to synthesize simple sugars, although the pyridine nucleotide $NADP^+$ serves as an intermediate electron acceptor from the light reactions (see fig. 3.7). The oxygen-releasing photosynthesis of the cyanobacteria and plants is often summarized by the following equation, which emphasizes the source of the oxygen released during photosynthesis:

$$\overset{\text{light}}{\underset{\text{chloro}}{nH_2O + nCO_2 \rightarrow (CH_2O)_n + nO_2}}$$

In this equation, n is typically designated as 6 to correspond with the biosynthesis of simple sugars, like hexose ($C_6H_{12}O_6$). In contrast, anoxygenic photosynthetic bacteria do not use water as the electron-donor

Light Reactions

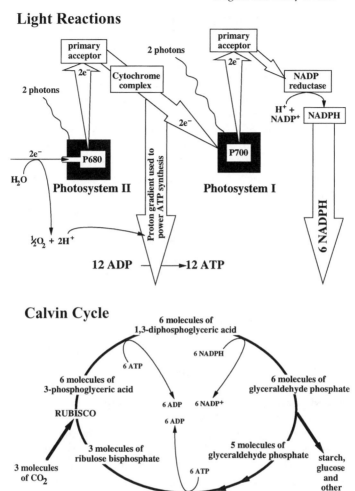

Figure 3.7. Summary diagrams of the light reactions and the Calvin cycle.

molecule. Some use hydrogen sulfide, thiosulfate, or hydrogen gas instead. For example, the sulfur bacteria use hydrogen sulfide:

$$2H_2S + nCO_2 \xrightarrow[\text{chloro}]{\text{light}} (CH_2O) + H_2O + 2S$$

As a consequence, the sulfur bacteria produce globules of sulfur instead of oxygen as their photosynthetic "waste product."

As noted, the evolutionary route by which the cyanobacteria acquired the ability to "split water" and release oxygen is not yet known. One possibility is that a genetically chimeric ancestor possessing the rudiments of photosystem I and photosystem II acquired the ability to use reductants weaker than compounds like hydrogen sulfide but stronger than water (Olson and Pierson 1987). Such an organism may have used Fe^{2+} as a reductant, much like some species of cyanobacteria that have the capacity to oxidize Fe^{2+} to produce the reduced Fe^{3+} ion (Cohen 1984). The electrons released from ions like Fe^{2+} could have been used by a cyanobacteria-like ancestor to oxidize carbon dioxide, and the electron flow could have been used to pump protons across membranes to establish the energy gradients necessary to make ATP.

Much more needs to be known about the evolutionary origin and development of both anoxygenic and oxygenic photosynthetic prokaryotes. At present there are only tantalizing bits of information relating to their origin. What can be said with assurance is that photosynthesis had dramatic biological and physical effects that shaped the course of subsequent evolutionary history. It is clear from geochemical and paleontological data that, after their initial appearance, oxygen-releasing prokaryotes very similar to current cyanobacteria prospered, multiplied, and diversified during the Archean. As a consequence, the primary global productivity of these organisms increased, and, by the end of the Archean and during the early Proterozoic the rate of oxygen production eventually exceeded the rate at which oxygen was consumed by the oxidation of reducing metals in soils and rocks. Somewhere between 2.4 and 2.8 billion years ago, Earth's atmosphere attained 1%–2% of the present-day atmospheric level of oxygen (Knoll 1992). The physical result was the emergence of a stable, oxygen-enriched atmosphere that undoubtedly restricted the habitats available to strict anaerobes and that favored the radiation of aerobic prokaryotes. Clearly, aerobic metabolic pathways identical to and differing from that of classical respiration (e.g., the flavin oxidase-based respiration of *Thermoplasma;* see Searcy, Stein, and Searcy 1981) have evolved independently in a great many different lineages (Fox et al. 1980). Thus oxygen, the waste product of cyanobacteria-like Archean organisms, supplied the selection pressure favoring varying degrees of oxygen tolerance in diverse lineages.

The Evolution of the Eukaryotes

The "oxygenation" of the early Proterozoic atmosphere by early photosynthetic organisms resulted in a physical environment that did not biochemically favor bacteria incapable of metabolically tolerating free oxygen. Some aerobic prokaryotes, particularly representatives of the purple nonsulfur bacteria, possess all the key features in the respiratory chain found in the mitochondria of eukaryotic cells. These aerobic prokaryotes are physiologically similar to the mitochondrion, which is the site of cellular respiration. Cellular respiration is the most prevalent and energy efficient catabolic pathway during which oxygen is consumed (as a reactant) and glucose is broken down (to generate ATP) (fig. 3.8). In this sense the mitochondrion uses the waste product of photosynthesis, oxygen, and releases the essential raw materials, carbon dioxide and water, that cyanobacteria require for photosynthesis. Conversely, the chloroplast produces carbohydrates and releases oxygen as a by-product of photosynthesis. Thus the Earth's first ecosystems may have comprised assemblages of oxygenic photosynthetic prokaryotes (chloroplastlike prokaryotes) and oxygen-consuming bacteria (mitochondrialike prokaryotes). These prokaryotic confederations would have derived energy from sunlight and beautifully recycled the chemical compounds essential to photosynthesis and cellular respiration (fig. 3.9).

Such an ecosystem may have prefaced the evolutionary origin and development of the first eukaryotic cells as envisioned by the endosymbiotic hypothesis. First proposed in the late nineteenth and early twentieth centuries (Schimper 1883; Mereschkowsky 1905, 1920) and subsequently brought to intellectual fruition by Lynn Margulis (1970), the endosymbiotic hypothesis states that eukaryotic cells are the evolutionary consequence of ancient confederations of different kinds of prokaryotes that merged to produce the eukaryotic cell (fig. 3.10). Specifically, the endosymbiotic hypothesis argues that mitochondria evolved from once autonomous aerobic bacteria and that chloroplasts evolved from once autonomous cyanobacteria-like bacteria. Once protomitochondria and protochloroplasts gained entry to a host cell, they continued to function within the cell. It is reasonable to assume that the host was either an anaerobic, albeit aerotolerant, prokaryote or an inefficient aerobic prokaryote (Margulis 1970, 1981).

A number of ways a host cell could garner endosymbionts have been proposed. One way may have been endocytosis, the ability to engulf an extracellular particle by surrounding it with a portion of the external cell membrane and then invaginating the particle encapsulated within a microvacuole. The endocyctotic method of "capturing" prokaryotic protomitochondria and protochloroplasts gains partial support from the fact that mitochondria and chloroplasts are double-membrane-bound organelles that are topologically external to the eukaryotic cell. Perhaps the

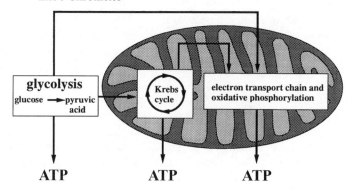

Figure 3.8. Locations of glycolysis, the Krebs cycle, and the electron transport chain and oxidative phosphorylation relative to the structure of the mitochondrion.

"host" prokaryote endocytotically ingested prokaryotes with differing metabolic capabilities, failed to digest them, and acquired the ability to work in partnership with its internal occupants. (In this sense contemporary mitochondria and chloroplasts are the consequences of an undigested dinner.) However, the ability to flex, fuse, and pinch off membranes to form microvacuoles is poorly developed within living prokaryotes as a group. In part this is due to the general absence of the sterols and polyunsaturated fatty acids found in the cells of eukaryotes that confer membrane flexibility and fusability, and so the capacity for endocytosis may have been the limiting factor in the evolution of endosymbiosis. As an alternative to endocytosis, which requires membrane flexure, it is possible that protomitochondria and protochloroplasts evolved from internal prokaryotic parasites. This hypothesis receives some support from modern parasitic prokaryotes, such as the aerobic bacterium *Bdellovibrio*, which enters and takes up residency in the periplasmic space of its host cell (Margulis 1981). Perhaps in the distant past a similar parasite entered an anaerobic but aerotolerant host cell that provided it with some nutrition. In turn the endoparasite could have benefited its host by consuming oxygen through aerobic respiration, thereby detoxifying its host while simultaneously supplying limited quantities of ATP. (Thus the first eukaryotes may have been the product of a disease cured to mutual benefit.)

Regardless of how they gained entry, the first endosymbionts likely were not entirely benign tenants. Because they replicated themselves within its cytoplasm, the host cell had to gain some genetic control of them. This may account for the loss of portions of the chloroplast and mitochondrion genome in modern eukaryotes (see chapter 8). Endosymbionts also released "exotic" materials that may have initially been toxic to their host cells (e.g., modern mitochondria release abundant oxidants

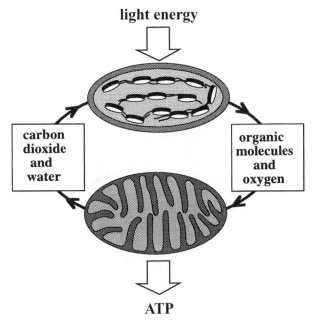

light energy

carbon dioxide and water

organic molecules and oxygen

ATP

Figure 3.9. Reciprocal relationship between the physiological requirements of the chloroplast (above) and the mitochondrion (below).

that can be harmful to prokaryotic organisms). Consequently many genetic and physiological relations had to be "ritualized" before ancient prokaryotic confederations evolved into the first eukaryotic cells (Blackstone 1995).

The endosymbiotic hypothesis, which is an old concept, had many detractors who argued that organelles evolved through the differentiation and invagination of functionally different membrane-bounded areas that eventually pinched off into the cell (fig. 3.10). This alternative to the endosymbiosis hypothesis is called the autogenous hypothesis. It is particularly attractive for the evolution of the nucleus and endomembrane system (consisting of the endoplasmic reticulum, Golgi apparatus, lysosomes, and other single-membrane-bound organelles). But the autogenous hypothesis encounters the same difficulty as the endocytotic mechanism for the incorporation of protomitochondria and protochloroplasts within a host cell—that is, the generally poor membrane flexibility and fusibility of prokaryotic membranes. Indeed, for either hypothesis to be correct, the flexibility and fusibility of the membranes of the first cells must have differed from the membranes of present-day cells. But the autogenous hypothesis entirely fails to account for the dramatically different genomes

Endosymbiotic hypothesis

Photoautotroph

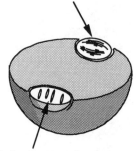

Aerobic heterotroph

Chloroplast

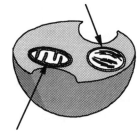

Mitochondrion

Autogenous hypothesis

Cytoplasmic invagination

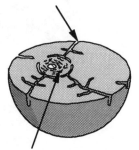

Prokaryotic "host" DNA

Endoplasmic reticulum

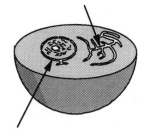

Nuclear envelope

Eukaryotic plant cell

found in current chloroplasts and mitochondria, nor can it account for the remarkably conservative nature of the DNA sequences in chloroplasts (see chapter 8).

Clearly, the autogenous and the endosymbiotic hypotheses are not mutually exclusive (fig. 3.10). Although the mitochondrion and chloroplast may have evolved through the lateral transfer among bacterial lineages envisioned by the endosymbiotic hypothesis, the nuclear envelope and endoplasmic reticulum may have evolved autogenically through membrane invagination. Increased cell size may have favored an amplification of membrane systems as a consequence of well-known area-volume ("law of similitude") relations that require disproportionate increases in surface area as cell size increases (Gould 1966; Niklas 1994).

The endosymbiotic origins of mitochondria and chloroplasts is unquestionably confirmed by a variety of data. As early as the late 1960s, Carl Woese began to assemble catalogs of oligonucleotide sequences released by the in vivo digestion of rRNA isolated from living prokaryotes and eukaryotes. By analyzing shared and unshared oligonucleotide sequences, Woese and his colleagues constructed a phyletic hypothesis for the origins of the eukaryotes and the phyletic relations among the prokaryotes. Although no single molecule can define an evolutionary lineage once lateral gene transfer is admitted to occur, methodologically rRNA was and is the molecule of choice for an evolutionary chronometer because it occurs in all living cells and is functionally conservative, evolving very slowly (Doolittle and Brown 1994). The rRNA data amassed by Woese and others indicate that the chloroplast and the mitochondrion found in extant eukaryotes (plants, animals, etc.) descend from very different groups of free-living prokaryotes. Specifically, the oligonucleotide sequences of the chloroplast closely align with those from living cyanobacteria, while the oligonucleotide sequences from mitochondria align with those of a group of purple nonsulfur bacteria called the proteobacteria (fig. 3.11). This phyletic hypothesis continues to agree with new information about prokaryotic biochemistry and molecular biology. It is also compatible with the essential features of previous classification

Figure 3.10. Comparisons between the endosymbiotic theory and the autogenous theory for the origins of organelles. The endosymbiotic theory posits that the chloroplast and the mitochondrion evolved as a consequence of the incorporation of a photoautotrophic prokaryote and an aerobic heterotrophic prokaryote, respectively, by a "host" prokaryote cell. The autogenous theory states that chloroplasts, mitochondria, and all other organelles evolved by means of cytoplasmic invagination and subsequent physiological differentiation. An amalgam of the two theories may reflect the actual course of evolutionary events leading to the appearance of the first eukaryotes (see bottom of figure).

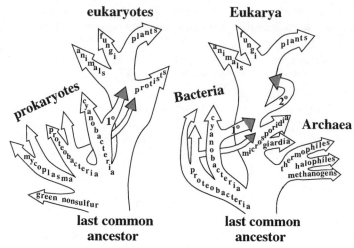

Figure 3.11. Two renditions of the evolutionary relations among prokaryotes and eukaryotes. In each rendition, mitochondria descend from endosymbiotic proteobacteria and double-membrane-bound chloroplasts descend from a primary endosymbiotic event (denoted by 1°) involving a cyanobacteria-like organism. The relations shown in the diagram on the left are those suggested by Woese and Fox 1977. This rendition reflects the five-kingdom system (the prokaryotes, protista, animals, fungi, and plants). The relationships shown in the diagram on the right are those more recently suggested by Woese, Kandler, and Wheelis 1990. In this rendition the bacteria are formally recognized to consist of two major groups, the Bacteria and the Archaea, and all eukaryotes are placed in the Eukarya. Secondary endosymbiotic events, involving photosynthetic eukaryotes engulfed by non-photosynthetic eukaryotes (denoted by 2°), are believed to have led to the evolution of some algal lineages whose chloroplasts have more than two bounding membranes.

schemes for bacteria. Additionally, chloroplasts and mitochondria are the appropriate size to have descended from prokaryotes; their inner membranes have several enzymes and transport systems resembling those found in the external membrane encapsulating modern bacteria; both organelles duplicate during the division of the eukaryotic cell by a splitting process that resembles the binary fission of bacteria; and chloroplasts and mitochondria contain DNA in the form of a circular molecule that is not associated with histones or other proteins as it is in eukaryotes (see chapter 8).

The modern version of the "tree of life" is reticulated, and not invariably branched as Charles Darwin believed. All the available evidence indicates that the eukaryotes, or formally the "Eukarya" (see Woese, Kandler, and Wheelis 1990), are the result of the lateral transfer of separate prokaryotic evolutionary lineages (fig. 3.11). But the "reticulate tree of life" is even more curious because there is mounting evidence that some photosynthetic eukaryotic lineages may be derived from secondary endosymbiotic events (Lee 1989). The primary endosymbiotic event was a partnership between cyanobacteria-like and nonphotosynthetic prokaryotic organisms, during which the photosynthetic endosymbiont relinquished some of its autonomy to the host cell by giving up part of its genome (see chapter 8). This event ultimately led to the red and the green algal lineages. In each of these lineages, the chloroplast is a double-membrane-bound organelle whose inner membrane is derived from the outer cell membrane of the photosynthetic endosymbiont and whose outer membrane is derived from a vesicle created by the prokaryotic host cell. However, the chloroplasts of other algal groups have more than two bounding membranes (e.g., cryptomonad and chlorarachniophyte chloroplasts have four membranes). These "supernumerary" bounding membranes likely resulted when a nonphotosynthetic eukaryote engulfed a photosynthetic eukaryote and acquired its chloroplast as a secondary endosymbiotic event (see McFadden and Gilson 1995). The nucleomorph, a nucleuslike organelle sandwiched between the second and third chloroplast membranes, may be the vestigial remains of the nucleus of the engulfed photosynthetic eukaryote. Thus, even if the double-membrane-bound chloroplast evolved only once as a consequence of primary endosymbiosis, the organisms traditionally called the "algae" likely are polyphyletic because some lineages have chloroplasts that are drastically reduced remnants of photosynthetic eukaryotes. The current view of algal phylogeny gives new meaning to "something old, something new / something borrowed, something blue."

But what of the ancestral genome of the Eukarya? Both the autogenous hypothesis and the endosymbiotic hypothesis suggest that the eukaryotic genome arose from some type of bacterial genome. Therefore the last common ancestor to all organisms alive today (the "LCA" of Cairns-Smith 1982 and the "cenancestor" of Fitch and Upper 1987) must be some type of prokaryote whose nuclear genome bears an imposing similarity to that of the basal eukaryotic organism. If a single nuclear prokaryotic genome underlies the host cell garnering chloroplasts and mitochondria, then the Eukarya can be thought of as a monophyletic group, albeit highly diversified. However, attempts to identify the LCA based on molecular data have led to equivocal results despite analysis of representative species of every known prokaryotic and eukaryotic lineage. That no current prokaryotic lineage can be positively identified as the source of the "host" nuclear genome suggests that the split, separating all living things into two camps (the prokaryotic/eukaryotic dichotomy), may extend back vertically to the very beginning of organic evolution. Alternatively, the failure to identify the LCA from among extant groups may reflect the still scanty molecular data currently available.

Comparatively few species in each of the many prokaryotic and eukaryotic lineages have been examined. Another possibility is that lateral gene transfer has occurred so many times and so far in the past that the phyletic memory of molecules used to render phyletic hypotheses (rRNA, proteins, and so on) has decayed beyond recognition.

Sexual Reproduction and Meiosis

The evolution of the eukaryotic nucleus presaged a dramatic departure from the previous way cells divided and reproduced. Prokaryotes reproduce by binary fission, during which the single bacterial chromosome replicates and one copy is allocated to each of the two new cells (fig. 3.12). Both copies of the duplicated chromosome are bound to the inner surface of the cell membrane and then are drawn apart by the subsequent growth of the intervening portion of the membrane. During the final stages of binary fission the bacterial cell "pinches inward" when the cell reaches roughly twice its original size. In contrast, the division of eukaryotic cells is far more complex, although it involves the duplication of the chromosomes and the production of two new cells, just as in binary fission (fig. 3.13). Unlike the duplication of the prokaryotic chromosome, the duplicates of each eukaryotic chromosome, or "sister chromatids," are attached to one another by a structure called the centromere. Each centromere becomes attached to a spindle fiber composed of microtubules and their associated proteins radiating from two opposing spindle poles. Eventually the chromosomes in a cell align on a single plane, called the metaphase plate, at the equator of the spindle. Because each chromosome arrives at the metaphase plate with an apparently random orientation relative to the two spindle poles, chance determines the segregation of the chromatids of each chromosome into the two new cells.

The events resulting in the duplication of eukaryotic chromosomes and the division of the nucleus into two new nuclei constitute mitosis, the process that obtains two genetically identical new nuclei. During the later stages of mitosis, at about the stage called anaphase, cytokinesis begins to occur. In cytokinesis, literally "movement of the cytoplasm," a parent cell divides into two new cells. At the stage called telophase, when the chromatids arrive at the two opposing poles and new nuclear membranes form around them, materials are deposited along the plane of the old metaphase plate. These materials are contained in vesicles amassed along the middle of the cell. The coalescence of the vesicles results in the formation of a double membrane that extends from the center of the old metaphase plate and eventually unites with the preexisting cell membrane of the original cell. Once the new membranes unite with the old

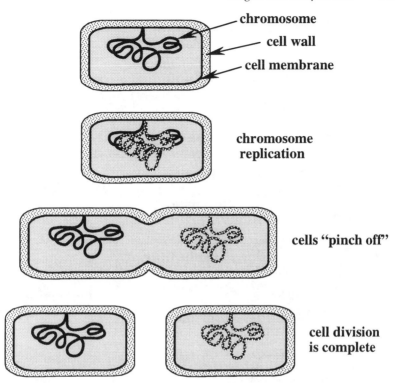

chromosome

cell wall

cell membrane

chromosome
replication

cells "pinch off"

cell division
is complete

Figure 3.12. Bacterial cell division by binary fission. The circular bacterial chromosome is attached to the cell membrane, it replicates, and its two copies are drawn into two future cells resulting from the invagination and "pinching in" of the dividing cell membrane.

cell membrane, cytokinesis is complete—each new cell is completely separated from its twin by its own cell membrane. Among land plants, a new cell wall forms within the double membrane along the old metaphase plate. The formation of the double membrane and the new cell wall is also associated with the phragmoplast, a complex structure composed of microtubules (see fig. 3.13). The distinction between mitosis, the division of the original nucleus into two new nuclei, and cytokinesis, the separation of the original cell's living material into two new cells with their own cell membranes, is important because mitosis is not always followed by cytokinesis. When cytokinesis does not occur, a cell increases in size but doubles the number of its nuclei. Repeated mitosis in the absence of cytokinesis produces a multinucleated cell.

Space does not permit a detailed discussion of the evolution of mitotic cell division. The following references, however, provide provocative reading on this subject: Cavalier-Smith (1978); Carlile (1980); Dodge

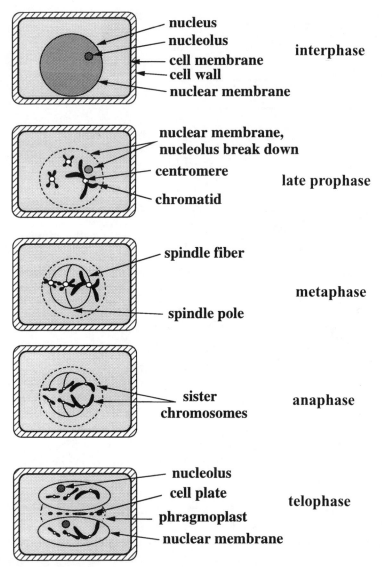

Figure 3.13. Diagrammatic rendition of the mitotic cell division of a land plant cell. The cell division cycle is divided into five phases (interphase, prophase, metaphase, anaphase, and telophase). Each phase is distinguishable in terms of the condition of the nuclear contents. During mitosis, chromosomes replicate to form sister chromatids that normally are genetically identical (see fig. 1.9). Sister chromatids are drawn toward opposing poles that will later be the sites of two nuclei, each containing the same number of chromosomes as the original cell.

and Vickerman (1980); Margulis (1981). Nevertheless, attention must be drawn to the provocative suggestion that the terminal replication site on the bacterial chromosome and the centromere of eukaryotic chromosomes are evolutionarily homologous (Maynard Smith and Szathmáry 1995). This suggestion is based on a comparison between bacterial chromosome replication and an atypical form of mitosis called pleuromitosis. Recall that the bacterial chromosome is a single circular strand of DNA. Chromosome replication begins at a single site, proceeds in both directions around the circular DNA strand, and ends at a single point called the terminus. In contrast, chromosome replication in most eukaryotic cells involves a bipolar spindle apparatus that emerges from the two opposing ends of the dividing cell. Each eukaryotic chromosome is attached to the equator of the spindle apparatus by its centromere, the structure involved in pulling a chromosome to one of the two spindle poles (see fig. 3.13). At first glance it is hard to see how the eukaryotic pattern of chromosome replication could evolve from that of a bacterial antecedent. However, some presumably ancient eukaryotes have an unusual form of mitosis called pleuromitosis (see Raikov 1982). In these organisms the two halves of the spindle apparatus of the dividing cell lie side by side rather than across from one another. In pleuromitosis the centromeres of chromosomes are attached to the inside of the nuclear membrane very much the way the single chromosome in a bacterial cell remains attached to the rigid bacterial cell membrane. Thus it is possible that the centromere of the linear eukaryotic chromosome is evolutionarily derived from the terminus of the prokaryotic circular chromosome that somehow split to acquire the familiar eukaryotic linear form. If so, then it is equally likely that the site of attachment of the bacterial chromosome to the rigid cell wall and the pole of the eukaryotic spindle apparatus are also homologous.

The selective advantages of the transformation from a circular chromosome to a linear chromosome are not immediately obvious. However, if there are limits to the rate at which chromosome replication proceeds, the single site of bacterial chromosome replication may have placed an upper limit on the size of the prokaryotic genome, because genome size tends to correlate positively with the rate of cell division across broad categories of organisms. Thus one advantage to the linear eukaryotic chromosome is the manifold sites for replication that could have permitted an increase in genome size without a concomitant increase in the duration of cell division.

Naturally, all scenarios for the evolution of mitosis beg the question of the origin of chromosomes, which regardless of their shape or size are collections of linked genes. Why and how this linkage happened remains a puzzle because connected genes would theoretically take longer to replicate than their disconnected counterparts, while the molecular mechanisms responsible for linking ancient genetic molecules to form "proto-chromosomes" are complex and problematic (see Szathmáry and Maynard Smith 1993). However, there are a few clear selective advantages to linking genes (Maynard Smith and Szathmáry 1995). First, because a complement of unlinked genes can be unequally partitioned during cell division, so that the two derivative cells may have vastly different genetic compositions, linked genes are certain to leave copies of themselves in each of the derivatives of division. Second, whereas unlinked genes can potentially replicate at very different rates and thus compete with one another for molecular resources in the same cell, linked genes must replicate synchronously. By means of chromosomes, all linked genes observe the same tempo of cell replication. A third advantage to linked genes is that rapid "selfish" gene replication can be effectively suppressed by mutations at other loci. In this sense gene linkage serves to democratize originally ancient and very loose confederations of potentially competing genetic molecules. How genes were linked to form the first chromosomes and managed to function in a orderly way remains a matter of conjecture, although it is obvious that this happened well before the advent of the evolution of mitosis and the first eukaryotes.

Barring random mutation or accidental damage to chromosomes, mitotic cell division normally results in no genetic variation among cells tracing their lineage to a single cell (all the cells are "chips off the same genetic block"). Genetically identical individuals may be advantageous to an organism particularly well adapted to its environment, provided the environment does not change, or to an organism that is capable of migrating and tracking the environment to which it is well adapted, or to an organism whose physiology or morphology is sufficiently "plastic" to deal with a wide range of environmental conditions. However, over long periods environments change; the ability of many types of organisms to migrate or disperse over great distances to the habitat that suits them best has limits. Thus some genetic variation is desirable in a population because it affords the opportunity to produce physiological or morphological variants that may survive when the environment changes or that may colonize new environments. Sexual reproduction and its resulting genetic recombination has evolved independently in many lineages, probably because it produces significant genetic variation among the individuals within a biparental population. Sexual and asexual reproduction are not mutually exclusive. Many unicellular eukaryotes reproduce asexually by means of mitosis. These types of organisms benefit from the genetic variation resulting from sexual reproduction when environmental conditions change, and they benefit from asexual reproduction whenever the environment is stable and amenable to growth and survival.

Sexual reproduction has evolved only among eukaryotic organisms because its precondition is an organism that has perfected meiosis and therefore possesses a nucleus (fig. 3.14). To understand meiosis, first appreciate that within the life cycle of all sexually reproducing organisms there exists at least one cell whose nucleus contains homologous pairs of chromosomes, or simply "homologues." The cell containing the pair of homologues is called a diploid cell, and the number of chromosomes within the diploid cell is the diploid number, denoted as $2n$. During meiosis the diploid number is reduced by half. Typically, four cells are produced when meiosis is complete. These cells are called haploid cells, and the number of chromosomes within each haploid cell is the haploid number, denoted as n. The importance of reducing the chromosome number by half is readily apparent because two haploid cells must fuse to produce a new diploid cell during sexual reproduction. If the chromosome number were not reduced to the haploid number, then the number of chromosomes would double endlessly each time sexual reproduction occurred.

As we have seen (chapter 1), although each pair of homologues within the nucleus carries the genes that control the same inherited traits, the genes controlling the same trait are not necessarily identical. They can have different forms, called alleles, each of which may evoke a different phenotypic condition. For example, consider the color of a flower's petals. A gene on one homologous chromosome may code for a plant whose flowers have white petals. In contrast, the gene on the other homologue may result in an individual with red petals. Since the nuclei of plants with flowers contain both homologues, a plant whose nuclei have the "white petal" gene and also the "red petal" gene may have pink petals, while the flowers of an individual whose nuclei contain homologues with exactly the same form of the gene will have either white or red petals. These differences among the individuals within a population of flowering plants result from meiosis, because during meiosis the homologous chromosomes within a diploid cell eventually come to reside in four haploid cells (see fig. 3.14). The number of possible permutations of different homologues in haploid cells depends on the number of chromosomes within the diploid nucleus from which they arose. For example, if the diploid nucleus has twenty chromosomes, the possible number of genetically different haploid cells is 2^{20} or 1,048,576. Each diploid human cell has twenty-three pairs of chromosomes, and so each diploid cell is theoretically capable of producing 2^{23} or 8,388,608 genetically different

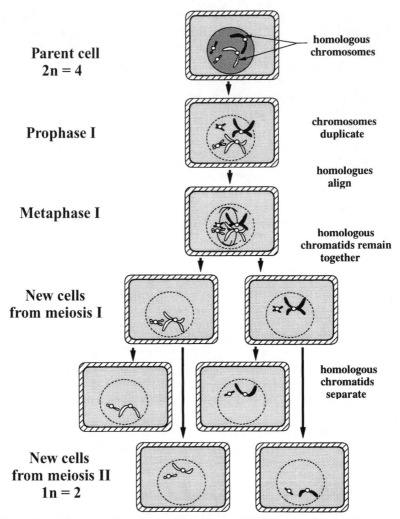

Parent cell
2n = 4

homologous
chromosomes

Prophase I

chromosomes
duplicate

homologues
align

Metaphase I

homologous
chromatids remain
together

New cells
from meiosis I

homologous
chromatids
separate

New cells
from meiosis II
1n = 2

Figure 3.14. Diagrammatic rendition of the meiotic cell division of a plant cell. During meiosis, homologous chromosomes pair and replicate to form sister chromatids. During metaphase I, homologous chromosomes are drawn to opposing cell poles; during metaphase II, sister chromatids are drawn to opposing cell poles. Meiosis normally results in the formation of four haploid nuclei that subsequently reside in four separate cells as a consequence of cytokinesis.

haploid cells. The genetic variation that can be introduced into a population is compounded because, during sexual reproduction, two haploid cells (called gametes) must fuse to produce a new diploid cell (called the zygote). Even in the absence of mutation or other types of genetic alterations, the number of genetically different permutations of the zygote equals the square of the number of possible genetically different gametes. Thus the human zygote has $(2^{23})^2$ or 70,368,744,177,664 different genetic combinations, which can be increased by mutations.

But is the genetic variation obtained from sexual reproduction always evolutionarily beneficial? Is sexual reproduction evolutionarily favored because it obtains new, potentially adaptive gene combinations? Consider that whenever a well-adapted individual reproduces sexually it conveys only 50% of its genes to its offspring. Thus some of the offspring may not be as well adapted to their environment as the most fit parent. This consequence is called the "50% fitness cost" to sexual reproduction (see Williams 1975; Maynard Smith 1978). Clearly, we cannot invoke the argument that "sexual reproduction evolves because it confers long-term evolutionary advantages." Because natural selection cannot foresee the future, it cannot have a "plan." So how do we resolve the issue of sex?

First, recognize that the "50% fitness cost" of sexual reproduction holds true only when the relative fitness of mating individuals within the population differs significantly and when asexual reproduction cannot occur. If the former is true, then the precondition that "a hypothetical population . . . is particularly well adapted to its environment" is not fulfilled—the population may be in the *process* of becoming adapted, but great differences among the fitnesses of individuals arguably are not the hallmark of a population that has *become* adapted to its particular environment. Second, note that the preclusion of asexual reproduction is biologically unreasonable. There is every reason to assume that the first unicellular eukaryotes could reproduce asexually (Knoll 1983). Barring high mutation rates, asexual reproduction through mitosis can serve to maintain a well-adapted population in its particular environment. The bonus of sexual reproduction is that it produces virtually endless genetic variants that can supply the raw materials with which to explore new and different habitats or persist in the same habitat whenever environmental conditions change. Thus the "50% fitness cost" of sexual reproduction is illusory in the sense that asexual and sexual reproduction are not mutually exclusive (Williams 1975).

Because sexual reproduction requires diploid cells and the perfection of meiotic cellular division, the critical questions are, What was the initial advantage to diploidy? and How did diploidy first evolve? The most plausible answer to the first question is that diploidy enriches the diversity of a population's gene pool. Each diploid cell contains pairs of homologous chromosomes that can bear different alleles, and so a population of diploid cells consists of heterozygous and homozygous individuals. And in the case of codominance or incomplete dominance, the interactions of alleles in the same cell can establish subtle variations in biochemistry, physiology, morphology, and behavior. Indeed, a high correlation exists between diploidy and the complexity of developmental interactions (as first noted by Stebbins 1960; Raper and Flexer 1970).

Another advantage of diploidy is that damaged chromosomes can use their undamaged homologues as templates to repair damaged DNA. Recall that during meiosis prophase I, homologous chromosomes pair. At this juncture, gene conversion can change a defective chromosome back to its original, undamaged state. This phenomenon has inspired the DNA-repair hypothesis arguing that meiosis and sex evolved in response to selection pressures for DNA repair. The hypothesis implies that the diploid state need be only transient, existing just long enough to bring homologous chromosomes together to effect DNA repair. The hypothesis gains credibility in that many extant unicellular plants belonging to extremely ancient lineages have a sexual life cycle involving a long-lived haploid cell and a very short-lived diploid cell.

An alternative hypothesis is the selfish DNA or transposon hypothesis. (A transposon is a fragment of DNA with the ability to multiply itself and move spontaneously within the genome of an organism.) It is called the selfish DNA hypothesis because a transposon's main function seems to be simply to make more copies of itself. In this sense transposons are genomic parasites. The hypothesis argues that the driving force for the origin of sex was parasitic DNA sequences that enhanced their own fitness by promoting the cycles of chromosome fusion and segregation that occur during meiosis (see Michod and Levin 1988).

How diploidy and meiosis evolved is not known with any assurance, but the first diploid cells resulting from the fusion of two haploid cells (syngamy) most likely benefited from the equivalent of hybrid vigor (heterosis). Most deleterious mutations are partially or wholly recessive and thus negatively affect the heterozygote to a lesser extent than the homozygote. The first diploid cells may have evolved when haploid cells accidentally fused and remained so for a time as a single, functionally proficient cell. An additional advantage to maintaining the diploid condition is the ability to repair any damage to double-stranded DNA, because repairs can be effected only if a second undamaged DNA molecule on a homologous chromosome is available as a template. Assuming that the first eukaryotes were haploid, diploid cells may also have evolved as a consequence of mitotic nondisjunction. That is, the first diploid eukaryotic cells may have been the first autopolyploids (see fig. 2.6). The phenomenon called endomitosis, in which the chromosome number of haploid cells spontaneously doubles, occurs in a variety of presumably ancient unicellular plants and animals. In opposition to the benefits of heterosis, the immediate adaptive advantage to this form of "diploidiza-tion" may have been the conferral of a smaller ratio of surface area to volume relative to the haploid condition. This may have afforded diploid cells a higher metabolic efficiency than their haploid counterparts (Cleveland 1947; Margulis and Sagan 1986). In either case meiosis may have become perfected through natural selection as a mode of cell division to reacquire the ancestral haploid condition, which would have been an advantage when nutrients were limited because haploid cells tend to grow faster than diploids (see Weiss, Kukora, and Adams 1975). It has been suggested that ancient cell cycles alternating between haploid and diploid phases in asexually reproductive organisms may have promoted the origin of sex by providing a preexisting mechanism for a regular reduction in chromosome number (Kondrashov 1994).

The fossil record helps very little in adducing the "ploidy" of the most ancient fossils of eukaryotic cells. The chromosome number of an ancient eukaryotic cell cannot be judged with certainty from fossil remains. Indeed, the "nucleus" identified for some Proterozoic fossil cells may be nothing more than the remains of the shriveled, once living cytoplasm that pulled away from its cell wall as it died and became fossilized. Nor does the fossil record really shed much direct light on the initial evolution of meiosis and sexual reproduction, except perhaps to identify the time when the first sexual organisms evolved. Although some fossil eukaryotic cells appear to be preserved in the act of "fusing" (in prelude to subsequent meiosis?), they could just as conceivably have been preserved in the act of "dividing" (mitotically?). These and other ambiguities constantly frustrate the professional paleontologist who tries to piece together the complex history of ancient life forms from their scattered fossil remains, although numerous areas of agreement exist among active workers in this field of paleontology (see Knoll 1983; Schopf 1983).

Multicellularity and the Plant Life Cycle

When is an organism multicellular? A generic answer that simultaneously encompasses animal and plant life is that multicellularity results when two or more neighboring cells adhere, interact, and physiologically communicate. Among multicellular organisms, direct physical contact among neighboring cells is achieved principally in one of four nonexclusive ways: tight junctions, desmosomes, gap junctions, and plasmodesmata (fig. 3.15). Tight junctions occur where proteins built into the external membrane of one cell bond with like proteins in the cell membrane of a neighboring cell. Where they occur, tight protein junctions cause the two neighboring cell membranes to adhere. Desmosomes are regions of intracellular filaments that extend through the external spaces between adjoining cell membranes. The filaments, which cross into the intercellular space, permit substances to pass freely between adjoining cells. Gap junctions are open pores surrounded by transmembrane proteins that pass through the spaces between adjoining cells. Like desmosomes, gap junctions permit materials to move between cells. Plasmodesmata are open channels through the plant cell wall that connect neighboring cells. In multicellular plants these channels are required because each cell is typically surrounded by a cell wall composed of comparatively densely packed materials that may present a physical obstacle to the passage of modestly sized molecules from one adjoining cell to another. Plasmodesmata permit plant cells to interact and intercommunicate physiologically. They are small open channels within the cell walls through which strands of cytoplasm interconnect the living protoplasts of adjoining cells. The cell membranes of neighboring plant cells, which line each of the plasmodesmatal channels, are continuous from one cell to the next. A multicellular plant thus consists of a single living protoplast incompletely subdivided by an infrastructure of cell walls. Throughout this continuous living mass, water and small molecules may pass with comparative ease. One of the exquisite features of the multicellular plant body is that the flow pattern of water and other small molecules can be modified by altering the number and location of plasmodesmata among neighboring cells. Adjoining cells sharing many plasmodesmata may serve as preferred routes of transport for nutrients. Preferential transport of water and nutrients is a requisite for the survival and growth of large plants (see chapter 4).

In contrast to eukaryotes, which include unicellular and multicellular representatives, no prokaryote is known to produce plasmodesmata, tight junctions, desmosomes, or gap junctions. Colonial and filamentous prokaryotes exist (e.g., *Coelosphaerium* and *Oscillatoria*, respectively), but these growth forms do not involve cytologically connected cells. Nevertheless, narrow cytoplasmic channels, called microplasmodesmata, that adjoin the cells of some prokaryotes have been reported (e.g., Westermann et al. 1994). Whether microplasmodesmata serve as cytoplasmic conduits for the transport of metabolites between adjoining prokaryotic

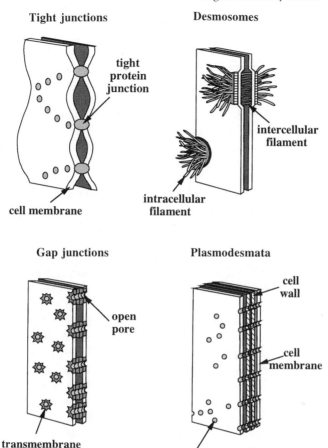

Tight junctions

tight
protein
junction

cell membrane

Desmosomes

intercellular
filament

intracellular
filament

Gap junctions

open
pore

transmembrane
protein

Plasmodesmata

cell
wall

cell
membrane

plasmodesma

Figure 3.15. The four ways the cells of multicellular eukaryotic organisms are physically and cytolologically interconnected. Each diagram depicts a small section of two adjoining cells. The space between the adjoining cells has been greatly exaggerated to emphasize the structures interconnecting cells.

cells remains unclear. If they do, then multicellularity, broadly defined, is not restricted to the eukaryotes and finds its equivalent in some prokaryotes. If not, then the absence of multicellular prokaryotic organisms suggests that the evolution of nuclei and other organelles was either a prerequisite for multicellularity or a collateral condition.

It is clear that multicellularity evolved independently in different eukaryotic lineages and so may have directly or indirectly conferred some advantage in survival or reproductive success. One of the more provocative explanations for multicellularity is offered by Lynn Margulis (1981), who notes that cells can divide and move, but not at the same time. Margulis notes that this presents a problem for unicellular organisms suspended in water, because dividing cells can sink to depths that exclude photosynthesis; likewise, they cannot flee from predators. Some organisms, like the ciliates, have evolved elaborate nonstandard mechanisms for mitotic cell division. Others have permanently retracted their cilia or flagella and abandoned motility altogether. Still others divide and move in different stages of their cell division cycle. As Margulis points out, however, one solution of great evolutionary significance is to keep the two cells produced by mitotic cell division together and to specialize one of these cells for locomotion and the other for cellular division. According to this model for multicellularity, the evolution of mitosis created a dilemma that was resolved by multicellularity. To be sure, although unicellular species cannot divide and move at the same time, many of these organisms are eminently successful, and therefore the failure to perform both biological tasks simultaneously has not imposed any obvious "universal" selection pressure. Thus we are obliged to consider the possibility that the evolution of multicellularity has occurred for different reasons in different lineages.

In ideal circumstances, one would like to explain each separate origin of multicellularity, but this is not possible because living transitional forms between the ancestral unicellular and the derived multicellular condition are absent in some lineages. No unicellular brown algal species currently exists. All are multicellular, although few scientists doubt that the multicellular species of the brown algae trace their origins to unicellular brown algal ancestors. Perhaps some transitional forms were too rare or too fragile to be preserved in the fossil record, or perhaps they are simply not immediately recognizable as belonging to the brown algal lineage. In the absence of detailed chemical or cellular preservation, it would be virtually impossible to distinguish a fossil unicellular brown alga from a fossil unicellular green or red alga. Alternatively, multicellularity may have evolved suddenly in some lineages as a consequence of a genetic revolution.

Even though the origins of multicellular plants are shrouded, some of the selective advantages conferred by partitioning the plant body into "cells" are immediately obvious. For example, a multicellular plant continues to survive after it has sexually reproduced. In contrast, a unicellular plant that sexually reproduces must convert itself into a gamete, or sex cell. From the perspective of its population it ceases to exist—quite literally giving its all to the next generation! The consequences of this self-sacrifice are quickly seen by comparing the growth of populations of multicellular and unicellular species. In both cases, assume that each adult requires a "mate" to complete the sexual reproductive cycle; two haploid gametes fuse, and the resulting diploid zygote divides meiotically to produce four functional haploid plants. The only salient difference between these two species is that the multicellular parents continue to exist after sexual reproduction whereas the unicellular parents essentially vanish from the population (fig. 3.16). Note that each pair of multicellular parents produces six individuals: the two original parents and four offspring. In the more general case, the total number of multicellular plants in the population increases over successive generations according to the progression N $3N$ 3^2N, 3^3N, 3^4N, 3^5N, and so forth, where N is the original number of parents in the population. For example, starting with two multicellular parents of which only one cell per parent forms a gamete, the population will increase as 2, 6, 18, 54, 162, 486 . . . assuming that no offspring or parents die. In contrast, the number of unicellular plants merely doubles with each successive generation. That is, starting with any N unicellular adults, the population increases according to the procession N^1, N^2, N^3, N^4, N^5, N^6, . . . When $N = 2$, the population will increase as 2, 4, 8, 16, 32, 64, . . . once again assuming none of the offspring or parents die along the way. By dividing the number of multicellular individuals by the number of unicellular individuals produced in successive generations, one quickly sees that, for the successive generations, N/N, $3N/N^2$, $3^2N/N^3$, $3^3N/N^4$, $3^4N/N^5$, $3^5N/N^6$, . . . Thus, when $N = 2$, these quotients are 1, 6/4, 18/8, 54/16, 162/32, 486/64, . . . (see fig. 3.16). By the sixth generation the population of multicellular plant species is more than seven times that of unicellular plant species. Notice that, because it was assumed each multicellular plant produces only one gamete at a time, this is the "best case" scenario for the unicellular species. These calculations ignore that reproductive rate, on the average, declines as the 3/4 power of an organism's size. But the difference in the rate of reproduction between a unicellular plant and a multicellular one composed of, say, two or three cells is trivial, whereas the slightly larger size of the two-cell plant provides an immediate selective advantage perhaps sufficient to favor an even larger multicellular body size.

Perhaps because of the selective advantage of "persistence," multicellularity has arisen in either one or both phases in the life cycle of the same plant species (fig. 3.17). When both phases in the same life cycle are multicellular, each species consists of two individual organisms, and both phases have the opportunity to diverge morphologically and anatomically to cope with the particular environmental context facilitating its reproductive function. To understand this, consider a life cycle in which the haploid and the diploid phases are unicellular. In this life cycle, two haploid cells fuse to produce a diploid cell that then divides meiotically

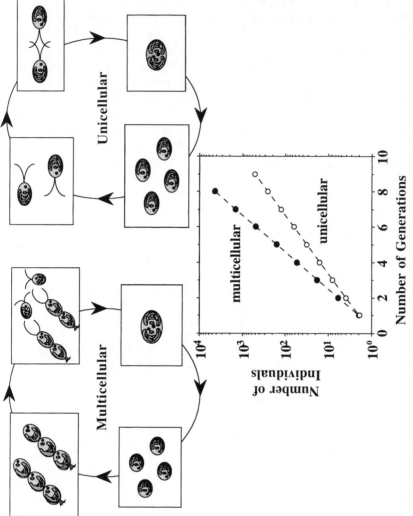

Figure 3.16. Comparison between the number of individuals produced by successive generations of a multicellular and a unicellular plant (lower graph). The life cycles of the multicellular plant (left) and the unicellular plant (right) are identical. The number of individuals composing the successive populations of the multicellular plant increases at a greater rate than the number composing the successive populations of the unicellular plant.

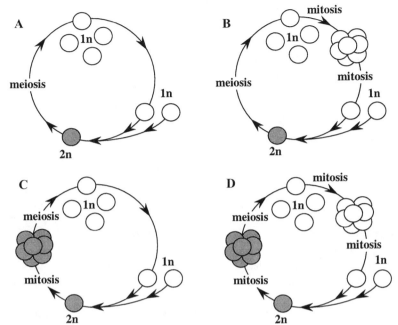

Figure 3.17. Comparisons among life cycles, indicating where multicellularity has occurred as a consequence of mitotic cellular divisions. Shaded circles indicate diploid individuals; unshaded circles indicate haploid individuals: connected circles indicate multicellular individuals. (A) life cycle with no multicellular individual (e.g., the green alga *Chlamydomonas*); (B) haplobiontic-haploid life cycle (e.g., the green alga *Stigeclonium*); (C) haplobiontic-diploid life cycle (e.g., the brown alga *Fucus* and humans); (D) diplobiontic life cycle (all embryophytes). A comparison among the four life cycles shows that they are identical except where multicellularity has been "intercalated" in the life cycle (see A).

to produce haploid cells (e.g., *Chlamydomonas*). A life cycle with a multicellular haploid individual capable of producing many gametes results by intercalating mitotic cell divisions after meiosis (e.g., *Spirogyra*, *Chara*, *Coleochaete*). Conversely, a life cycle with a multicellular diploid individual results by intercalating mitotic cell divisions before zygotic meiosis (e.g., *Fucus*). And intercalating mitotic cell divisions before and after zygotic meiosis produces a life cycle that alternates multicellular haploid and multicellular diploid individuals (e.g., *Polytrichum*, *Pinus*; *Prunus*). Although there are significant differences among them, the four life cycles just reviewed share a very basic feature—an alternation between a diploid cell and a haploid cell (Niklas 1994). This similarity appears to have gone unnoticed in early debates about the origins of the plant life cycles (Celakowsky 1874; Church 1919; Bower 1935).

Aside from conferring the potential for morphological and anatomical divergence between the diploid and haploid phases in the same life cycle, another advantage of multicellularity is the ability to specialize cells into tissues and tissues into organs. Recall that plant cells are bonded by their cell walls, which are perforated by plasmodesmata (see fig. 3.15). Small molecules like glucose and water can pass through these perforations, and the more plasmodesmata there are between two adjoining cells, the greater the rate of molecular transport. Modifications in the location and number of plasmodesmata provide a way to direct the flow of metabolites and water along preferred pathways (Biao et al. 1988) and occur throughout the plant kingdom. For example, some large marine kelps produce "trumpet cells," elongated cells with numerous interconnecting plasmodesmata at their end walls. Very much like the phloem cells in vascular plants and the leptoids found in some mosses, the trumpet cells transport cell sap loaded with sugar from sunlit fronds to more deeply submerged portions of the kelp plant. By the same token, numerous and different plants have evolved functionally analogous preferred routes for water transport (e.g., hydroids in mosses and xylem elements in vascular plants).

Multicellularity confers other advantages. Partitioning the plant body into cells dramatically increases the volume and total surface area of cell membranes, enhancing the ability to exchange chemicals between cells and the external environment. The "endoskeletonlike" infrastructure of cell walls resulting from multicellularity also provides a rigid system of internal struts and beams and columns that mechanically reinforces the plant body against the effects of moving water or air. Yet another advantage of multicellularity is that portions of the plant can be isolated from the rest of the plant body when they are attacked, damaged, or invaded by an herbivore or a pathogen. Also, multicellularity allows for large size, and large organisms are less affected by small changes in temperature, humidity, or nutrient availability than their smaller counterparts (multicellularity enhances homeostasis). Finally, large size, which generally requires some degree of cellularization, is correlated with life span. In general terms, larger multicellular organisms live longer than their smaller counterparts and so may have a competitive advantage in terms of occupying space and acquiring nutrients and sunlight, in addition to producing more reproductive structures. In the next chapter we shall see that multicellularity is a precondition for survival and successful reproduction on land (or more properly speaking, in air).

The question remains why different variants of the same basic life cycle

diploid phase

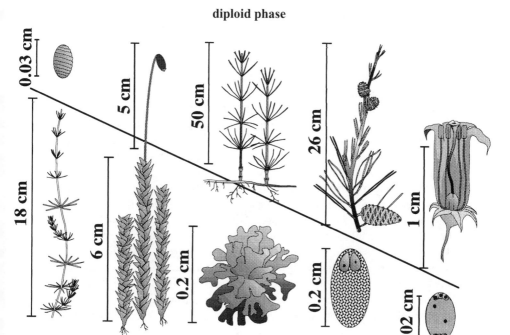

haploid phase

Figure 3.18. Comparisons between the relative size of the diploid and haploid phases in the life cycles of representative plants from progressively more recent lineages (from left to right). The diploid phase in each life cycle is shown above its corresponding haploid phase. The reduction in the size of the haploid phase relative to the size of the complementary diploid phase broadly corresponds to a reduction in the duration of the phase in the life cycle. From left to right: *Chara,* a green alga; *Polytrichum,* a moss; *Equisetum,* a horsetail; *Pinus,* a gymnosperm; *Lycopersicum,* a flowering plant.

dominate different plant lineages and why the diploid and haploid phases in the same life cycle have different durations. The relative duration of the haploid and diploid phases during the sexual life cycle of species varies greatly among organisms (Bell 1982, 1994; Klinger 1993), yet no one has offered a good evolutionary explanation for this variation (Maynard Smith 1978; Kirkpatrick 1994). It has been suggested that a prolonged diploid phase evolved because the diploid genome can mask the deleterious effects of mutations (Muller 1932; Crow and Kimura 1965;

Bernstein, Byers, and Michod 1981). Others point out that this advantage may be counterbalanced by the disadvantage that diploids have twice the mutation rate of haploids. Theoretically, diploidy would reduce the mean fitness of the population when the duration of the diploid phase equals or exceeds the duration of the haploid phase (Crow and Kimura 1965).

Cheryl Jenkins and Mark Kirkpatrick (1995) have modeled the evolutionary consequences of the relative duration of the haploid and diploid phases based on different assumptions about the rate of recurrent deleterious mutations, genetic linkage, and epistasis. Their mathematical models predict that a life cycle dominated by the diploid phase is evolutionarily stable when allelic recombination is entirely free (no linkage) even though prolonging the diploid phase reduces the mean fitness of a population because the mean number of mutations carried per genome is independent of the dominant phase in the life cycle. Their models predict that individuals with a prolonged diploid phase have, on average, the same number of mutations per genome as do more haploid individuals. Because reverting to haploidy exposes deleterious mutations that were previously masked by diploidy, the spread of individuals with long haploid durations is predicted to be very difficult. However, the models predict that the haploid phase will dominate when complete linkage occurs or when mutations are highly deleterious.

Jenkins and Kirkpatrick (1995) found that life cycles retaining both the haploid and diploid phases are evolutionarily unstable. Thus their mathematical models do not account for the life cycle of land plants or many different kinds of algae even under conditions where selection pressures on the viability of the diploid and haploid phases overwhelm the force of selection against deleterious mutations. The models of Jenkins and Kirkpatrick predict that the phase with the lowest mortality rate will evolve in duration and become the dominant phase in the life cycle provided the differences in the mortality rates of the haploid and the diploid phases are very large. That is, pronounced phase-specific mortality, perhaps resulting from morphological or ecological differences in the two phases, can override the effect of selection against deleterious mutations in the evolution of stable life cycles. A consistent feature of these models is that life cycles containing both the haploid and diploid phases are evolutionarily unstable. This may shed light on the fact that most green algae have life cycles dominated by either the haploid or the diploid phase and explain why, over the course of land plant evolution, the diploid phase in the life cycle has progressively increased in duration as well as size relative to the duration and size of the haploid phase (fig. 3.18). Technically, all land plants have mixed life cycles in which the multicellular haploid phase alternates with the multicellular diploid phase, but there is nevertheless an evolutionary trend toward reduction of the haploid phase.

4   The Invasion of Land and Air

> *The history of life is to be studied by a great variety of means, among which special importance attaches to the actual historical record in rocks and the fossils contained in them.*
>
> George Gaylord Simpson

Compared with the early events in plant history discussed in chapter 3, the four adaptive evolutionary events treated here may appear less significant (the invasion of land, the evolution of conducting tissues, the appearance of the first seed plants, and the evolution of the flowering plants). But consider that the first organisms to live and reproduce on land were plants, and that by providing shelter and food they paved the way for invasion of the land by animals. Therefore the first zoological landfall was contingent on the greening of the terrestrial landscape by plant life, which was really more an invasion of air than of land. Many organisms can live in the wet interstices of soil and rock, but the first to challenge and later subdue the air were plants. Life's onslaught on air involved the evolution of vascular tissues, which provide low-resistance conduits for the flow of water and food. Without these plant tissues all terrestrial life, animals as well as plants, would remain physiologically anchored to a semiaquatic existence.

The final greening of the land required seeds. Seed plants can successfully reproduce as well as survive vegetatively even in arid habitats, and so they have radiated into the drier uplands and deserts of the Earth. Eating their way through this green world, animals did not politely follow: throughout most of life's long history, animals were the mortal enemies of plants. With the evolution of flowering plants this changed into a truly symbiotic partnership between some plants and animals. Like their other terrestrial plant counterparts, flowering plants are sedentary organisms, providing animals with fodder and shelter. But most flowering plants require or benefit from the transfer of pollen from one individual to another. Wind and water can serve in this capacity, but many pollen grains fail to connect with flowers and so are wasted in a reproductive sense. In contrast, animals are effective pollinators because they can recognize flowers belonging to the same species. Flowering plants have evolved intricate rewards and morphological subterfuges to attract animals for this purpose. As noted by Bruce Tiffney, it is not hyperbole to say that the flowering plants have trained animals to become their surrogate brain and muscle.

The Invasion of Land: The Embryophytes

As noted, the botanical invasion of "land" was really an invasion of "air," an evolutionary assault on the dehydrating effects of the atmosphere (Raven 1985). Plants survive, grow, and reproduce on land because they are able to resist or tolerate desiccation. This ability involves numerous morphological and physiological departures from the ancestral aquatic condition of plants. For example, many land plants, like the mosses and liverworts, hug the moist substrates they grow on. This growth posture maximizes the surface area through which water can be absorbed and minimizes exposure to rapidly moving and drier air (Niklas 1994). Although many land plants embrace this water-conservation tactic, hugging a moist substrate is not a dependable method for survival on land. Even short, prostrate plants dehydrate when exposed to high wind speeds. Many moss species can withstand extreme dehydration, but at the expense of suspending growth and reproduction. A far more dependable method of reducing water loss, one that permits the true invasion of air, is to coat external surfaces with a pellicle impenetrable to water molecules. Unfortunately the cuticle of land plants, which resists the diffusion of water, is also impermeable to carbon dioxide and oxygen. The solution was to pierce the cuticle with holes through which these gas molecules can pass. The stomata that perforate the cuticles of vascular plants provide microscopic avenues for gaseous diffusion, and the guard cells flanking stomata adroitly limit the rate at which water is lost through external surface areas while simultaneously ensuring a direct communication between the external atmosphere and the plant's moist interior. Life in the air was further assisted in most land plants by elaborating internal tissues with labyrinthine intricacy, thereby providing a large moist surface area through which gas can diffuse. Much like the alveoli of the mammalian lung, the aerenchyma of land plants reflects one of the overarching morphological themes of terrestrial life—the topological invagination of external into internal surface area.

Nevertheless, the persistence of plants on land was not ensured by mere vegetative acclimatization to air. The appellation "land plant" must be reserved for photosynthetic eukaryotes capable of successful reproduction on land. The distinction between two fundamental classes of adaptation—those concerned with the survival of the individual and those concerned with genetic continuity among individuals in successive generations—is important. Although reproduction is not essential to the survival of the individual, it is the route by which every species historically persists and evolves. Just like their ancestors, the first land plants undoubtedly were capable of some form of asexual reproduction, multiplying by throwing forth genetically identical individuals into the aerial realm. But asexual reproduction cannot account for the first land plants, which by definition were a significant evolutionary departure from their aquatic ancestors, nor can it account for the dramatic evolutionary changes that followed the first vegetative landfall. Seen in its entirety, the history of land plants could not conceivably have happened in the absence of sexual reproduction.

If it is true that colonization of the land required reproductive and vegetative adaptations, then a comparatively straightforward ecological definition of "land plant" emerges: *a land plant is any photosynthetic eukaryote that can survive and sexually reproduce on land.* This definition combines the two fundamental classes of adaptation, those that permit the survival and growth of the individual and those that ensure the genetic continuity of its species through time. It also emphasizes that two lines of evidence are required to prove an organism is a land plant—the organism must be capable of surviving on land, and its species must be shown to persist evolutionarily on the terrestrial landscape by virtue of sustained reproductive success. By the same token, this definition excludes any type of organism that attempted to colonize the land but ultimately failed. In this regard the fossil record indicates that the terrestrial landscape may initially have been occupied by very different phyletic groups of plants, but it also shows that these forays onto land were largely unsuccessful. In this respect they provide a dim glimpse into eclectic but ultimately unsuccessful ways different types of plants coped with the desiccating effects of air. In the final analysis, these biological experiments are curiosities that tend to detract from two important conclusions: first, the only bona fide land plants are the embryophytes, and second, the embryophytes are a monophyletic group of plants. That is, all embryophytes ultimately descend from a common ancestor. The juxtaposition of these two conclusions reveals that the ecological definition of "land plant" and the taxonomic definition are synonymous—the land plants are the embryophytes (see Gray 1985; Graham 1993).

Surprisingly, the term embryophyte may be unfamiliar even to some professional biologists. Yet all modern land-dwelling plants are embryophytes. Specifically, the embryophytes include the mosses, liverworts, hornworts, ferns, horsetails, lycopods, gymnosperms, and angiosperms. Despite the tremendous diversity in size and appearance among these plants, all modern embryophytes share two important life cycle features: an alternation of generations involving a multicellular haploid generation and a multicellular diploid generation, and the retention of eggs and their diploid embryos within a structure called the archegonium. The haploid generation is called the gametophyte because this multicellular plant produces gametes—sperm cells or egg cells or both. The diploid generation is called the sporophyte because it produces spores by meiosis, each encased in a spore wall composed of the highly chemically resistant biopolymer sporopollenin. The term embryophyte literally means "embryo-bearing plant," alluding to the fact that the gametophyte generation of every embryophyte retains its eggs and its juvenile diploid embryos within archegonia.

Unlike the egg- and sperm-bearing reproductive structures of most algae, embryophytes enclose their gamete-producing cells within a sterile jacket of cells. The egg is enclosed within the archegonium; the sperm are enclosed within the antheridium (fig. 4.1). Although the evolution of the antheridium is important, the evolutionary significance of the archegonium is the focus because this structure serves to protect, nourish, and developmentally influence the growth of the juvenile sporophyte embryo.

Technically, the archegonium is a multicellular organ ontogenetically derived from a single haploid cell on the gametophyte. Although superficially simple in appearance, the archegonium is an extremely sophisticated reproductive device. In addition to protecting the egg and the diploid embryo that will eventually develop within it, the archegonium serves as a physiological interface shuttling nutrients produced by the gametophyte to the developing diploid embryo (Thomas et al. 1978; Ligrone and Gambardella 1988; Renault et al. 1992; Graham 1993). This interface is physically manifested in the form of specialized transfer cells at the base of the archegonium, called the venter. These cells have highly branched or convoluted walls that dramatically increase the surface area of the cell membranes through which nutrients pass. The cells of the venter are physiologically active, transporting sugars from the gametophyte into the fertilized egg and developing embryo. In this respect the sterile cells of the archegonium function in a way analogous to the "placental transfer cells" occurring at the junction between the female placental mammal and its embryo.

How the embryophyte life cycle and archegonium evolved has been the subject of intensive research and speculation (see Graham 1993), but it is clear that, despite their complex appearance and physiology, the embryophytes evolved from some type of alga. We can say this with absolute, albeit ironic, confidence because by definition the "algae" are all plants lacking archegonia. Thus the algae represent the only possible types

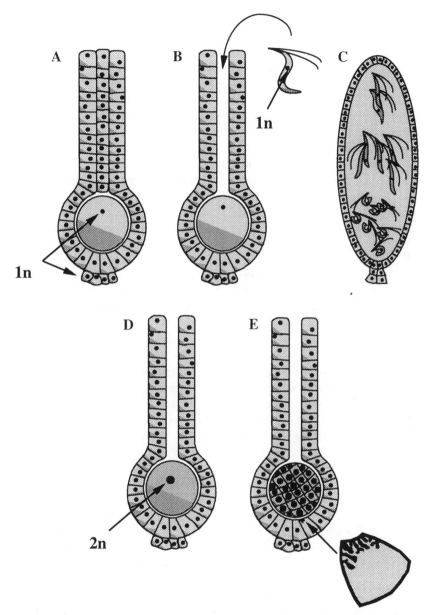

Figure 4.1. Fertilization and embryogenesis in the archegonium: (A) archegonium consisting of a tubular-shaped neck and a basal globose venter containing a single haploid egg; (B–C) fertilization of the egg (B) by sperm cell released by antheridium (C); (D–E) diploid zygote (fertilized egg) in archegonium (D) develops into multicellular diploid embryo (E); (E) amplified view of a venter cell showing the convoluted wall in contact with the embryo.

of plants from which the embryophytes could have evolved. Nonetheless, there are about 70,000 known fossil and extant species of algae, and as previously noted, the algae are polyphyletic in origin. Indeed, some comparative studies suggest that the algae constitute at least six distinctly different eukaryotic lineages (see Bold and Wynne 1978). Within this century this broad field of candidates for the embryophyte "cenancestor" has been dramatically narrowed. Only the green algae, the Chlorophyta, share with the embryophytes the same cell wall chemical composition (a variety of polysaccharides including cellulose), the same complement of photosynthetic pigments (chlorophyll *a* and *b*; carotenoids), the same principal carbohydrate food reserve (starch), and the same number, location, and type of flagella on motile cells (two apically or laterally inserted whiplash flagella of equal length). Further, among the green algae, only the class Charophyceae shares with the embryophytes the same mode of cell division, which involves a persistent mitotic spindle (Pickett-Heaps 1975, 1976). And within the Charophyceae, the biological features characterizing the order Coleochaetales most completely align with those of the embryophytes (Graham 1993). Among these features are the disintegration of the nuclear envelope and the formation of the phragmoplast during mitotic cell division, the presence of a very particular photorespiratory enzyme called glycolate oxidase, found in organelles called peroxisomes, and the presence of a multilayered cellular structure at the base of the flagella of reproductive cells. Thus all the evidence indicates that, of the 70,000 known species of algae, only the 19 or so currently assigned to the order Coleochaetales share the greatest number of biological features with the over 250,000 living species of embryophytes.

Additional comparisons between the Coleochaetales and the embryophytes help with the reconstruction of the evolutionary steps leading to the first land plants. Specifically, there are only two genera currently placed in the Coleochaetales, *Chaetospheridium* and *Coleochaete*. Because all these species have large, nonmobile eggs just like those of the embryophytes, the ancestor to the embryophytes was undoubtedly oogamous. And though *Chaetospheridium* releases its eggs once they are fertilized, all the species of *Coleochaete* retain their fertilized eggs just as the embryophytes do. In *Coleochaete* this is accomplished by the growth of cortical cells around the zygote, and in at least one species, *Coleochaete orbicularis*, these cells develop localized wall ingrowths similar to those produced in the venter cells of archegonia. There is some evidence that cortical cells may actively transport sugars to fertilized eggs—*Coleochaete* zygotes continue to accumulate large stores of lipids and starch after the egg is fertilized, and overall vegetative growth is stimulated by adding glucose to the medium in which *Coleochaete* is cultivated (Graham et al. 1994). Thus, although they are purely circumstantial, there are good reasons to believe that vegetative and reproductive cells have the ability to actively transport sugars. Additionally, distinctive acetolysis-resistant, autofluorescent materials, which bear a striking similarity to those found in the placental cells of the hornwort genus *Anthoceros*, are produced in the cortical cell walls of all *Coleochaete* species examined. All these similarities may be the result of convergent evolution, but viewed in the broader context of other shared features it seems reasonable to conclude that the common ancestor to the embryophytes and *Coleochaete* had the ability to produce a jacket of sterile cells around its zygotes, and that these cells functioned physiologically in a way analogous to the venter cells of the archegonium.

Yet it is obvious that the Coleochaetales as a group lack two very important features—the archegonium itself and a multicellular diploid phase in the life cycle. In marked contrast to the archegonium and its egg, which develop from a single cell, the cell that gives rise to the egg in *Coleochaete* does not invest the fertilized egg with cortical cells. And among all the nineteen or so species currently placed in the order Coleochaetales, the only diploid cell found in the life cycle is the fertilized egg (which subsequently divides meiotically and then mitotically to produce up to thirty-two flagellated cells). These differences make it unwise to speculate on whether the multicellular diploid generation of the embryophytes evolved before, after, or concurrently with the evolution of the archegonium. Notice that all extant plants with archegonia have a multicellular diploid individual in their life cycle. But this correlation comes from a curious intellectual hindsight. Just because all *modern* embryophytes have archegonia and a multicellular sporophyte, there is no reason to suppose that these two features evolved concurrently.

Indeed, because there are numerous advantages to each of the biological innovations that collectively define an embryophyte, there is no reason to assume that "archegoniate plants" and "embryophytes" are synonymous. Likewise there is no a priori reason to assume that all past embryophytes were land plants. Consider a hypothetical sequence of evolutionary events beginning with an organism very like *Coleochaete* that lived in shallow bodies of freshwater or in moist microhabitats, perhaps near streams, rivers, or lakes (fig. 4.2). Based on previous discussions, this hypothetical organism is assumed to have large, nonmobile eggs protected by cortical cells produced after fertilization. The cortical cells would have reduced the probability of microbial attack on zygotes, in addition to reducing the chances that zygotes would be washed away or dehydrate when the habitat flooded or the water level dropped precipitously. We will assume that this organism already had some type of active

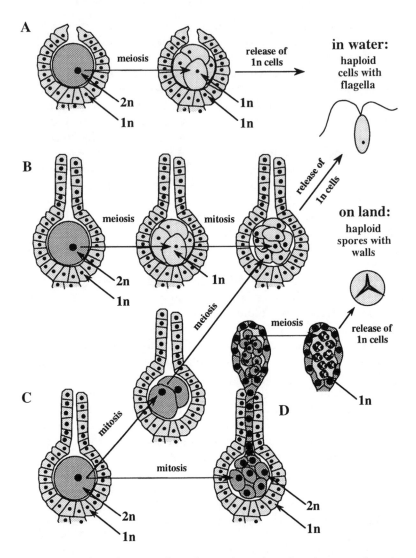

Figure 4.2. Hypothetical events in the evolution of an embryophyte from an archegoniate nonembryophyte ancestor. Events are arranged to emphasize that an archegoniate plant need be neither terrestrial nor an embryophyte.

sugar transport system, because the large, nonmobile eggs of most green algae, even those produced by algae unrelated to the Coleochaetales, continue to accumulate nutrients after the egg cell severs its cytoplasmic connections with other adjoining cells. However modestly developed, an active nutrient transport system would confer an immediate benefit to flagellated cells resulting from the meiotic divisions of the zygote—greater vigor and possibly larger size, and therefore greater success at establishing the next generation of gametophytes.

The next stage in the hypothetical sequence is an organism whose eggs were retained within a true archegonium capable of transporting nutrients to its egg. The advantages are obvious. The egg would be supplied with nutrients and protected from inimical physical and biological environmental conditions *before* fertilization, thereby fostering a healthy population of eggs and subsequently assisting the establishment of a healthy population of zygotes. The next step in the hypothetical sequence is the evolution of delayed zygotic meiosis, which would result in a multicellular diploid phase in the life cycle. The advantage of this innovation is that even a few mitotic cellular divisions would amplify the reproductive benefits of each fertilization event because every cell of the diploid phase would be functionally equivalent to a single fertilized egg. Thus even very rare fertilization events could potentially yield a population of new adults. The last step in the hypothetical sequence is the evolution of a multicellular diploid sporophyte composed of vegetative (somatic) and reproductive (generative) cells. The growth and division of somatic cells could have elevated the reproductive portion of the sporophyte, thereby enhancing the probability of long-distance dispersal of spores and the establishment of new populations of gametophytes far away from the parent population. Even for an aquatic organism, the ability to form desiccation-resistant spores would be advantageous in habitats prone to seasonal drying and wetting. Many freshwater algae encase their zygotes with walls composed of sporopollenin. It is possible that the genetic modifications attending the evolution of delayed zygotic meiosis in the first embryophytes also delayed the initiation of the biosynthetic pathway leading to the formation of sporopollenin-encased spores.

There is no good reason to believe that this hypothetical "gradualistic" scenario reflects the actual course of events leading to the appearance of the first embryophytes. Conceivably the archegonium and delayed zygotic meiosis could have simultaneously and suddenly appeared as a consequence of some type of macromutation. Nonetheless, the gradualist scenario serves a pedagogical function, showing that we should not assume the first archegoniate plants were embryophytes, or that all past embryophytes were land plants.

I said earlier that the invasion of land required features permitting the survival of the individual (e.g., cuticles and stomata) in addition to features that ensure successful reproduction on land (e.g., archegonium and sporopollenin-walled spores). Although the sequence of historical events that procured these adaptive innovations may never be known with absolute assurance, the fossil record indicates that land plants most likely evolved by the close of the Ordovician period and certainly no later than the Silurian (fig. 4.3). Spores with sporopollenin walls, cuticlelike sheets bearing the impressions of cells, and microfossils called "banded tubes" are reported from the Ordovician, specifically from the Caradoc epoch, dated as 458 million years ago (Gray, Massa, and Boucot 1982; Gray 1985; Gray and Shear 1992). Morphologically similar microfossils are also found in early Silurian rocks (see Pratt, Phillips, and Dennison 1978). Although spores with sporopollenin walls are not prima facie evidence for the existence of multicellular terrestrial sporophytes (see Banks 1975), spores with resistant walls suggest the existence of plants with reproductive cells that could survive limited exposure to dry conditions. That some of the Caradoc spores are the products of meiosis, and therefore derived from diploid plants (i.e., sporophytes), is indicated by the suggestion of a trilete marking on spore walls. The trilete mark is a Y-shaped ridged surface on the spore wall that occurs when a diploid cell divides to produce four pyramidially arranged, closely abutting spores. The co-occurrence of cuticlelike sheets bearing cellular impressions is also suggestive of terrestrial plants. A true cuticle is totally useless (even detrimental) to an aquatic plant because it impedes the passage of carbon dioxide and oxygen from the water into living plant cells (Raven 1984). And finally, the microfossils called banded tubes could represent extremely ancient water-conducting cell types, signifying a level of anatomical sophistication expected for a bona fide land plant.

Unfortunately, the biological affinities of cuticular sheets and banded tubes are far from unambiguous. Although some paleontologists contend these microfossils are the remains of ancient plants (Gray and Boucot 1977; Gray and Shear 1992), others find troubling morphological parallels with the remains of aquatic and semiaquatic animals (Banks 1975). The arguments marshaled on either side of this debate are cogent. Those who believe that some Ordovician cuticular sheets and banded tubes belong to plants rather than animals point out that these microfossils are found alongside plant spores and therefore most likely all belong to the same kind of organism. Perhaps in frustration, these experts also point out that the botanical affinities of the Ordovician remains would not be contentious if "cuticles" and "banded tubes" had been found in Silurian rocks. In turn, skeptics argue that the burden of proving the botanical

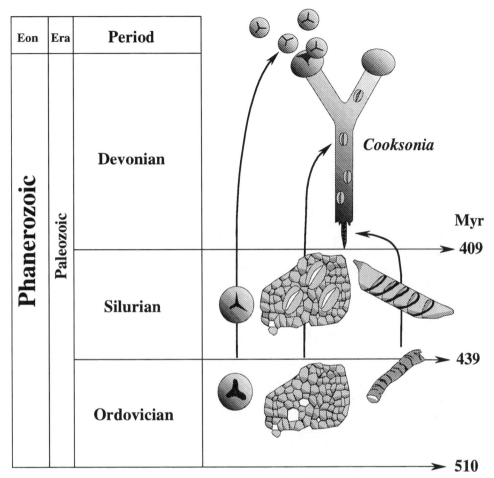

Figure 4.3. Chronology of appearance of various micro- and megafossils pertaining to the evolution of vascular land plants. Spores, cuticular membranous sheets, and banded tubes appear in the Ordovician. Spores with trilete markings, plant cuticles bearing the impressions of guard cells, and tracheids appear in the Silurian. *Cooksonia*-like megafossils with sporangia containing trilete spores, cuticles with stomata, and axes containing water-conducting cells appear in the Upper Silurian and Lower Devonian.

affinities of these remains lies with those who would push the origin of land plants into the Ordovician. Few doubt that the first land plants evolved during early Silurian times and that the vascular land plants had evolved by the end of the Silurian (see Edwards and Fanning 1985; Edwards, Davies, and Axe 1992; Taylor and Taylor 1993), but the Ordovician microfossils are more dubious in terms of what may be said about their biology. Nonetheless, I take the position that, although a skeptical attitude is warranted, particularly when the implications of a hypothesis rub against traditional thinking, land plants most likely evolved in the Ordovician. Whether all these Ordovician "land plants" were embryophytes or whether some were only transient explorers of the terrestrial domain, derived from algal lineages unrelated to the charophycean algae, remains to be seen.

The Evolution of Vascular Tissues

When juxtaposed, the evolutionary history of the nonvascular embryophytes (the "bryophytes") and the vascular plants (the "tracheophytes") shows that the embryophytes as a group evolutionarily explored one of two largely mutually exclusive routes, either the amplification of the size of the gametophyte generation or the amplification of the size of the sporophyte generation. The bryophytes followed the former route, and the tracheophytes followed the latter.

The bryophytes are the nonvascular embryophytes and include the mosses, liverworts, and hornworts and the enigmatic genus *Takakia* (Crandall-Stotler 1980; Scagel et al. 1982). As a group, the bryophytes have elaborate gametophytes and small "ephemeral" sporophytes. The tracheophytes include ferns, horsetails, lycopods, and the seed plants (gymnosperms and angiosperms) as well as a host of extinct plant groups. These plants have elaborate sporophytes and comparatively small, short-lived gametophytes. The either/or choice in embryophyte evolutionary history reflects the fact that the sporophyte generation is physically attached to the gametophyte generation and is incapable of an independent existence during its early ontogeny. Because the sporophyte is physically attached to its subtending gametophyte and, at least initially, physiologically dependent on it, indeterminate growth of the sporophyte generation necessarily has destructive consequences for the continued survival and growth of the gametophyte. Conversely, the elaboration of the size of the gametophyte generation can be entertained only when the sporophyte generation is comparatively small or short-lived.

The elaboration in size of the sporophyte or the gametophyte generation literally takes two very different "directions," or more precisely orientations, because the two multicellular generations of every embryophyte fulfill very different reproductive roles that in turn have different biological and physical requirements. The reproductive functions of the gametophyte are to produce gametes, to ensure the fertilization of eggs, and to maintain at least for a time the developing sporophyte. These functions require continued access to liquid water for the survival of the gametophyte, the transfer of sperm to egg, and the initial growth of the sporophyte embryo. For this reason the free-living gametophytes of embryophytes tend to be short, prostrate organisms that hug their moist substrates, thereby reducing the distance sperm must travel to reach eggs and minimizing the rate of water loss from growing tissues. In this sense indeterminate vertical growth of the gametophyte generation is not favored, whereas continued horizontal growth against a moist substrate is possible and even beneficial. In contrast, the reproductive roles of the sporophyte generation are to produce and eventually disseminate meiospores. In the terrestrial landscape, wind provides a cheap and more or less dependable abiotic vector for spore dispersal. Because wind velocity diminishes toward ground level and because the time it takes for spores to settle out of the moving air increases as spores are released higher into the air, the reproductive roles of the sporophyte generation of land plants are favored by vertical rather than horizontal growth.

The elaboration in the size of a prostrate gametophyte or a vertical sporophyte requires water- and food-conducting tissues once either organism reaches a critical size. The functional significance of conducting tissues can be explained in terms of a basic physical principle: the time for even small molecules like water and glucose to passively diffuse through living cells increases with the square of the distance to be traversed. This principle follows directly from Fick's second law of diffusion (Nobel 1983; Niklas 1994). In the absence of specialized conducting tissues, height is physiologically limited by the time it takes molecules to passively diffuse through cells and reach aerial portions of the plant. Simple calculations show that it takes about one second for water molecules to passively diffuse through a living cell measuring 50 μm in length but about one year for the same number of molecules to diffuse through a string of five hundred of these cells placed end to end (a total distance of roughly 25 cm). Clearly, passive diffusion of water and glucose molecules is rapid enough to maintain the metabolic requirements of plants over short distances but woefully insufficient to sustain even modestly elevated aerial portions of plants. Thus Fick's law shows that some kind of conducting tissue is a physiological necessity for the continued vertical growth of land plants.

Physical laws and engineering principles further dictate the nature of conducting tissues. The efficient passage of water requires tubular conduits offering low resistance to fluid flow along their length, and the volume of water passing through a conduit is proportional to the fourth power of the radius of the conduit. Therefore broad hollow tubes are the best water-conducting devices. Water conduction through plant tissues is accomplished by removal of the protoplasts of living cells whose principal (longitudinal) axes are aligned parallel to the direction of preferred flow (e.g., hydroids, tracheids, and vessel members). In some plants the end walls of these cells remain after the death of the protoplasts (e.g., hydroids and tracheids). In other plants, where the demand for water is high, the end walls of water-conducting cells are partially or entirely removed by protoplasts before their death (e.g., vessel members), thereby minimizing resistance to water flow from one cell to another and increasing the rate of flow. A characteristic of many water-conducting cells is a cell wall that is differentially thickened internally (e.g., tracheids and vessel members). These thickenings mechanically reinforce the cell wall from mechanical implosion as a consequence of the negative internal pressure created when a fluid moves rapidly through a tube. Thus these thickenings are important when the demand for water is great and flow rates through conducting tissues are high. They are of less importance when the demand for water is small. It is hardly surprising, therefore, that the water- and food-conducting cells of embryophytes—bryophytes and tracheophytes alike—tend to be tubular, stacked end to end along the principal dimension of organs like stems, leaves, and roots. Nor is it surprising that the water-conducting cells of mosses lack internally thickened cell walls and that taller plants possess them.

Physical principles also reveal the magnitude of the critical size mandating water- and food-conducting tissues. To be sure, the idea of "critical size" is vague because it depends on environmental conditions (ambient wind speed, temperature, and so forth), the extent to which the plant is elevated or reclines against a moist substrate (and therefore is exposed to the drier atmosphere), and the extent to which the surface of the plant is protected from water loss through evaporation (e.g., presence or absence of a cuticle and stomata). However, calculations indicate that in the absence of a cuticle a cylindrically shaped plant 2 cm tall growing in a still atmosphere at 70% relative humidity requires a water-conducting tissue even if the plant has free and unlimited access to liquid water at its base (see Niklas 1994). These calculations are necessarily crude, but they help show the order of magnitude of "critical size"—the threshold of vertical height expected to impose selection pressures favoring conducting tissues.

It should not escape attention that the size threshold mandating a conducting tissue is totally indifferent to whether a plant is a sporophyte or a gametophyte. Any selection pressure favoring tall gametophytes or tall sporophytes is expected to also favor organisms with conducting tissues. In this regard it is hardly surprising that tall moss gametophytes have water- and food-conducting tissues—the hydrome and the leptome, respectively—that are strikingly similar to the xylem and phloem tissues found in the sporophytes of vascular plants. By the same token, any

selection pressure favoring a reduction in the vertical size of either the gametophyte or the sporophyte would likely reduce anatomical complexity
to a parallel degree. Conducting tissues are highly reduced or wholly absent in short and small moss species phylogenetically related to tall and
large species with conducting tissues (Hébant 1977).

The relation between critical size and anatomical complexity for land
plants is embedded within the duality of the embryophyte life cycle. Because the embryophyte life history presents an either/or choice of whether
to elaborate the size of its sporophyte or that of its gametophyte generation, an evolutionary amplification of the size and anatomy of one
generation will be mirrored by an evolutionary reduction in the size and
anatomical complexity of the other. Crudely put, the embryophyte life
cycle permits one "degree of freedom," although the "choice" is not irreversible. A lineage that has aggrandized its sporophyte generation in the
past could conceivably reverse this trend. The only constancy is in the
consequences of the choice—beyond a critical size threshold, increasing
the size of one generation mandates the presence of conducting tissues
and a diminution in the size and anatomy of the other generation.

This either/or choice and its anatomical consequences make the "bryophytes" and the "tracheophytes" a foregone conclusion in the sense that
they are the only two broad categories of biological organization available to the earliest terrestrial embryophytes. By virtue of the presence and
absence of vascular tissues, the tracheophytes and the bryophytes were
also considered two distinctly different levels of anatomical organization.
Largely for these reasons, the bryophytes and the tracheophytes are traditionally viewed as two distinct lineages, though they trace their evolutionary history back to a common ancestor. Thus the bryophytes and the
tracheophytes were simultaneously thought of as clades (monophyletic
groups of plants) and grades (different levels of biological organization).
These phyletic and anatomical commonalties were formally codified by
placing all the bryophytes into the division Bryophyta and all the vascular
plants into the Tracheophyta. Additionally, Bryophyta was positioned
between Thallophyta (collectively, all the algae) and Tracheophyta, thereby
conveying the notion that the nonvascular embryophytes were ancestral
to the vascular land plants.

This traditional view of the bryophytes and tracheophytes is now seriously challenged. Recent comparative and cladistic studies argue that the
bryophytes are a paraphyletic group of plants and that the mosses, alone
among the bryophytes, share a last common ancestor with the tracheophytes (fig. 4.4). If the mosses, liverworts, and hornworts have separate
evolutionary origins (Crandall-Stotler 1980; Mishler and Churchill 1984,

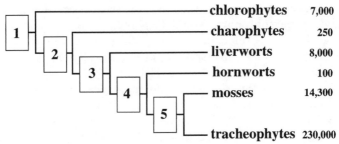

Figure 4.4. Phyletic hypothesis purporting to show the evolutionary relationships among green algae, bryophytes (liverworts, hornworts, and mosses), and tracheophytes (pteridophytes, gymnosperms, and angiosperms). The approximate number of extant species in each group is shown on the right. Some of the characters used to construct the phyletic hypothesis are as follows: 1 = chlorophylls *a* and *b*, carotenoids, starch; 2 = phragmoplastic cell division, glycolate oxidase; 3 = archegonium and antheridium; 4 = indeterminant growth of sporophyte; 5 = water and cell sap conducting tissue system. Adapted from Mishler and Churchill 1984, 1985.

1985; Duckett and Renzaglia 1988; Mishler et al. 1994), then the "bryophytes" cannot be considered a clade, although they still reflect a collection of lineages sharing the same grade of biological organization (i.e., nonvascular archegoniates). By the same token, the monophyletic status of the vascular plants has also been recently challenged. Paleontological studies based on the ultrastructural details of water-conducting cells once believed to be bona fide tracheids suggest that the vascular plants may have at least two separate evolutionary origins (Edwards 1986; Kenrick and Edwards 1988; Kenrick, Edwards, and Dales 1991). If so, then the "tracheophytes" are also a paraphyletic group of organisms that share the same anatomical grade of organization but through evolutionarily independent means. It is important that none of these studies challenge the monophyletic status of the embryophytes as a group. The plants traditionally called "bryophytes" and "tracheophytes" share too many features to reject the hypothesis that they ultimately trace their origin to a common algalike ancestor from which the modern Charophyceae also descended (Pickett-Heaps 1976; Mattox and Stewart 1984; Graham 1993; Mishler et al. 1994). Rather, the issue is whether the bryophytes and the tracheophytes represent two distinct clades as formerly believed or whether they are two dissimilar grades of biological organization achieved at different times by different lineages *within* the embryophytes.

Much of this issue centers on whether certain biological features shared among different lineages are homologous features or are simply analogous (see chapter 7). The cladistic hypothesis aligning the mosses with the tracheophytes is a good case in point because it is based principally, albeit not exclusively, on the presumed homologies between the water- and food-conducting tissues found in both groups of plants (Hébant 1977; Scheirer 1980; Mishler and Churchill 1984, 1985; Mishler et al. 1994). Specifically, the water-conducting tissue of the mosses (the hydrom) is stated to be homologous with the xylem tissue of vascular plants, while the food-conducting tissue of the mosses (the leptom) is stated to be homologous with the phloem tissue (Scheirer 1980; Mishler and Churchill 1984, 1985; Mishler et al. 1994). These homologies have led some to speculate that the water- and food-conducting tissues of the mosses are highly reduced vascular tissues. Hébant (1977) suggests neoteny as a possible developmental mechanism for the reduction (and in some cases total loss) of conducting tissues in the mosses. Neoteny refers to the precocious appearance of reproductive features attended by a delay in somatic growth. In this respect many features of the moss hydrom appear to be "juvenile" features of developing xylem cells (e.g., absence of secondary cell walls and internal lignified thickenings or cell wall perforations). Along these lines, neoteny could also explain the small size of the moss sporophyte. Unlike the branched sporophytes of vascular plants that can bear numerous sporangia, the moss sporophyte bears a single terminal sporangium. At least in theory, a developmental delay in somatic (vegetative) growth of a polysporangiate sporophyte could give rise to a short, unbranched monosporangiate sporophyte. Clearly, differences between the hydrom and the xylem tissues are developmentally recondite, as are the ways moss and tracheophyte sporophytes grow in length. For example, the hydrom differentiates before the leptom, whereas the xylem differentiates after the phloem, and whereas the hydrom differentiates after the elongation of the moss axis, xylem differentiation attends axial elongation (Hébant 1977). In terms of sporophytes, no homologue to the apical meristem of the tracheophyte sporophyte is known for any bryophyte sporophyte. Without exception, the sporophytes of bryophytes grow in length as a consequence of the meristematic activity of subapical rather than apical cells (Crandall-Stotler 1980).

What can be said of these presumed homologies? At the outset, it is important to bear in mind that the liverworts and hornworts cladistically disalign from the mosses and tracheophytes based in part on the presumption that the conducting tissues found in the liverworts evolved independently and in parallel with the conducting tissues of the moss-tracheophyte clade and that the intercalary meristem, nonsynchronous spore production, and pseudoelaters of the hornworts have no homologues in other embryophyte groups. Space does not permit a detailed discussion of these presumptions, but it is clear that cladistic analyses reinforce the impression that the bryophytes as a group have undergone numerous character reversals and evolutionary episodes of morphological and anatomical reduction as well as convergent and divergent evolution. As a consequence, the roughly 22,400 extant bryophyte species are highly specialized organisms, making inferences among character states very difficult (Schuster 1966, 1981; Hébant 1977; Crandall-Stotler 1980; Crosby 1980; Scagel et al. 1982).

The juxtaposition of the fossil record of early vascular land plants with the biology of current bryophytes supports the alignment of the mosses with the tracheophytes but not the belief that the mosses are highly "reduced" vascular plants. Some fossil plants formerly believed to be tracheophytes are now known to have water-conducting cells very like hydroids, and a number of fossil plants have been shown to contain "tracheids" whose wall structures differ significantly from the tracheids of extant plants and share some morphological features with the hydroids of mosses (see Taylor and Taylor 1993). These fossil plants may represent intermediary stages in the evolution of tracheophytes from an ancestral stock of plants that diverged to give rise to the bryophytes as well. One such "intermediary" organism is the Devonian plant *Aglaophyton* (*Rhynia*) *major* (fig. 4.5). Historically this plant was treated as a tracheophyte, called *Rhynia major,* because it had a central strand of conducting tissue consisting of hollow tubular cells believed to be tracheids (Kidston and Lang 1920, 1921). Recent detailed microscopic studies suggest, however, that the water-conducting cells of this plant fossil lack secondary walls and are more similar to the hydroids of modern mosses. For this reason the fossil formerly called *Rhynia major* has been removed from the genus *Rhynia* and assigned to the new genus *Aglaophyton,* in recognition of

Figure 4.5. The early Devonian nonvascular sporophyte *Aglaophyton* (*Rhynia*) *major* (A) and its presumed gametophyte *Lyonophyton rhyniensis* (C). Anatomical similarities between the conducting tissues occurring in the center of the cylindrical axes of both fossil plants and similar epidermal features provide some of the evidence that *Aglaophyton* and *Lyonophyton* are parts of the life cycle of the same organism. (A) The sporangia of *Aglaophyton* were borne aloft at the tips of dichotomous axes (about 18 cm high) attached to horizontanially growing axes that absorbed water through tufts of water plant hairs (called rhizoids). (B) A slender strand of conducting tissue transported water to the nonphotosynthetic inner cortex and the photosynthetic outer cortex of the sporophyte. (C) The aerial axes of *Lyonophyton* were at least 2 cm long and terminated in a deeply lobed bowl-like shield with numerous sperm-bearing antheridia on its inner surfaces. (D) A longitudinal section through the bowl-like tip of *Lyonophyton* typically reveals a central strand of conducting tissues that ramifies beneath the spherically shaped antheridia. A and B adapted from Edwards 1986; C and D adapted from Remy, Gensel, and Hass 1993.

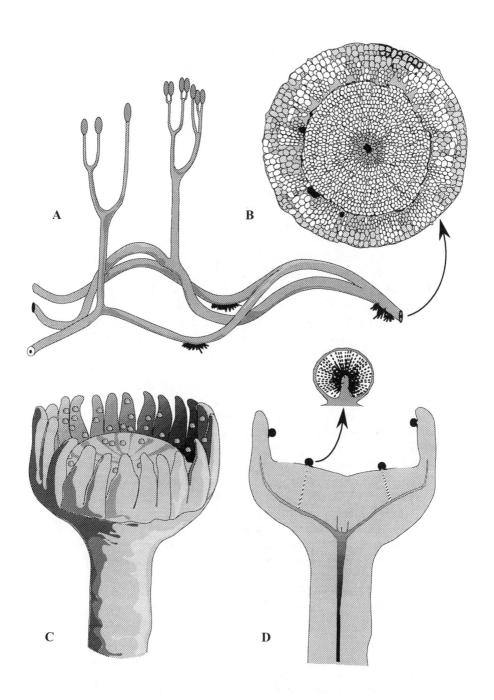

its putative nontracheophyte status (Edwards 1980, 1986). (Whether *Aglaophyton* should be considered a nonvascular plant is arguable. Conceivably this organism may have evolved from a vascular ancestor through reduction, in which case its mosslike water-conducting cells are derived from tracheids.) Similarly, tracheid-like cells in the Devonian plant *Sennicaulis hippocrepiformis* possess helically thickened cells that have a spongy organization unlike that found in the tracheids of current vascular plants, and the Devonian plant *Gosslingia breconensis* has "tracheids" with discontinuous inner cell walls that are not found in the tracheids of any known extant tracheophyte (fig. 4.6) (Kenrick and Edwards 1988; Kenrick, Edwards, and Dales 1991).

Although the peculiar cell walls seen in the conducting tissues of Devonian plants like *A. major, S. hippocrepiformis,* and *G. breconensis* may represent unusual states of fossil preservation, it is more likely that they reflect preliminary stages in the evolution of xylem tissue from the hydroidlike cells of some common ancestor. If so, then the hydroids of modern mosses more likely reflect the ancestral condition of the moss-tracheophyte clade rather than highly reduced tracheids. This hypothesis also resolves the developmental differences between the moss and tracheophyte sporophyte—the former need not be considered a "neotenous" version of the former, but rather may be seen as a monosporangiate counterpart of the polysporangiate vascular sporophyte. The most parsimonious description of the early history of the land-dwelling embryophytes and the evolution of conducting tissues involves the early evolutionary divergence and subsequent parallel evolution between two groups of embryophytes sharing a nonvascular, archegoniate ancestor with a life cycle alternating between a multicellular gametophyte generation and a multicellular sporophyte generation sharing much of the same appearance (fig. 4.7). Here the hypothetical common ancestor is envisioned to have a bryophyte-like grade of biological organization.

The presumption that the gametophyte and sporophyte generations had a similar morphological and anatomical repertory resulting in haploid and diploid plants with very similar appearance seems reasonable. The life cycle of all embryophytes consists of a regular alternation of generations that is obligatory in the sense that a zygote produced by the union of sperm and egg always gives rise to a sporophyte, and the spores

Figure 4.6. Anatomical variations of tracheids from Devonian fossil plants. Gross morphology of plants shown above a median longitudinal section through a representative tracheid: (A) *Sennicaulis hippocrepiformis;* (B) *Gosslingia breconensis.* Adapted from Kenrick and Edwards 1988:, Kenrick, Edwards, and Dales 1991; and Taylor and Taylor 1993.

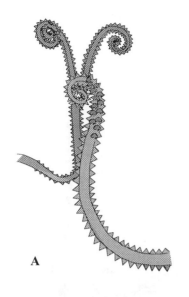

A

B

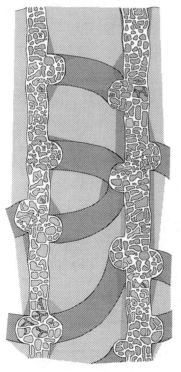

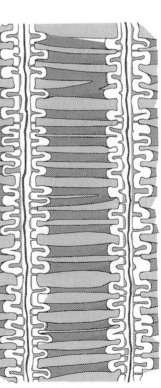

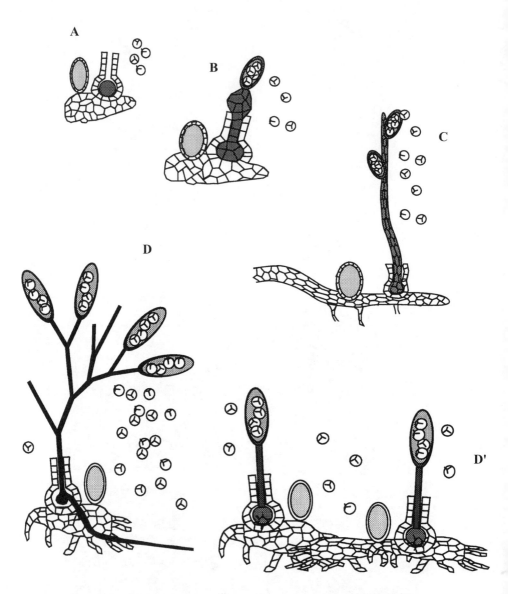

Figure 4.7. Hypothetical sequence showing the evolutionary origins of tracheophytes (D) and bryophytes (D′) from a *Coleochaete*-like ancestor (A). Divergence between the bryophyte and tracheophyte lineages is posited to have followed the acquisition of a diplobiontic life cycle involving an alternation of more or less isomorphic generations (B–C).

produced by a sporophyte always develop into gametophytes. However, the alternation of generations is not obligatory in the sense that the gametophyte and sporophyte must always alternate with each other. The gametophytes of embryophytes have the capacity to bud off cells that can develop into a plant that morphologically looks like a sporophyte (a phenomenon called apogamy); the sporophyte has the ability to bud off cells that can develop into a plant that morphologically looks like a gametophyte (a phenomenon called apospory). Apogamy and apospory occur normally among the bryophytes as a group and are not infrequent among ferns and lycopods (Smith 1955; Scagel et al. 1982; Gifford and Foster 1989). Likewise, the gametophytes of seed plants can develop into plants that are morphologically and anatomically indistinguishable from sporophytes. Thus apogamy and apospory must be considered shared primitive features among the embryophytes as a group (Bell 1992). More important, these phenomena (together with genetic considerations) indicate that the developmental repertory of the gametophyte and sporophyte generations of embryophytes can achieve very similar morphological and anatomical results. This may account for the ability of both generations to elaborate conducting tissues in current embryophytes, which in turn favors the hypothesis that the first land-dwelling embryophytes had morphologically similar, though not identical, sporophyte and gametophyte generations. This hypothesis is also consistent with recently discovered fossil gametophytes, called *Lyonophyton rhyniensis* (see fig. 4.5), which are morphologically and anatomically very similar to the fossil remains of *A. major* as well as to early Devonian vascular plants preserved in the same rock layers (Remy and Hass 1991; Remy, Gensel, and Hass 1993).

Subsequent evolution involved morphological and anatomical divergence between the two generations in the life cycle of individual clades. Those lineages that by chance elaborated the gametophyte generation did so at the morphological and anatomical expense of their sporophyte generation. Collectively these lineages are the "bryophytes." Within the bryophyte grade of organization, further diversification and ecological specialization resulted in the anatomical and developmental dissimilarities seen among current mosses, liverworts, and hornworts, while parallel evolution among some lineages produced analogues of conducting tissue, elaters, and pseudoelaters as well as "intercalary" sporophyte meristems. Those lineages that elaborated and aggrandized the sporophyte generation did so at the expense of the morphological and anatomical elaboration of their gametophyte generation. Collectively these lineages are the "tracheophytes"; one of their characteristics is the polysporangiate condition. The initial divergence of the mosses and the tracheophytes may have involved a common ancestor very much like the fossil *Cookso-nia,* long believed to be a vascular plant. Tracheid-like cells have only recently been isolated from some fossil remains believed to be *Cooksonia* (Edwards, Davies, and Axe 1992), but these conducting cells differ in no substantive way from the "banded tubes" found in older geological strata and are assumed to be tracheids only because *Cooksonia* is tradi-tionally believed to be a vascular plant. The conducting tissue found in plants like *Cooksonia* may be neither xylem nor hydrome, and rather than being the benchmark for the first appearance of vascular plants, *Cooksonia* may represent an organism that is anatomically intermedi-ate between the mosses and tracheophytes. The stratigraphic position of *Cooksonia* (Silurian–Early Devonian) is compatible with this phyletic hypothesis because unequivocal tracheophytes and bryophytes coexist by Middle Devonian times. Based on the available data, it seems entirely reasonable to believe that the mosses and tracheophytes had a common ancestor. The tracheophytes may be biphyletic, one lineage giving rise to modern lycopods through the zosterophyllophytes, an ancient group of lycopod-like Devonian plants (see Taylor and Taylor 1993; Stewart and Rothwell 1993), and another lineage, the rhyniophyte-trimerophyte clade, eventually diversifying into present-day ferns, horsetails, and seed plants (fig. 4.8).

Heterospory and the Seed Habit

Most Devonian vascular land plants had sporophytes that freely shed their spores into the air (Taylor and Taylor 1993). The gametophytes of these Devonian species grew independent of the sporophyte generation that produced them. Vascular plants that freely shed their spores are called pteridophytes. The term literally means "fernlike plant" and al-ludes to the fact that the sporophytes of plants like ferns shed their spores and therefore have "free-living" gametophytes. The independent lifestyle of the free-living gametophytes in part dictates the ecological distribution of pteridophyte species. Although the sporophyte may be well adapted to survival and growth in a dry habitat, the gametophyte generation of the same species typically requires moist conditions for vegetative growth,

Figure 4.8. Chronology of appearance and relative species diversities of major vascular plant lineages beginning with the appearance of archegoniate plants (possibly in the Or-dovician). Inserted diagrams depict representative plants or plant organs from each lin-eage. Width of shaded areas for each lineage renders relative species diversity. L = lyco-pods, H = horsetails, F = ferns, G = gymnosperms, A = angiosperms. All lineages trace their ancestry to a Devonian plexus of early archegoniate land plants.

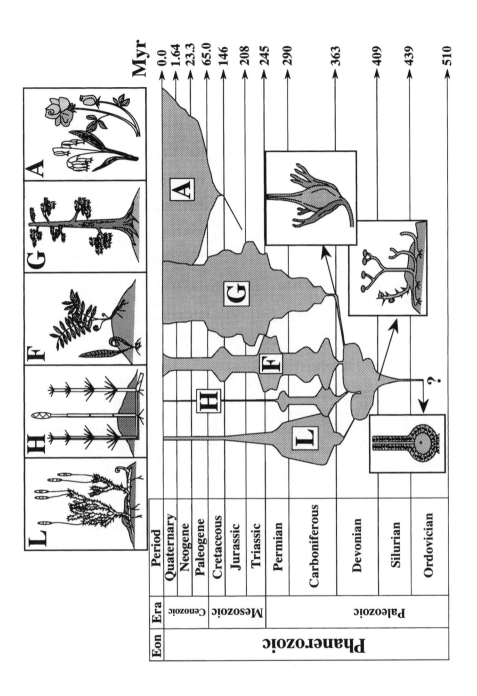

the transfer of sperm cells to archegonia, the fertilization of eggs, and the survival of the sporophyte embryo. In a very crude sense, pteridophytes may be thought of as the amphibians of the plant kingdom in the sense that the completion of their sexual life cycle requires a "return to water." Although this simile is somewhat whimsical, pteridophyte sexual reproduction is ecologically confined either to comparatively moist habitats or to those that have a wet season during which free-living gametophytes can grow and develop eggs and new sporophyte embryos.

The colonization of new sites by early vascular land plants was likely aided by their comparatively small spores that developed into bisexual gametophytes. Small spores have a greater probability of being carried long distances by wind currents than larger spores because, other things being equal, the rate at which any particulate settles out of the air is proportional to the size and density of the particles (Vogel 1988). This simple physical relation also helps to explain one of the major trends in early vascular land plant evolution—the comparatively rapid evolutionary increase in the size and height of sporophytes (Chaloner and Sheerin 1979). If it is true that most early vascular land plants relied on wind currents to transport and deposit spores in microhabitats conducive to the growth and sexual function of gametophytes, then the elevation of sporangia far above ground level undoubtedly enhanced the long-distance dispersal of spores regardless of their size. This elevation likely increased the probability that at least some spores would reach a moist refuge where they could grow and develop into sexually mature gametophytes. It seems reasonable to conclude, therefore, that species with tall sporophytes would have had a selective advantage over their shorter counterparts, and that with time the size and height of sporophytes would tend to increase.

Regardless of the size and density of spores or the size and height of sporophytes, spores developing into bisexual gametophytes would have initially conferred an advantage in colonizing new sites. Provided a gametophyte can produce both egg and sperm and that the gametes produced by the same gametophyte are sexually compatible and can form viable zygotes, only one gametophyte is required to colonize a new site by establishing a sporophyte (Tryon and Lugardon 1991). On the other side of the life cycle coin, the sporophytes of many modern pteridophytes are capable of asexual reproduction and may live for many years. Assuming that the sporophytes of ancient pteridophytes were also capable of asexual reproduction, once a sporophyte was established in a dry habitat (by virtue of the felicitous survival of a bisexual gametophyte in a temporarily moist microhabitat) it could establish a "population" of genetically identical individuals (a clone) and simultaneously colonize distant sites by discharging literally millions of spores.

But what is the basis for assuming that the earliest vascular land plants had bisexual gametophytes? The answer lies in two characteristics of pteridophytes that produce bisexual gametophytes—their spores tend to be comparatively small, and all the spores produced by a particular species have the same general external appearance (Chaloner 1967; Pettitt 1970; Taylor and Millay 1979). Plants that produce only one kind of spore are called homosporous, meaning "same spores." Because the spores of the most ancient vascular land plants were comparatively small and because the spores found in the fossil remains of these plants tend to all look alike, these plants are generally assumed to have produced bisexual gametophytes. This assumption is somewhat problematic, however, because some modern mosses produce spores that look alike but develop into sexually dimorphic gametophytes (e.g., *Macromitrium comatum*). This condition is called anisospory, literally "unequal spores." Here we see the distinction between "spores of unequal size" and "spores that develop into unisexual gametophytes." The former reflect morphological heterospory; the latter presage sexual dimorphism.

Many modern species of ferns, lycopods, and horsetails have retained the presumably ancient homosporous condition and therefore produce bisexual gametophytes. However, a significant number of living and extinct pteridophyte species are heterosporous; that is, they produce two kinds of spores: small spores that develop into sperm-producing gametophytes and comparatively larger spores that develop into egg-bearing gametophytes. In recognition of their relative size, the small spores of heterosporous species are called microspores and the sperm-bearing gametophytes they develop into are called microgametophytes. The larger spores produced by heterosporous species are called megaspores, and the egg-bearing gametophytes they develop into are called megagametophytes. Even though "micro" and "mega" imply that size is the salient feature of heterosporous plants, this is not the case. The essential feature of heterosporous plants is that they produce unisexual gametophytes.

Heterospory appears to confer a selective advantage, because it evolved independently in at least four very different plant groups: ferns, lycopods, horsetails, and seed plants. Likewise, the first heterosporous vascular plants make their appearance fairly early in the fossil record. By plotting the frequency distribution of fossil spore sizes during the Devonian, William Chaloner (1967) showed that lower Devonian spores tend to be uniformly small, typically less than 100 μm in diameter, but that toward the middle of the Devonian the frequency distribution of spore size begins to expand toward larger spores, some much more than 200 μm in diameter. By the end of the Devonian, some spores measure 2,200 μm in diameter (e.g., *Cystosporites devonicus*). This trend in the distribution of spore size, which incidentally reinforces the view that the earliest Devonian vascular plants were homosporous, supports the hypothesis that heterosporous plants evolved about 386 million years ago.

Another line of evidence supporting the hypothesis that heterospory evolved early in the history of plant life are fossil sporangia containing both small and large spores ("mixed" sporangia) and sporangia attached to the same plant containing either small or large spores (micro- and megasporangia). Both situations occur by Devonian times, and both fall in line with the logical expectation that heterosporous plants would eventually evolve sporangia dedicated to the exclusive production of microspores or megaspores (fig. 4.9). For example, the Devonian plant *Barinophyton citrulliforme* had fairly large mixed sporangia containing several thousand small spores (each measuring 50 μm in diameter) and approximately thirty large spores (each measuring up to 900 μm in diameter). In addition to bearing mixed sporangia, the Devonian plant *Chaleuria cirrosa* produced microsporangia (containing only small spores measuring from 30 to 48 μm in diameter) and megasporangia (containing spores measuring 60 to 156 μm in diameter). The presence of mixed sporangia on *Barinophyton* and *Chaleuria* plants suggests that the sporophyte generation had gained some physiological control over the sexuality of the spores it produced. The presence of micro- and megasporangia on *Chaleuria* plants suggests that this control was tighter, perhaps as a consequence of regulating the flow of nutrients to cells destined to undergo meiosis and produce spores. In contrast, heterospory among bryophytes seems to be a consequence of sex chromosomes (Sussex 1966; Bell 1979).

Although the "when and how" of vascular plant heterospory can be traced with some precision by virtue of the fossil record, no one is quite sure why heterospory evolved. If the colonization of new sites was important and favored by producing bisexual gametophytes, then it seems counterintuitive for plants to abandon homospory in favor of heterospory. Nonetheless some did so, and rather quickly in terms of the evolutionary time scale. The reason may lie in the benefits that come from reapportioning metabolic resources in favor of the spores that will develop into megagametophytes. Consider that heterosporous plants tend

Figure 4.9. Diagrams of micro- and megasporangia. (A) The Devonian plant *Barinophyton citrulliforme* bore sporangia containing small and large spores (mixed sporangia). (B) The Devonian plant *Chaleuria cirrosa* produced three different types of sporangia; one type was mixed, another contained only small spores (microsporangia), and the other contained only large spores (megasporangia). (C) Many fossil and living plants produce sporangia exclusively containing microspores or megaspores (e.g., *Archaeopteris* and *Selaginella*).

A

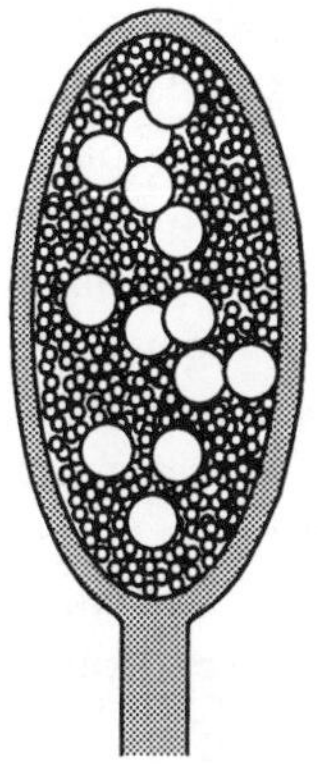

B

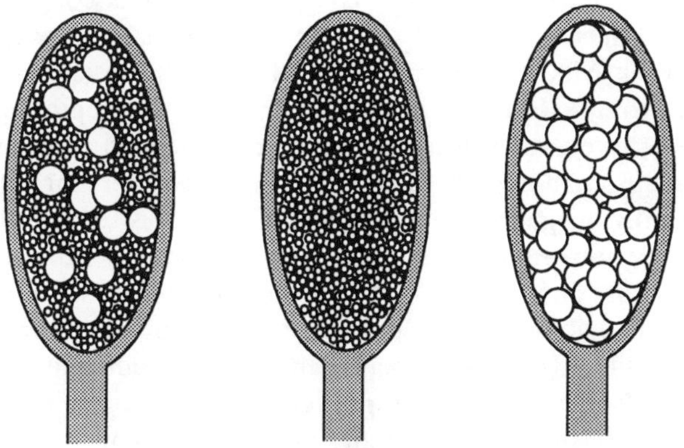

C

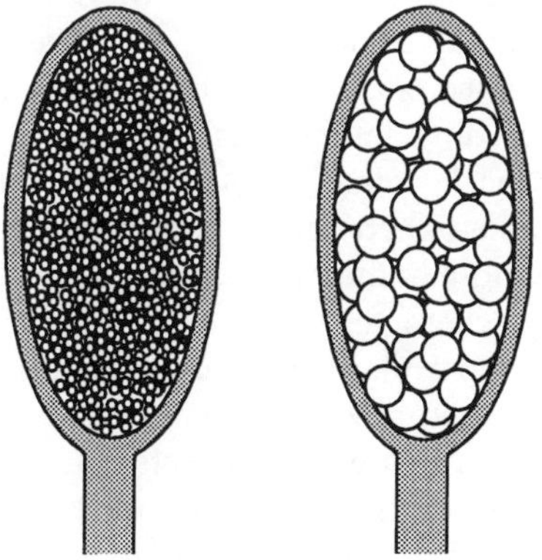

to produce many more microspores than megaspores and that the microspores of each species are generally much smaller than megaspores. Assuming that each sporangium is invested by its sporophyte with roughly the same amount of metabolic resources, then partitioning this resource into "many small microspores" reduces the amount of metabolites allocated to each microspore, which in turn may reduce the vigor of the microgametophyte. Although this may appear "biologically foolish," recognize that once sperm are released, the microgametophyte has fulfilled its reproductive function and is expendable. By producing many small microspores, heterosporous plants increase the chances that many sperm-bearing plants will be available for sexual reproduction even if the gametophytes that produce them die shortly thereafter. In contrast, consider the other side of the reproductive coin—producing "few large" megaspores means that each is invested with a comparatively large amount of metabolic reserve that the megagametophyte can use to initially establish itself (Tryon and Tryon 1982). The megagametophyte must continue to survive after it produces eggs if the developing embryo within it is to survive and eventually take up an independent existence. In this sense many small microgametophytes and a few very large (and presumably healthy) megagametophytes ensure a high probability of fertilization and the recruitment of at least a few new sporophytes in the next generation. By loose analogy, the heterosporous condition is a biological extension of the logic underwriting oogamy: eggs (and egg-bearing plants) are metabolically favored over sperm (and sperm-producing plants).

One consequence of this reproductive allocation is that large spores tend to settle out of air currents faster than small spores, and so the megagametophytes and microgametophytes produced by an individual sporophyte are likely to part ways aerodynamically when wind carries them a great distance from the parent plant. Curiously, aerodynamic sorting based on spore size may confer a reproductive advantage in terms of outbreeding and heterosis (hybrid vigor). Because they share the same genetic background, eggs fertilized by sperm produced by the same microgametophyte invariably produce homozygous sporophytes, diploid plants whose homologous chromosomes have genes encoding for exactly the same character states. Homozygous sporophytes can be at a disadvantage, particularly when deleterious mutations occur. Thus, inbreeding among gametophytes produced by the same sporophyte can have long-term deleterious genetic effects on a population. In contrast, the production of very small microspores, some of which can be carried a long distance, has the positive effect of ensuring the exchange of genetic information among genetically different gametophytes, thereby increasing the frequency of heterozygous sporophytes in the populations of hetero-sporous species (see chapter 7).

In light of the resource allocation model for the evolution of hetero-spory, we would expect an inverse relation between the number and the size of the megaspores produced within a megasporangium, ultimately resulting in some species with a single megaspore in each megaspo-rangium (fig. 4.10). These predictions appear to be borne out. For exam-ple, among the extant species of the lycopod genus *Selaginella,* some pro-duce sixteen megaspores within each megasporangium, others produce eight megaspores per megasporangium, and in one species only four megaspores are found in a megasporangium. Among these species, megaspore size proportionally increases as the number of megaspores per sporangium decreases, showing that the contents of sporangia are partitioned more or less evenly among the cells destined to divide meiot-ically to form megaspores. Much more dramatically, the Devonian fossil *Cystosporites devonicus* consists of a single quartet of spores; one mea-sures 2.2 mm in diameter while the other three are aborted megaspores only 100 μm in diameter. All four spores are surrounded by a delicate membrane, indicating that the megaspore tetrad was the meiotic product of a single diploid cell that occupied virtually the entire contents of its megasporangium.

The partitioning of resources among the spores produced within a spo-rangium may have set the stage for the evolution of the seed habit. In contrast to heterosporous pteridophytes, which freely shed their micro-spores and megaspores and have free-living gametophytes, seed plants retain megaspores within their megasporangia. Importantly, seed plants are unanimous in reducing the number of functional megaspores per megasporangium to one. The retention of the megaspore (and subse-quently the megagametophyte) within sporophytic tissues confers con-siderable benefits. It eliminates the requirement for an external supply of water to fertilize eggs, and it affords the opportunity to protect and nour-ish megagametophytes, their eggs, and eventually the diploid embryos that develop from fertilized eggs. Seed plants are sometimes spoken of as the "amniotes of the plant kingdom" because their sporophyte genera-tion retains the egg and embryo just as the female amniotic animal does. Of course this fanciful simile ignores the fact that the egg of the seed plant is produced by a multicellular haploid organism, the megagameto-phyte, which is a parasitic organism living within the host sporophyte and deriving its nourishment from it. It also ignores the fact that the sporo-phyte embryo is contained within the megagametophyte and not directly attached to the diploid parent as it is in amniotic animals. Nevertheless,

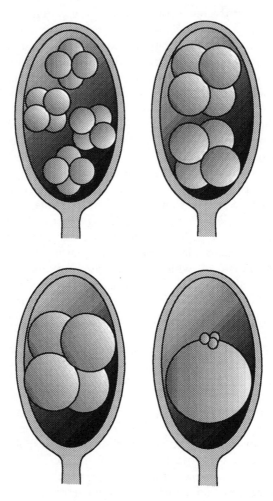

Figure 4.10. Hypothetical sequence of reduction in the number of megaspores per mega-sporangium leading to the formation of megasporangia containing only one functional megaspore and three aborted smaller spores (lower right).

we are seeing precisely the same bio-logic in the sense that seed plants and amniotic animals have evolved reproductive syndromes that eliminate the dependency on freestanding water for fertilization. Naturally, plants did this first.

The evolution of seed plants involved the acquisition of a *syndrome* of reproductive features that goes well beyond the evolution of the structure called the seed. To understand this syndrome, it is necessary to briefly examine how the seed develops and define a few technical botanical terms along the way.

Strictly speaking, the seed is a mature ovule that typically contains at least one viable sporophyte embryo (fig. 4.11). The ovule consists of a single megasporangium (whose wall, among seed plants, is called the nucellus) invested with one or two ensheathing structures called integuments (i.e., the ovule is an integumented megasporangium). Among modern seed plants, the integument has a clearly defined opening at its tip called the micropyle, through which the microgametophyte gains entry to ultimately deliver its sperm. In contrast, the integuments of the oldest known seed plants either were deeply lobed or consisted of a girdling truss of sterile branchlike axes. Strictly speaking, many of these ovules lacked micropyles and integuments. The integument of modern seed plants is believed to have evolved by the fusion of these lobes (Long 1977). Another hypothesis is that the integument is an evolutionary modification of the nucellus (Andrews 1961). Regardless of the origins of the integument, among all current seed plants meiosis occurs in the nucellus and results in a single functional megaspore that subsequently develops into a megagametophyte that bears at least one egg (Gifford and Foster 1989). (Among some seed plants eggs are borne in archegonia, just as they are in pteridophytes and bryophytes. However, among flowering plants, the distinguishable features of the archegonium are absent.) The death and breakdown of nucellar cells expose the megagametophyte and its egg(s), previously buried within the nucellus. In some species, notably gymnosperms, the disintegration of these nucellar cells simultaneously provides a droplet of fluid, called the pollination droplet, that fosters the entry of sperm cells into the ovule. Under any conditions, the fertilization of the egg requires the receipt of microspores and the development of sperm-bearing microgametophytes. The immature microgametophyte of seed plants is called the pollen grain. The microgametophyte initially develops within the microspore wall of the pollen grain, whose wall opens to release sperm cells. Among many seed plants, the microgametophyte develops a pollen tube that delivers sperm cells to the megagametophyte. After fertilization, the ovule continues to develop. Its

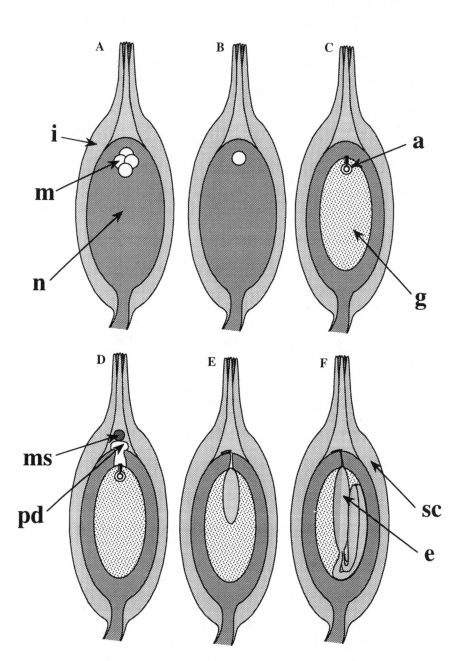

integument(s) mature(s) to form the seed coat, and the sporophyte embryo grows and matures within the megagametophyte (fig. 4.11).

This brief review of the ovule and seed shows that the reproductive syndrome of all seed plants involves the retention of megaspores within megasporangia; the reduction of the number of megaspores per megasporangium to a single functional megaspore; the enclosure of the megasporangium by an integument; some kind of modification of the megasporangium to receive microspores and their sperm-bearing microgametophytes; and the elaboration of microspores, enabling them to deliver sperm cells to eggs. Clearly, a number of important features had to evolve before the seed habit was acquired. These features appear comparatively early in the history of vascular plants. The oldest known fossil seed plant is the late Devonian plant called *Elkinsia polymorpha* (Rothwell, Scheckler, and Gillespie 1989). The ovules of *Elkinsia* were produced at the tips of slender branches whose tips were modified into a structure called a cupule (fig. 4.12A–B). Each ovule had a single integument consisting of four or five vascularized lobes fused at the base. The tip of the nucellus was extended to receive microspores and contained a single large megaspore as well as three smaller aborted megaspores. A slightly younger Devonian plant, *Archaeosperma arnoldii* (Pettitt and Beck 1968), had a similar morphology (fig. 4.12C–D), but the cupules of this plant, which were borne in pairs, had extremely long vascularized extensions fused at the base. Two ovules were attached to each cupule, and each had an integument serrated into a number of lobes that formed a rudimentary micropyle. Because the ovules are poorly preserved, it is not possible to determine how the tip of the nucellus may have been modified to receive pollen grains, but the nucellus contained one large megaspore and three aborted spores toward the nucellar tip, just like *Elkinsia*. Curiously, unlike the megaspores of modern seed plants, which lack spore walls, the megaspores of *Archaeosperma* had recognizable spore walls. This distinction likely was important from a functional point of view. Thick spore walls can be an obstacle to the transfer of nutrients from the nucellus to the developing megaspore and subsequently to the megagametophyte. The megaspores of modern seed plants have reduced spore walls that are very thin and membranous, permitting the passage of

Figure 4.11. Diagrams rendering the stages in the development of an ovule (A) into a gymnosperm seed (F). (A–B) Three of the four megaspores (m) that develop within the nucellus (n) abort. (C) The surviving megaspore develops into a megagametophyte (g) that produces one or more archegonia (a). (D–F) A microspore (ms) attached to a pollination droplet (pd) fertilizes an egg within an archegonium, the resulting zygote develops within gametophytic tissues into a sporophyte embryo (e), and the integument (i, see A) develops into the seed coat (sc). Compare with fig. 4.12.

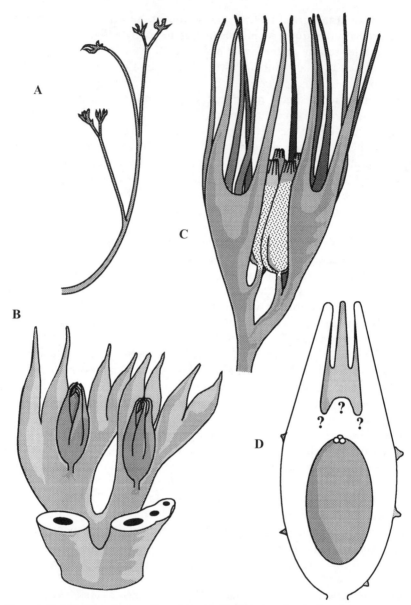

Figure 4.12. Late Devonian cupulate ovules: (A–B) *Elkinsia polymorpha*; (C–D) *Archaeosperma arnoldii*. Adapted from Pettitt and Beck 1968; Rothwell, Scheckler, and Gillespie 1989; and Serbet and Rothwell 1992.

nutrients. Some early seed plants, like *Archaeosperma*, may not have been able to supply nutrients to their developing megagametophytes as efficiently as modern seed plants, but the large size of *Archaeosperma* megaspores suggests that the rudiments of differential nutrient allocation existed early in the history of seed plants.

The earliest seed plants, like *Elkinsia* and *Archaeosperma*, were gymnosperms ("naked seed" plants): just like the ovules of modern species of pine and spruce, the ovules of *Elkinsia* and *Archaeosperma* were directly exposed to the air when they were receptive to pollination rather than buried within sporophytic tissues as they are in flowering plants. This is not to say that gymnosperm ovules and seeds are totally unprotected at all stages in their development. The term gymnosperm refers to the *topology* of ovules. The ovules of modern pine and spruce species are borne on the upper surfaces of scales attached to a stem. These scales flex backward when ovules are receptive to pollination, reflex toward the stem once pollination has occurred, and subsequently flex backward once again to permit the release of mature seeds. The choreography of the scale ensures that airborne pollen grains can reach receptive ovules, that developing seeds are later provided a modicum of protection from animals and microbes, and that seeds are eventually released when mature.

A good deal of evolutionary innovation and "experimentation" occurred among ancient gymnosperms. One overriding theme in this experimentation, at least during the Devonian period, was the extension of the nucellar tip into a funnel-like tube called the salpinx (Taylor and Millay 1979). This tubular extension appears to have directly received pollen grains. In this sense the salpinx was analogous to the micropyle of modern gymnosperms. As noted, the integuments of the most ancient seed plants were variously lobed, and a distinct micropyle was not present. It has been suggested that the deeply lobed integument may have functioned aerodynamically to direct airborne pollen grains to enter the salpinx, thereby increasing the efficiency of pollination. If so, then lobed integuments functioned much like a snow fence that directs windswept snowflakes to accumulate in a preferred site (Niklas 1981). This hypothesis does not discount the role of the integument in protecting the nucellus and the megagametophyte from predatory animals seeking nutrition, nor does it exclude the possibility that the integument of some fossil species may have attracted ancient pollinating animals (Taylor and Millay 1979). Indeed, the structure of some fossil ovules suggests that the integument may have served in a variety of ways. What can be said with assurance is that gymnosperms whose ovules had a salpinx eventually became extinct, and species whose ovules had a true micropyle gained ascendancy.

By Carboniferous times, some gymnosperms evolved microgametophytes that formed pollen tubes that grew within and through the nucellus (Rothwell 1972). Among living seed plants, pollen tubes have been experimentally shown to draw nutrients from the nucellus so as to grow and develop. In this sense microgametophytes with pollen tubes are parasitic plants that subsist for a time by drawing metabolites from their host sporophytes. The nourishment of the microgametophyte by the nucellus may have been an important mechanism by which the sporophyte generation biochemically and immunologically recognized pollen produced by its own species. It also would have provided a trigger to kill or retard the growth of microgametophytes from other species. The presence of fossil microgametophytes with pollen tubes suggests that some Carboniferous gymnosperms may have evolved an extremely important and sophisticated pollen recognition systems, perhaps prefiguring the pollination biology of flowering plants, many of which either retard the growth of "exotic" pollen grains that accidentally land on their stigmata or kill them.

The Flower

Horticulturists have modified flowers into such beatific essays in color, shape, and size that it is easy to forget the flower's raison d'être—seed production. From a functional perspective, the flower differs not a jot from the less affected pinecone. Likewise, it is easy to forget that the flower is nothing more than a determinate shoot bearing sterile and reproductive appendages. A shoot is a stem bearing leaves; "determinate" means that the shoot stops growing after it has reached a developmentally predetermined size and morphology. Although they may not look like leaves, developmentally all the parts of the flower—sepals, petals, stamens, and carpels—are highly modified leaves (Gifford and Foster 1989). Some of these leaves, like the sepals, may be scalelike in appearance and function to protect the inner parts of the flower as they mature. In some species, however, the sepals can be large and brightly colored, supplementing or entirely substituting for the petals that attract pollinators in most flowering plants. The sepals and petals collectively compose the perianth, which literally means "around the flower." Within the perianth, the stamens and carpels develop. The stamen bears microsporangia (the anthers) in which pollen grains develop from microspores. The carpel is a modified angiosperm leaf bearing one or more integumented megasporangia, that is, ovules. Just as in gymnosperms, the megagametophyte typically develops from a single functional megaspore. Unlike gymnosperms, whose ovules typically have one integument, the ovules of the flowering plants are bitegmatic—each ovule is surrounded by two integuments.

Because the carpel surrounds the ovules within it and therefore hides seeds from direct view, the flowering plants are called angiosperms, literally "seeds are contained" (*angos* is the diminutive of the Greek word for container, referring in this case to the carpel). More important, the eggs within the ovules of flowering plants are secluded from pollen grains in the sense that once a pollen grain attaches to the stigma of a carpel it must grow through the tissues of the carpel before it can deliver its sperm cells to an egg. The pollen grains of all living angiosperms invariably produce pollen tubes that grow through the style and deliver two sperm cells to each egg. Elements of this delivery system had already evolved in some Carboniferous gymnosperm species lacking carpels, perhaps because it afforded a mechanism whereby the sporophyte could immunologically identify pollen from its own species and conversely reject or kill foreign pollen grains.

The passage of the pollen grain from one flower to another is a painstakingly investigated affair among angiosperms and accounts in part for the stunning diversity in flower size, shape, and number of parts (Faegri and van der Pijl 1979; Crepet 1984). Some flowers, like the common duckweed *Lemna minor*, measure less than 1 mm in length and have waterborne pollen; other flowers, like those of the parasitic Sumatran plant *Rafflesia arnoldii*, approach 3 m in diameter and are pollinated by flies. Some are bilaterally symmetrical with fused parts that, like many orchids, mimic the appearance of a pollinator's mate or prey, thereby attracting an enterprising animal and dusting it with pollen grains. Other flowers, like *Magnolia*, have spirally arranged parts so numerous and fleshy that portions of the flower may be gratuitously tendered as food to attract pollinators. Although most flowers are "perfect" in the sense that they have both stamens and carpels, some are self-sterile or configured to release their pollen before their own carpels mature, and many species have evolved "imperfect" flowers bearing exclusively either stamens or carpels. Many plants are wind pollinated and have evolved floral traits that reduce the chances of capturing pollen from the same plant (e.g., grasses). These and other floral contrivances can decrease the probability that a flower will pollinate itself and so suffer the deleterious genetic effects of inbreeding.

The intricate relation between floral size and shape and the biology of pollinators offers a potential explanation for the extraordinarily large number of angiosperm species and the apparent rapid rise of the angiosperms compared with other plant groups (Grant 1963, 1971; Crepet,

Friis, and Nixon 1991). Any mutation altering the pollinator attracted to the flowers of a species could invoke a reproductive barrier between the mutant and its parent population. Subsequent genetic and phenotypic divergence could lead to speciation. Likewise, species hybrids that attract a novel pollination syndrome would be reproductively isolated from both parent species (Grant 1949). Nevertheless, it is highly problematic that the success of the flowering plants is entirely due to the evolution of the flower per se. A constellation of morphological, anatomical, developmental, and ecological features coevolved with the flower and likely contributes to the great success of the angiosperms (Tiffney 1981). The taxonomic separation of the flowering plants from the gymnosperms is technically vouchsafed not merely by the enclosure of ovules within carpels. (The term "angiosperm" inadequately describes the biological features that truly set this group of seed plants apart from all other groups.) The angiosperms differ from the gymnosperms not because they have "hidden seeds" but because their biology entails a large number of additional features such as the ability to propagate asexually, grow rapidly, and produce specialized xylem elements called vessel members and phloem cells called sieve tube members in addition to the features collectively defining the flower. When this constellation of vegetative and reproductive features is viewed in its entirety, the adaptive breakthrough of the flowering plants is seen to be not the evolution of the flower but the embedding of this reproductive organ within a backdrop of many other adaptations.

"Biological success" is not easy to measure in an unambiguous way, but by almost any yardstick the angiosperms appear to be the most successful seed plant group. Consider the number of living angiosperm species and families compared with the number of surviving bryophyte, pteridophyte, and gymnosperm species. Depending on the taxonomic authority consulted, there are anywhere between 300 and 400 distinct angiosperm families containing nearly 220,000 species. In contrast, there are currently only 22,400 bryophyte, 9,000 pteridophyte, and roughly 750 gymnosperm species. Thus there are about 7 flowering plant species for every extant nonflowering embryophyte. In fact, the number of species within a single flowering plant family can far outnumber all extant pteridophyte and gymnosperm species combined! The bean family alone has 14,000 species, compared with the roughly 10,000 surviving pteridophyte and gymnosperm species. Yet another yardstick of biological success is ecological diversity. Here too the flowering plants are the most successful. Aside from occupying polar regions, deserts, and bodies of freshwater, the angiosperms have also returned to the sea (e.g., the "seagrass," *Zosterophyllum*), a habitat totally devoid of any bryophyte, pteridophyte, or gymnosperm species. Yet another gauge of evolutionary success is how rapidly one plant group assumes dominance over previously successful ones. The fossil record shows that the angiosperms rose to taxonomic dominance over their gymnosperm contemporaries within 40 million years or so. It is not clear whether this was a consequence of competitive displacement or of the radiation of angiosperms into niches vacated when earlier gymnosperm and pteridophyte species became extinct just before the Cretaceous (see chapter 8). Nevertheless, in terms of their current species richness, their ability to survive and prosper in many different habitats, and their historical rise as the dominant element in the flora, the flowering plants are the most successful type of land plant.

Attempts to lay the success of the angiosperms squarely on the reproductive virtuosity of the flower are easily thwarted. Some of the most successful flowering species have totally abandoned sexual reproduction. For example, although it has all the accouterments for sexual reproduction, the common dandelion produces seeds with viable embryos not from the fusion of sperm and egg but from the asexual division of cells within the nucellus. The flowers of the lemon and mango equally eschew the zygote as the means of reproduction. By the same token, duckweed crowds the surfaces of lakes and the water hyacinth clogs lowland waterways with genetically identical copies produced by asexual reproduction through fragmentation. On land the kudzu and poplar, like the weedy dandelion, duckweed, and water hyacinth, illustrate the same theme—rapid colonization and occupancy of a particular site by means of vegetative growth and propagation. Even among those species that normally reproduce sexually, many rely on rapid vegetative growth to establish individuals in new sites and to hold on to a previously occupied habitat when populations are disturbed either physically or biologically. The rapidity with which some angiosperm species grow vegetatively relates to their herbaceous growth habit, which is totally absent in other extant seed plants. The seeds of the mouse-ear cress *Arabidopsis* grow into a flowering adult within twenty-four days. Even the most rapidly growing gymnosperms, like *Ephedra*, typically take years to reach sexual maturity, and the seeds of most gymnosperms take months or years to develop and mature from fertilized ovules.

Even though the evolutionary success of the angiosperms cannot be ascribed solely to benefits conferred by possessing flowers, the features of the flower are typically used to trace the evolutionary origin of the angiosperms to some type of gymnosperm ancestor. Tantalizingly, many of the morphological and developmental features that characterize the "flower" overlap with those found in the reproductive organs of fossil gymnosperms. For example, the ovules of the Jurassic gymnosperm

Caytonia were surrounded by carpel-like cupules, and the cupules of another group of Jurassic gymnosperms, the Czekanowskiales, had papillate flanges that superficially resembled the stigmatic surfaces of modern flowers. Although the pollen grains of these fossil gymnosperms appear to have landed directly on the micropyles of the ovule, many primitive extant angiosperms have carpels that are not completely closed around their ovules (e.g., *Drimys, Degeneria,* and *Exospermum*). Clearly, a number of ancient gymnosperm lineages appear to have explored the benefits of enclosing their ovules with carpel-like structures.

Another morphological feature, the bitegmatic ovules of angiosperms, may have been explored well before the angiosperms made their first appearance. The seeds of the Pentoxylales, a group of Jurassic and Cretaceous gymnosperms, have two integuments. Likewise the seeds of the Bennettitales, another gymnosperm group extending from the Triassic into the Cretaceous period, are interpreted to be bitegmatic. The Bennettitales are of interest for another reason. Some of these organisms had bisporangiate reproductive organs; that is, pollen and ovules were produced in the same structure. This is unusual among gymnosperms, which typically produce their pollen and ovules in separate organs. In sum, familiarity with the diversity of fossil and living gymnosperms quickly shows that some of the most salient "floral" features appear in ancient gymnosperms before the first angiosperms made their evolutionary entrance.

Yet one feature of the flowering plants remains unique: the formation of polyploid endosperm, the storage tissue within the seed. The endosperm is a consequence of the biological phenomenon called double fertilization. After the angiosperm pollen tube grows through the style of a flower, it delivers two sperm cells to the megagametophyte (fig. 4.13). The haploid nucleus of one sperm cell fuses with the haploid nucleus of the egg cell to produce a diploid zygote that subsequently develops into a sporophyte embryo. The nucleus of the second sperm cell enters the central

Figure 4.13. Diagrams rendering one way an angiosperm ovule (A) can develop into a seed (F). Alternative developmental routes exist. The one depicted here is called monosporic megagametogenesis (i.e., the megagametophyte develops from one megaspore. (A) Meiosis results in one functional haploid megaspore (m) within the diploid nucellus (n) of an ovule invested with two integuments (i). (B–E) The haploid nucleus of the megaspore undergoes three mitotic division cycles to produce eight haploid nuclei. (F) Cytokinesis apportions eight haploid megagametophyte nuclei into seven cells, one of which is the egg (e), located opposite three antipodal cells (ac), and the largest of which is the binuclear central cell (cc) with its two polar nuclei (pn). (G–H) The pollen tube cell delivers two sperm cells (sc) to the megagametophyte, and one sperm fuses with the egg while the second sperm fuses the central cell (whose two nuclei may have fused to form a diploid nucleus before double fertilization; see G). (I) Double fertilization is complete with the formation of a diploid zygote and, in this case, a triploid nucleus of the central cell.

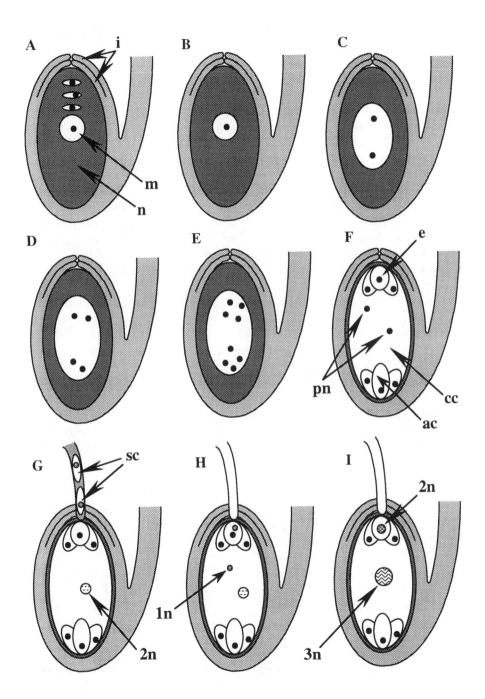

cell. Initially, the central cell is binucleate. Its two nuclei are called the polar nuclei; in the simplest case, the polar nuclei are haploid. As a consequence of double fertilization, a triploid nucleus is produced in the central cell. The triploid central cell subsequently divides mitotically and produces a triploid endosperm tissue. As it develops, the endosperm absorbs nutrients from the parent sporophyte and later relinquishes these storage materials to the embryo.

Double fertilization has been reported in the living gymnosperm *Ephedra* (Friedman 1994), although this claim is highly problematic, as may be seen from the following. The megagametophyte of *Ephedra* produces archegonia, each containing a binucleate cell at the time of fertilization. In *E. nevadensis* and *E. trifurca,* the pollen grain releases two sperm cells into this cell. Each sperm fuses with one of the two nuclei, resulting in two diploid nuclei (fig. 4.14). The development of these two diploid nuclei is identical—each divides mitotically to form several diploid cells, which in turn develop into manifold embryos. No polyploid endosperm is formed. The *Ephedra* megagametophyte provides the developing embryos' nutrition. In contrast, the concept of double fertilization posits the interaction of two sperm cells with two specialized cells within the megagametophyte—the egg cell and the central cell—whose fertilization fates differ profoundly (the fertilized egg develops into the embryo; the fertilized central cell develops into endosperm). Thus double fertilization sensu stricto cannot be said to occur in *Ephedra*.

That two sperm cells fuse with the same cell, and that a polyploid endosperm never forms, distances the reproductive biology of *Ephedra* from that of the flowering plants. The pollen grains of many gymnosperms produce two sperm cells, one of them "left with something to do." In pines the two sperm cells differ in size, and the larger one carries out fertilization while the smaller sperm cell disintegrates. In *Ephedra* the second sperm cell obtains supernumerary embryos within the same developing seed. *Ephedra* and the angiosperms may have shared a common ancestor; cladistic analyses typically place the Gnetales (a plant order containing the genera *Ephedra, Gnetum,* and *Welwitchia*) in close phyletic alignment with the angiosperms (Doyle, Donoghue, and Zimmer 1994; Nixon et al. 1994). Given the information currently available, however, this alignment does not provide proof that the "double fertilization" seen in some gnetalean gymnosperms is homologous with that seen among the angiosperms. It is just as likely—perhaps more so—that the gnetalean and angiosperm lineages evolved along a variety of nearly parallel reproductive avenues that in some cases produced analogous features.

Charles Darwin called the origin of the angiosperms an "abominable mystery." Things have not changed appreciably: the angiosperm ancestor

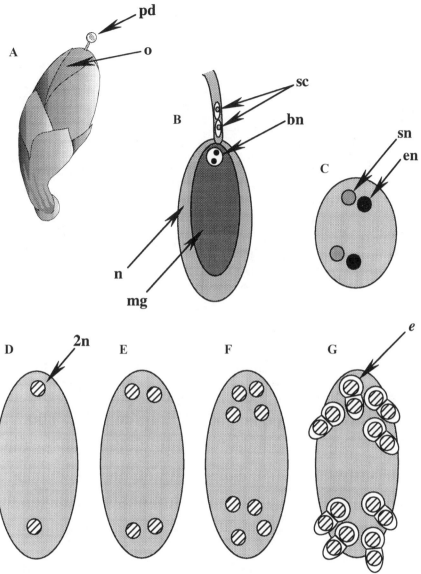

Figure 4.14. Fertilization and early stages of embryo development in *Ephedra*. (A) ovule (o) with extended micropyle exuding a viscous pollination droplet (pd); (B) pollen tube delivers two sperm cells (sc) to binucleate cell (bn) produced by megagametophyte (mg) within the nucellus (n) of the ovule; (C) sperm nuclei (sn) fuse with nuclei (en); (D–G) diploid nuclei within binucleate cell mitotically divide twice to produce a cell containing eight diploid nuclei (F), and cytokinesis results in eight diploid cells, each of which develops into a filamentous proembryo (e). Adapted from Friedman 1994.

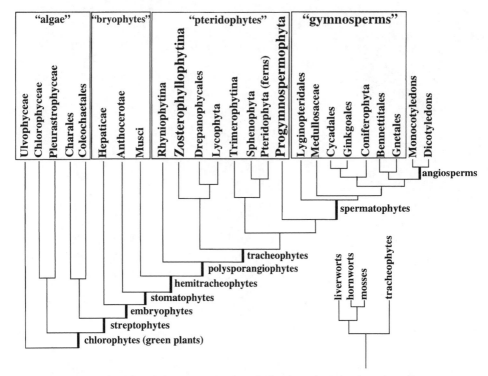

Figure 4.15. Simplified cladogram rendering a phyletic hypothesis for the embryophytes and evolutionarily related green algal lines. Different lineages sharing the same level of morphological or reproductive organization (i.e., evolutionary grades) are grouped in boxes. The informal names for these grades are given in parentheses (e.g., "algae"); formal taxonomic designations are given in boxes (e.g.., Ulvophyceae). Plant groups sharing the same ancestor are indicated by dark vertical links (e.g., chlorophytes). Note that not all the formal taxa are of equal taxonomic rank (e.g., "aceae" designates family rank, "ales" indicates an order, and "phyta" indicates division). According to this cladogram, "bryophytes" are a paraphyletic group (an assembly of organisms that excludes some species that share the same common ancestor with species included in the group). The small cladogram (bottom right) depicts the more traditional view that the bryophytes (Hepaticae = "liverworts," Anthocerotae = "hornworts"; Musci = "mosses") are monophyletic (i.e., Bryophyta) and shared a last common ancestor with the vascular plants (tracheophytes). Adapted from Mishler et al. 1994; Nixon et al. 1994; and Rothwell and Serbet 1994.

is still unknown, although it must be some type of gymnosperm. By its nature, evidence of angiosperm double fertilization has a low probability of being preserved in a fossil seed. And even when preserved, the endosperm can be distinguished only based on the ploidy of cells, a cytological feature not likely to be preserved. Much less mysterious is the first appearance of the flowering plants, roughly 125 million years ago during the Cretaceous period (see Doyle and Hickey 1976; Taylor and Taylor 1993).

Closing Remarks

The most expeditious vehicle for summing up the salient features of this chapter is a cladogram showing the hypothetical phylogenetic relations among the various plant groups discussed (fig. 4.15). This cladogram is "hypothetical" in the sense that all cladistic reconstructions are hypotheses—every cladogram is subject to further testing and may change as new information comes to light. Even though most of the relations shown in this cladogram may continue to hold under future scrutiny, their details must be used cautiously, particularly regarding the alignment of fossils with living groups of plants. The incorporation of paleobotanical data in cladistic analyses of living plants is very much in its infancy, and newly discovered fossils may dramatically alter prior perceptions of phylogenetic relations. With these caveats in mind, this cladogram will be used as a backdrop for the discussions in the following chapters.

Part 3 Adaptive Walks: A Hypothesis

Plate 3. Aspects of plant morphology, such as the arrangement of leaves on stems (called phyllotaxy), are often amenable to simple mathematical descriptions. The simplest of all the spiral phyllotactic patterns is the distichous, or alternate arrangement, in which successive leaves are 180° apart around the stem (e.g., *Hedera helix*, in the background, and the orchid *Dendrobium cumenatum*). Since one circuit around the stem axis is required to pass from one leaf to another directly below it, the distichous pattern can be expressed by the fraction 1/2. In other species, one circuit around the stem axis passes through three leaves of which the top and bottom are superimposed. This tristichous pattern can be represented by the fraction 1/3. These and other patterns are described by a series of fractions called the Fibonacci series, where the numbers in both the numerator and the denominator of each fraction are the sums of those in the two preceding fractions in the series. The higher fractions in this series become more and more uniform and approach as a limit the decimal fraction 0.38197, for which no two leaves on a stem would ever lie precisely under one another.

5 The Aquatic Landscape

*The mechanism is, however, one which shuffles the
species about in the general field as a whole. Since the
species will be shuffled out of low peaks more easily
than high ones, it should gradually find its way to
the higher general regions of the field as a whole.*

Sewall Wright

In the previous two chapters I reviewed the broad trends
preserved in the plant fossil record, starting with the
origin of life and the genetic code and closing with the evolution of flowering plants. My objective in this chapter and chapter 6 is to see whether
this outline can be emulated by computer generated adaptive walks on
fitness landscapes. The idea is not to redact history but to discover why
plant evolution proceeded the way it did. The modus operandi is to construct a universe of all conceivable plant phenotypes, assign a relative
fitness to each phenotype, and beginning with a hypothetical ancestor,
simulate transformations leading from the ancestor to progressively more
fit descendants. The hypothetical universe of all conceivable phenotypes
may be called a "morphospace," a domain containing all morphological
or anatomical possibilities (see Thomas and Reif 1993). By assigning the
location and a relative fitness to each phenotype this domain, we construct
a "fitness landscape," and the sequence of phenotypes leading from
ancestor to the most fit descendant is an "adaptive walk."

The sequence of adaptive walks presented in this chapter and the next
is dictated by the course of actual evolutionary events. Life began in the
water in the form of unicellular entities, achieved multicellularity and radiated into diverse forms in the aquatic habitat, and finally invaded the
land. Here my concern is with the physical appearance (morphology) or
internal structure (anatomy) of aquatic plants. The morphospace for terrestrial plants and adaptive walks through it are dealt with in chapter 6.

Relative Fitness and "Complexity"

Computer simulated adaptive walks present a number of philosophical
and practical challenges that are best discussed at the beginning. As I mentioned, relative fitness depends on the particular environment in which a

phenotype grows and reproduces. Clearly, the environment has changed significantly over evolutionary history (see Kauffman 1993; Gould 1980; Levin 1978). The Earth's atmosphere and oceans have changed chemically, mountain ranges have come and gone, continents have moved, fragmented, and rejoined, and as plants evolved, new species entered the arena, thereby changing the biological environment. Even during periods of environmental stability, plants have moved from one type of environment to another. The transition from aquatic to terrestrial plants is a case in point. For the first land plants, the terrestrial landscape was a totally new environment. Clearly, it is impossible to deal realistically with all of the temporal and spatial changes that have occurred in the physical and biological environment during the long history of plant evolution. Nonetheless, the fitness of plants is known to correlate closely with the operation of physical laws and processes governing the exchange of mass and energy between the plant and the external environment (Gates 1965; Brent 1973; Nobel 1983; Sultan 1987; Van Tienderen 1990; Niklas 1992, 1994; Schmid 1993). These physical laws and processes have remained constant through evolutionary time, and so the broad outlines of the physical environment defining the fitness landscape of plants have not changed. Although we cannot identify subtle differences in the relative fitness of phenotypes, we can paint the fitness landscape with broad brushstrokes in the primary colors of biophysics and biomechanics.

When assigning a relative fitness to each phenotype in a morphospace, one must recognize that the fitness contributed by performing one biological task depends on the ability to perform other tasks (epistatic fitness contributions; Ewens 1979; Franklin and Lewontin 1970; Lewontin 1974). A plant's ability to capture sunlight and garner the raw materials for photosynthesis (carbon dioxide and water) undoubtedly influences its ability to reproduce. Plants invest their progeny with some of their own energy in the form of metabolites stored in spores or seeds. Likewise, a plant's ability to elevate its reproductive and photosynthetic organs through a water column or above the ground depends on the mechanical properties of its tissues and organs. The "mechanical fitness" of a plant undoubtedly influences the contributions to total fitness made by photosynthesis and reproduction. Assigning a relative fitness to each phenotype in a morphospace is therefore far from simple. Throughout this chapter and the next we will have to make simplifying assumptions about the number and kinds of biological tasks that collectively define the relative fitness of phenotypes.

Epistatic fitness contributions highlight the issue of biological complexity. Paleontologists and theorists have long debated what, if anything, is meant by biological "complexity." Some think that adaptive evolution results in increasingly complex organisms; others believe complexity is a purely human concept that is ill defined and largely meaningless in the context of organisms other than ourselves. I hope to show, however, that "complexity" can be defined in terms of the number of biological tasks an organism must perform simultaneously in order to survive, grow, and reproduce. I also plan to show that adaptive walks through a hypothetical morphospace can be used to quantify the influence of biological complexity on evolutionary trends. With this definition and pedagogical tool, we will address the relation between the number of phenotypes in the morphospace that perform their biological tasks equally well and the number of functional tasks that collectively define "complexity."

Although there is no a priori reason to assume that the number of equally "most fit" phenotypes depends on the number of tasks a plant must perform, there are good reasons to believe that increasing the number of functional obligations increases the number of phenotypes with equivalent relative fitness. Engineering theory shows that the number of equally efficient designs for an artifact is generally proportional to the number of the tasks it must perform simultaneously (Meredith et al. 1973). The reason is twofold. First, the efficiency with which each of many tasks an artifact must perform has to be relaxed owing to unavoidable conflicting design specifications for individual tasks (Gill, Murray, and Wright 1981). And second, as the number of tasks increases, the number of configurations (= phenotypes) that achieve equivalent or nearly equivalent performance levels also increases (Brent 1973). If these features hold true for organisms, then the number of equally fit phenotypes may increase as organisms become more "complex."

Although the problematic analogy between engineered and biological systems may speak to the meaning of the term "complexity," it does not directly shed light on the "dynamics" of an adaptive walk. That is, it does not address the relation between the number of tasks a plant must perform simultaneously and the magnitude of the morphological transformations between neighboring variants required to reach the most fit phenotypes in a morphospace. Here the principal question is, Are walks confined to nearest-neighbor variants, or are they free to reach comparatively distant phenotypes? In some respects this question harks back to the two diametrically opposite portraits of speciation painted in chapter 2. According to the Darwinian view of speciation, phenotypic transformations between ancestor and descendant species are slow and incremental. According to the portrait of speciation painted by genetic revolution, the

birth of a species may be sudden, with few if any morphological intermediates between ancestor and descendant. The Darwinian view, called
phyletic gradualism, dictates stately adaptive walks over fitness landscapes driven by natural selection. In contrast, genetic revolutions may
inspire punctated equilibriums such that the course of morphological
evolution hopscotches over fitness landscapes from one fitness peak to
another without the requirement of phenotypic intermediates.

At some level, the number and magnitude of phenotypic transformations must depend on both the location of fitness peaks and how closely
the fitness of neighboring variants is correlated. That is, the morphological transformations defining an adaptive walk must depend on the topology of the fitness landscape. However, it is also clear that the dynamics of
an adaptive walk equally depend on the ability to alter the phenotype in
a heritable manner. Although the developmental repertoire of most organisms permits some latitude in external shape and internal structure
(i.e., every organism has a norm of reaction), an adaptive walk is undoubtedly governed by genetic or developmental mechanisms that establish barriers to phenotypic transformations among neighboring variants
on the fitness landscape (Alberch 1980, 1981, 1989; Odell et al. 1981;
Oster, Odell, and Alberch 1980). If so, then morphological transformations among phenotypes are not equiprobable, and walks cannot be
governed exclusively by the topology of the fitness landscape.

On the other hand, the extent to which a walk is genetically or developmentally fettered is relative rather than absolute. The barriers established by genetic and developmental factors undoubtedly vary among organisms and change over evolutionary time for each kind of organism. In
this regard, the "phenotypic plasticity" of plants appears to be extremely
high in comparison with that of most animals (Schmid 1993; Sultan 1987;
Van Tienderen 1990). Also, certain periods of evolutionary time are characterized by exceptionally high rates of phenotypic innovation, for example, the colonizing of the terrestrial landscape by the first vascular
plants (see fig. 4.8). Therefore one is left with the impression that plant
development permits large phenotypic transformations compared with
those occurring in most animals and that the adaptive walks of plants
typically involve morphological transformations large enough to reach
most of the morphological fitness peaks on most fitness landscapes. We
will evaluate this impression by comparing, once again, hypothetical
adaptive walks against the fossil record. In any circumstances, the assumption that walks are unimpeded over spatially stable fitness landscapes greatly simplifies attempts to explore the relation between landscape topology and the dynamics of walks.

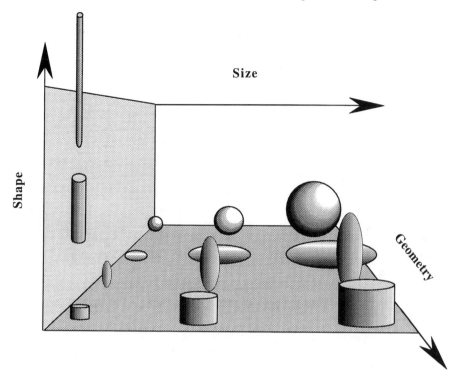

Figure 5.1. Hypothetical cell morphospace.

The Cell Morphospace

In many respects the cell morphospace is the simplest among those treated in this chapter and the next because it is a three-dimensional domain defined by "size," "geometry," and "shape" (fig. 5.1). Nevertheless, even though we may have an intuitive grasp of what is meant by size, geometry, and shape, these three variables are not easily defined or measured in nonarbitrary ways. Consider size, which is a substantial variable. That is, size must be expressed in units of a fundamental physical quantity, like mass, or of a derived physical quantity, like surface area or volume. The unit of measurement of size is arbitrary in the sense that any fundamental or derived physical quantity can be used, but it is not arbitrary in the sense that different units of measurement are not interchangeable (e.g., mass and volume). Also, the measurement of size should be convenient and also express biologically meaningful differences among the entities being compared. For example, cells often contain vacuoles or crystals and other inert materials that contribute to cell volume but are not necessarily physiologically active cell components. Therefore, although it is convenient and relatively easy to determine, cell volume can be a misleading measure of cell size, particularly for cells with large amounts of physiologically inert materials. In contrast, size measured in terms of the mass of organic carbon per cell may be elected as more representative of the physiological size of cells than volume. Yet the cell walls of some unicellular plants contain large amounts of carbon while the cell walls of others contain little or none (e.g., the silica cell walls of diatoms). Clearly, size comparisons among unicellular plant species differing in cell wall chemistry could easily go astray when size is measured in terms of the mass of carbon per cell. Yet another measurement of cell size, the quotient of cell surface S and cell volume V, is frequently used as the measure of cell size because the rates of many physiological processes depend on cell surface area or volume or both. Because this quotient has units of m^{-1} (i.e., $S/V = m^2/m^3 = 1/m = m^{-1}$), it is a more cumbersome measurement of cell size than a simple linear measurement, say cell length. Nonetheless, the quotient S/V is useful because it expresses size in terms of two variables that are embedded within many equations describing a number of physical and physiological processes. In the end, however, no single measure of cell size is perfect: each has its merits and debits.

If the measurement of cell size presents problems, cell geometry and shape are far more difficult to define or measure in useful ways. Though geometry and shape are often treated as synonymous terms, a sharp distinction must be drawn between these two variables. Geometry refers to the class of three-dimensional objects a solid body belongs to, whereas shape refers to the relative proportions of a solid body regardless of its geometric class. Consider a cylinder with a circular cross section. All circular cylinders belong to the same class of geometric objects; each is a solid body generated by translating a circle along the length of an axis. But not all cylinders have the same shape: they can be narrow or broad, depending on the diameter of their circular cross section, and they can be long or short, depending on the length of their principal axis. In general terms, shape is a natural variable. That is, unlike size, shape need not be measured in terms of a fundamental or derived physical quantity but can be described without reference to an external, essentially artificial standard of measurement. The shape of a circular cylinder can be described by the aspect ratio, the ratio of length L to diameter D (see fig. 5.1). Aspect ratios like L/D do not have units because the components of the ratio share the same units, but aspect ratios have magnitude without reference to absolute size. Thus $L/D = 100$ describes a very long or narrow cylinder, while $L/D = 0.01$ describes a very short or wide cylinder. Notice that the aspect ratio is independent of size because it describes the *proportionality* of an object's dimensions. Cylinders of comparable size can have very different shapes. Conversely, small and large cylinders can have the same shape.

Cell size, geometry, and shape are potentially independent variables in the sense that cell geometry or shape, or both, can be altered as a cell continues to increase in size. By the same token, different species of unicellular plants can be adapted to their particular environments by virtue of differences in cell geometry or shape. By way of illustration, it is well known that for any series of objects differing in size but sharing the same geometry, surface area is proportional to the 2/3 power of volume. This scaling relation results because surface area has the dimensions of the square of length while volume has the dimensions of the cube of length. Therefore when surface area is plotted against volume for objects with the same geometry but differing in size, the slope of S versus V equals 2/3 ≈ 0.667. What is generally not appreciated is that this "2/3-power law" holds true only if the series of objects differing in size maintains the same *shape* as well as the same geometry. If shape varies among the objects, then the slope of the relation between surface area and volume can be lower or higher than 2/3. Because unicellular plants may differ in size, geometry, and shape, the actual slope of the relation between their cell surface area and volume may be higher or lower than 2/3. For some collections of unicellular plants, the slope equals 0.699 (fig. 5.2). Although the difference between 0.667 and 0.699 superficially appears trivial, statistical analysis reveals that 0.699 actually is significantly different from 0.667. The difference may also be biologically significant because a greater cell surface area can expedite the rate of exchange of materials between a cell and its surroundings (Niklas 1994).

Comparisons across unicellular species of plants indicate that the geometries of progressively larger species confer disproportionately larger surface areas relative to their cell volumes than occur in smaller species. Specifically, small unicellular plant species tend to have a spherical geometry; larger species tend to be oblate or prolate (i.e., a "pancake" or a "cigar" geometry); and the largest species tend to have cylindrical cell geometries. Note that, with the exception of the sphere, all geometric categories (e.g., oblate and prolate spheroids, and cylinders) consist of things that can differ in shape as well as size. (The sphere cannot change shape because its geometry is defined by only one dimension, radius.) Indeed, within each class of cell geometry, progressively larger species tend to adopt a more slender shape that further increases the ratio of cell sur-

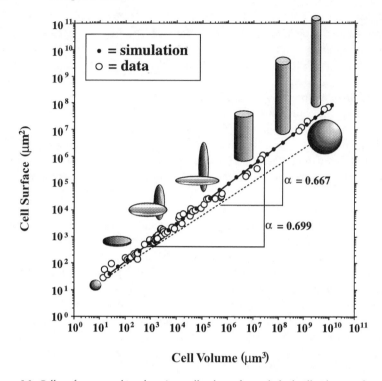

Figure 5.2. Cell surface area plotted against cell volume for real algal cells (data) and computer simulated geometries that maximize surface area with respect to volume as cell size increases (simulation). The changes in cell geometry required to maximize surface area with respect to cell volume are indicated by figures inserted above the dashed regression line. Data for real algal cells from Niklas 1994.

face area to volume. For example, larger cells with an oblate geometry are disproportionately flatter than smaller oblate cells, while larger cylindrical cells tend to be more slender than their smaller counterparts. Thus comparisons across unicellular plant species reveal that cell size, geometry, and shape vary in such a way that the classic "2/3-power law" is violated (Niklas 1994).

The relation between cell surface area and cell volume for unicellular plants is the maximum slope that can be achieved when cell size, geometry, and shape are simultaneously and independently varied. This can be shown by means of an "adaptive walk" through a cell morphospace instructed to find progressively larger cells with the highest possible ratios of cell surface area to volume (fig. 5.3). (In this morphospace, cell size is measured in terms of cell length; in spherical cells, size equals cell diame-

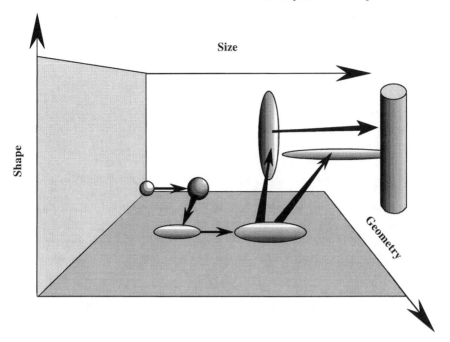

Figure 5.3. Simulated adaptive walk through the cell morphospace (shown in fig. 5.1). The walk begins with a small spherical cell (left rear) and bifurcates to end with a large flat oblate spheroid (right rear) and a large circular cylinder (right front).

ter.) With felicitous alacrity, the adaptive walk through the cell morphospace identifies not only the same sequence of cell geometries observed for progressively larger unicellular plant species (i.e., sphere → oblate spheroid → prolate spheroid → cylinder), but also a sequence of cell shapes within each geometric class that collectively obtains a slope of 0.699 for the scaling relation for surface area with respect to cell volume (see fig. 5.2).

This adaptive walk appears to hold true for bacteria as well as photosynthetic unicellular eukaryotes. Although most bacterial cells are nearly spheroidal and small (see fig. 5.5), some are surprisingly large, such as *Epulopiscium fishelsoni*, a gram-positive bacterium that can measure up to 600 μm in length and 80 μm in diameter (Montgomery and Pollak 1988; Angert, Clements, and Pace 1993). *E. fishelsoni* cells, which are perhaps the largest bacterial cells known, are very slender prolate spheroids (L/D $\approx$ 6–10). As predicted by the adaptive walk for cells, this shape maximizes cell surface area with respect to cell volume when cell size is very large. The ability to exchange materials between the *E. fishelsoni* cell and its external environment is enhanced by a complex of highly folded membranes appressed against the bacterial cell wall by a large central mass of more or less amorphous cytoplasm. The appearance of these membranes suggests that they are the principal site of metabolic activity within the cell. If so, their location is ideal because Fick's law shows that it minimizes the distance (and time) required for materials to passively diffuse into and out of the cell. The internal structure of *E. fishelsoni* may shed some light on how prokaryotes deal with large cell sizes rivaling those of photosynthetic unicellular eukaryotes. Its internal structure is functionally similar to that of most photosynthetic plant cells in which a large central vacuole appresses the cytoplasm containing chloroplasts and mitochondria against the plant cell wall, where organelles have greater access to light, carbon dioxide, oxygen, and nutrients.

That the adaptive walk for hypothetical cells mimics the real relation between cell surface area and volume for actual unicellular plants and bacteria may simply be fortuitous, or it may mean that cell geometry and shape have been adaptively varied among species to achieve the maximum surface area for any given cell volume. Undoubtedly, maximizing relative surface area enhances the rate at which a variety of substances passively diffuse into cells. For example, according to Fick's law for passive diffusion, the time it takes the cellular concentration of CO_2 to increase from zero to one-half the external ambient CO_2 concentration is inversely proportional to S/V. The same holds true for a variety of other substances essential for cellular photosynthesis and respiration (Nobel 1983; Niklas 1994). Thus, all other things being equal, unicellular plants with high ratios of surface area to volume acquire external metabolites like carbon dioxide faster than unicellular plants with smaller ratios. Small spherical cells have exceptionally high ratios of surface area to volume. As a spherical cell increases in size (diameter), however, its S/V progressively and unavoidably decreases in direct proportion to the increase in size. Beyond a certain size, therefore, it incurs the "physiological wrath of Fick's law." Other cell geometries are far more accommodating. The ratio of surface area to volume for a cylindrical cell is independent of cell length and varies only as a function of cell radius ($S/V = 2\pi R L / \pi R^2 L = 2/R$). Therefore a cylindrical cell can grow indefinitely in length without affecting the time needed for the passive diffusion of substances like carbon dioxide. For any cell size (cell length), slender cylindrical cells are more efficient in terms of Fick's law than their stouter counterparts.

It would be naive to claim that size-correlated differences in cell geometry and shape are a consequence of adaptive evolution in response to selection pressures favoring the maximization of cell surface area with respect to cell volume. Many other physical phenomena and physiological processes are influenced by cell size, geometry, and shape. For example, the ability of aqueous suspensions of unicellular plants to intercept sunlight also depends on cell size, geometry, and shape as a consequence of the "package effect," so named because photosynthetic pigments contained in discrete units or packages (e.g., chloroplasts or cells) are less efficient at harvesting light than equivalent amounts of pigments suspended in solution (Duysens 1956; Kirk 1975, 1976, 1983; see Niklas 1994, 66–72). Thus large spherical cells are less efficient at light capture than small spherical cells (fig. 5.4). Much more interesting is that mathematical models for the package effect show that moderate-sized cells with an oblate or prolate geometry are more efficient at harvesting sunlight than large spherical cells, and that large cylindrical cells are more efficient at intercepting sunlight than smaller cells with either an oblate or a prolate geometry (fig. 5.4). The consequences of cell shape on light interception are far more complex, but in general terms "slender" or "flat" cells tend to be more efficient at harvesting light than their stouter counterparts. Consequently, whenever cell size increases across unicellular plant species, the same sequence of cellular transformations that maximizes the ratio of cell surface area to volume is also predicted to maximize the ability of cells suspended in water to intercept sunlight. Because the package effect and Fick's law for passive diffusion operate in harmony (both independently obtain the same "adaptive walk" through the cell morphospace), either or both could account for the size-correlated differences in cell geometry and shape empirically observed among living unicellular species. By the same token, *neither* the package effect nor Fick's law may be the "governing principle," because correlation does not guarantee a cause-and-effect relation. Although computer simulated adaptive walks through a cell morphospace show that empirically observed relations among the size, geometry, and shape of unicellular plants achieve a near perfect adaptive "state of being" in terms of passive diffusion and light interception, the adaptive walks do not provide evidence that this adaptive state is the result of adaptive evolution, the consequence of the "process of becoming" by virtue of natural selection.

The adaptive walks predicted by the package effect and Fick's law were "driven" through the cell morphospace by stipulating a progressive in-

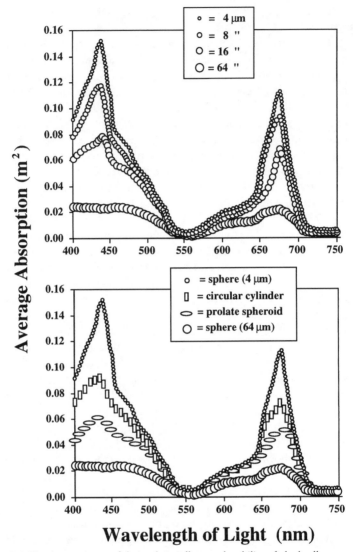

Figure 5.4. The consequences of the package effect on the ability of algal cell suspensions to absorb sunlight. Average absorption plotted against wavelength of light for hypothetical spheres differing in diameter (upper graph) and for various cell geometries (lower graph). Suspensions of very small spherical cells (4 μm in diameter) optimize light absorption (upper graph). The ability to absorb sunlight decreases as cells assume a cylindrical, prolate spheroid, or very large spherical geometry (lower graph). Adapted from Niklas 1994.

crease in cell size. This stipulation is arbitrary in the sense that there is no reason to assume that a progressive increase in cell size confers selective advantages. Indeed, the large cell size of some unicellular plant species poses something of a problem. Recall that the package effect predicts that very small cells are more efficient for light interception than larger cells (see fig. 5.4). By the same token, Fick's law reinforces the notion that unicellular photosynthetic organisms dependent on passive diffusion for external metabolites should be small rather than large because small cells have disproportionately higher ratios of surface area to volume than their larger counterparts. Thus small cells occupy the adaptive peak within the cell morphospace in terms of light harvesting and passive diffusion. The cyanobacteria as a group appear to "obey" the stricture against large cell size, whereas the photosynthetic eukaryotes, like the unicellular green algae, appear to flout the mandate of small cell size imposed by the package effect and Fick's law and often achieve cell sizes predicted to be less efficient. Indeed, the average size of unicellular photosynthetic eukaryotes is an order of magnitude greater than that of the bacteria (fig. 5.5). Thus the salient question is why the size of unicellular photosynthetic prokaryotes trespasses into regions of the cell morphospace predicted to confer less efficient light harvesting and passive diffusion.

The larger cell size of unicellular eukaryotes may be due to an "endosymbiotic package effect." Comparisons between photosynthetic prokaryotes and eukaryotes suggest this is a partial answer to the issue of large cell size in photosynthetic organisms possessing chloroplasts and other organelles. Recall that the endosymbiotic hypothesis states that chloroplasts and mitochondria are evolutionarily derived from ancient bacteria-like organisms that gained entry and took up permanent symbiotic residency within some ancient kind of host bacterial cell (see fig. 3.10). If so, then the eukaryotic cell is an "endosymbiotic package" whose total size logically must be larger than any of its constituent endosymbiotic parts. This observation is biologically consistent with the comparatively small cell sizes of photosynthetic prokaryotes and the larger cell sizes of photosynthetic unicellular eukaryotes, like the green algae (see fig. 5.5). The endosymbiont hypothesis further states that, even though they were most likely less efficient at light harvesting or passive diffusion than their endosymbionts, the first eukaryotic cells had an advantage in dealing with an increasingly oxygen-rich ambient environment and recycling the metabolic waste products of chloroplasts and mitochondria (oxygen and carbon dioxide, respectively). The first eukaryotic cells were barred from returning to the smaller size range occupied by their constituent endosymbionts. Metaphorically speaking, comparing the size

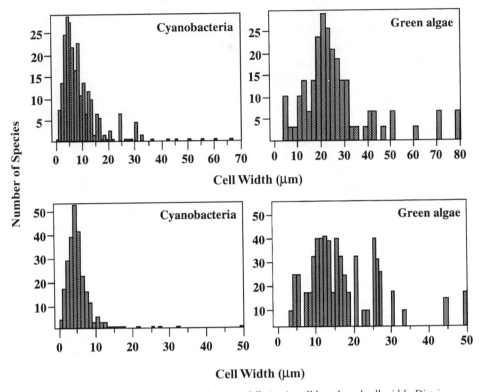

Figure 5.5. Frequency distributions of species differing in cell length and cell width. Distributions are for unicellular cyanobacteria (photoautotrophic prokaryotes) and unicellular green algae (photoautotrophic eukaryotes).

ranges of photosynthetic prokaryotes and unicellular photosynthetic eukaryotes suggests that the departure of eukaryotic cells from the optimal cell size defined by the package effect and Fick's law was jump-started by endosymbiosis.

Clearly this hypothesis begs a pivotal issue, the subsequent evolutionary appearance of progressively larger unicellular eukaryotes. That is, we are still left with the question why some *eukaryotic* species evolved larger cell sizes than their ancient eukaryotic ancestors. This question is answered in part by first recognizing that both the package effect and Fick's law *permit* further increases in cell size provided cell geometry or shape is modified to accord with the sequence of cellular transformations obtained by the adaptive walks through the cell morphospace. Large oblate, prolate, or cylindrical cells can be more efficient at light interception than spherical cells with equivalent cell volume (see fig. 5.4). Also, Fick's law shows that a cell can increase in overall size with no ill effects provided its ratio of surface area to volume remains the same. Cylindrical cells can grow indefinitely in length without altering their ratio of cell surface area to volume. This "strategy" has been adopted by a variety of unicellular organisms. The coenocytic green alga *Caulerpa* provides an excellent example (fig. 5.6). This unicellular plant may grow many meters long, yet its cell remains comparatively narrow in girth. *Caulerpa* shows that unicellular photosynthetic cells are not barred from being large organisms provided they adopt cell geometries and shapes that obtain and retain large surface areas with respect to cell volume.

Although physical processes and laws, like the package effect and Fick's law, permit large unicellular organisms, these same processes and laws do not *favor* the evolution of increasingly larger cell size. Indeed, depending on the environment where an organism lives, there are opposing advantages to small and large size. Large and tall terrestrial plants tend to have extensive root systems with which to garner water and soil nutrients efficiently. They also can cast shadows on their shorter neighbors, thereby depriving them of the sunlight essential for growth. Large plants typically produce many reproductive organs, while tall plants can disseminate their spores or seeds great distances by means of wind. Large land plants also tend to be insulated from rapid dehydration and changes in temperature, while larger aquatic plants suffer less rapid changes in external ion concentrations than their smaller counterparts. But there are many advantages to being small. For example, comparisons among many different plant species show that the rate of reproduction disproportionately decreases with increasing cell size (fig. 5.7). By the same token, larger cell size appears to correlate with a disproportionte reduction of the metabolically active ingredients of the cell. Larger cells tend to have lower intracellular concentrations of proteins, lipids, photosynthetic pigments, and so on (fig. 5.8). This metabolic "dilution" results from the fact that larger unicellular species have disproportionately larger cell vacuoles that contain very dilute chemical solutions. This feature may explain why, all other things being equal, a larger unicellular plant requires relatively more time to grow to maturity and reproduce than do its smaller counterparts. High growth and reproductive rates are advantageous in habitats where environmental conditions normally change rapidly. A small organism can grow to maturity and complete its life cycle in a relatively brief "window of opportunity" during which favorable ecological conditions prevail. In contrast, larger unicellular organisms, which require more time to grow and reproduce, run the risk of being exterminated when environmental

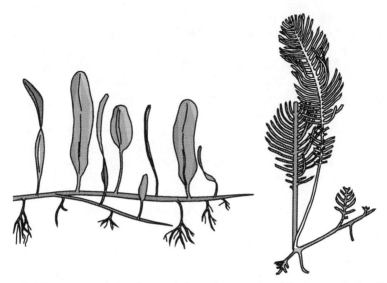

Figure 5.6. Diagrams rendering the morphology of two species of *Caulerpa,* a single-celled multinucleate green alga.

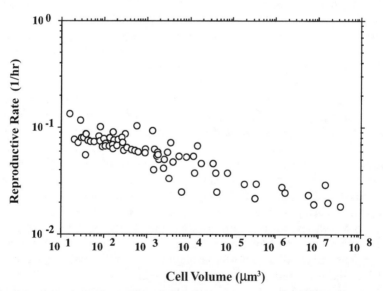

Figure 5.7. Reproductive rate of unicellular algae plotted against cell volume. The slope of the relation between these two variables is roughly −2/3, indicating that reproductive rate, on the average, decreases with increasing cell size. Adapted from Niklas 1994.

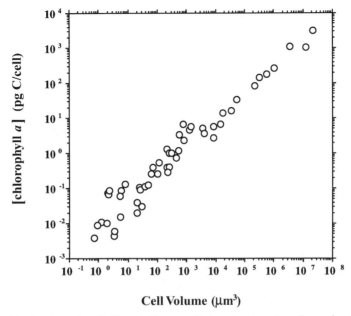

Figure 5.8. Concentration of chlorophyll *a* per cell plotted against the volume of unicellular algae. The slope of the relation between these two variables is roughly −2/3, indicating that concentration of the photosynthetic pigment disproportionately decreases with increasing cell size. Adapted from Niklas 1994.

conditions change drastically and unpredictably. Larger organisms tend to occupy habitats whose conditions remain relatively stable, whereas unstable conditions select against large organisms that grow and reproduce more slowly. These ecological generalities have macroevolutionary counterparts. During geological periods characterized by environmental instability, larger organisms tend to have higher probabilities of extinction than do smaller and presumably more rapidly growing and reproducing species, which tend to pass through periods of global inclemency relatively unscathed (Bakker 1977; Stanley 1979; Tiffney and Niklas 1985; Hallam 1990).

Nevertheless, a strict adaptationist explanation for the apparent trend toward larger eukaryotic cell size appears ill advised, particularly since an alternative model to natural selection exists and, in the case of increasing cell size, provides as cogent an explanation as does natural selection favoring larger size. This alternative model is inspired by the observation that the very first unicellular eukaryotes were very likely small and comparatively simple in shape and geometry compared with modern eukaryotic cells. The inaugural unicellular eukaryotes were undoubtedly poised on the edge of a potentially accessible, albeit initially totally unoccupied domain for larger cell size and more elaborate cell geometry and shape. A biological system like this one has at least one predictable property—it is expansive in nature in that the variance about the mean for each phenotypic feature (such as size, geometry, or shape) will tend to increase over time (Stanley 1973; Fisher 1991; Gould 1990). This kind of biological system is diffusive—over time, it will tend to invade and come to occupy progressively more regions within the theoretically accessible phenotypic domain. This expansive property is analogous to the way the molecules in a cube of sugar dissolve in a glass of water and eventually become equally distributed as a solute by means of passive diffusion. In like manner, an alternative to an adaptationist model is a "diffusive evolutionary model," which, simply put, argues that evolution abhors a phenotypic vacuum—whenever possible, organisms will evolve by chance into and come to occupy all regions in the domain of theoretically possible phenotypes that permit survival and reproductive success. (This model fits in with the theory of punctuated equilibrium, which suggests that selection can be weak at times; see chapter 2.) If true, then eukaryotes may have acquired larger size not because selection pressures favored larger cell size, but because selection pressures failed to act *against* its acquisition. According to this evolutionary model, most species are predicted to be small organisms, but mean cell size will nonetheless increase with time because a few new species will by chance evolve and elevate the maximum cell size (fig. 5.9). As a consequence, the mean, range, and variance of cell size are predicted to increase over time despite episodes of selection against large species that will reduce them. Over time, the frequency distribution of size will become progressively skewed toward larger cell size, but most species will remain "bunched" at the lower range of the accessible domain. Presumably the lower size range is defined by the minimal cell size required to contain the typical complement of eukaryotic organelles.

The frequency distributions of cell size for many unicellular organisms appear to comply with the predictions of a diffusive evolutionary model, particularly the "left-handed bunching" of small species and the "right-handed tail" reflecting a comparatively few larger species (see fig. 5.5). Likewise, the frequency distributions for fossil cell size through the Precambrian and Cambrian periods lend support to the diffusive model for the size of unicellular organisms, although the accuracy of some of these data is in doubt owing to poor preservation (see Schopf 1983).

If it is true that there exists no "intrinsic drive" toward larger cell size and that unicellular eukaryotes simply diffuse into unoccupied territory

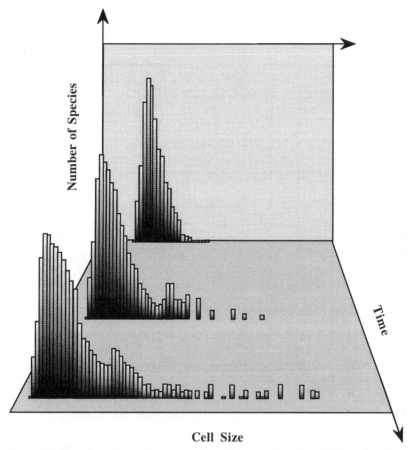

Figure 5.9. Hypothetical evolutionary trend in cell size resulting from "diffusive" rather than adaptive evolution (see fig. 5.5). Maximum cell size increases over time, but most species have a very small cell size.

within the domain of cell size and morphology, then the previous adaptive walks through the cell morphospace would appear, both literally and metaphorically, to have no power. But these walks were designed to predict only the adaptive transformations in cell geometry and shape required to explore the larger domain of the cell morphospace, regardless of why or how the domain was explored. Bear in mind that the diffusion model is random in the sense that any lineage may accidentally enter the domain of larger cell size, but it is also nonrandom in the sense that the cell geometry and shape of adventurously larger unicellular plants must abide by certain rules dictated by physical processes and laws like the package effect and Fick's law. Sewall Wright's metaphor for adaptive evolution is based on the simultaneous operation of the random processes of mutation and genetic drift and the nonrandom force of natural selection. Random genetic mutation, gene flow, genetic drift, and so on produce variation within populations and ultimately perhaps differences among species, but nonrandom selection pressures sort these variants so that the prevailing fitness of the population tends to progressively increase provided environmental conditions remain constant. The adaptive walks through the cell morphospace reflect this interplay of random and nonrandom evolutionary forces. They were inspired by mandating a progressive increase in size, regardless of whether large cells are relatively more fit than smaller cells. This mandate amounts to invoking random mutations and genetic drift that by pure chance cause some organisms to diffuse into the domain of larger cell size. Once these hypothetical variants enter the domain, however, nonrandom forces (selection pressures) come to dictate which cell geometries and shapes confer the highest fitness. These nonrandom forces are mathematically codified by physical laws and processes that have remained constant since the inception of life on Earth.

Fitness and the Multicellular Morphospace

Even the crudest morphospace for multicellular plants has three components or subdomains: cellular morphology, tissue organization, and overall plant size, geometry, and shape (fig. 5.10). The cell morphospace treated in the previous section is adopted here to define the subdomain for cellular morphology, but it is worth reiterating that this component of the multicellular morphospace is a simplistic rendition of real plant cells and does not distinguish among morphologically similar cells that serve very different biological functions. In turn, the cell morphospace is embedded within the second component of the multicellular morphospace required to describe tissue organization. As with cells, plant tissues can be very simple or extremely complex (e.g., pith parenchyma and wood, respectively), and so to be tractable the tissue subdomain must be greatly simplified. Here all plant tissues are assumed to be constructed in one of three ways, each relating to the planes in which cells normally are free to divide. Specifically, cell divisions confined to one plane give an unbranched filament of cells; cells free to divide in two planes but not three may grow into branched filaments, which can be used to fabricate a pseudoparenchymatous tissue; and finally, cells free to divide in all three planes may produce a true parenchymatous tissue construction. The focus here is on the *way* cells divide to produce tissues rather than on the actual anatomical details of how different types of tissues are constructed. In

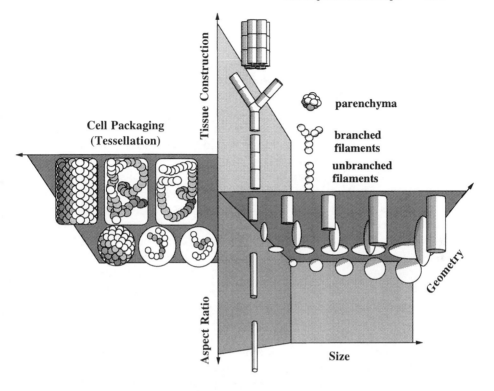

Figure 5.10. Multicellular plant morphospace consisting of three subdomains: cell morphospace (lower right), tissue construction (upper right), and tissue tessellation (left). Each of the three subdomains is a three-dimensional space, and so the entire morphospace is a multidimensional space that cannot be faithfully rendered by this diagram.

this manner the tissue subdomain is reduced to manageable proportions. The third component of the multicellular morphospace describes how tissues are packaged within the plant body. That is, the third subdomain defines the *tessellation* of the plant body with tissues—the rules by which small objects (cells) are packed within a larger object so as to minimize unoccupied space. Because any solid geometry can be constructed out of unbranched or branched filaments or parenchyma, tissue tessellation does not confine a multicellular organism to a particular morphology. In fact there is no a priori correlation between the external and internal appearance of a plant. Even though tissue construction and tessellation are independent of one another, the tessellation subdomain is defined in part by the overall appearance of a plant in addition to the size, geometry, and shape of the cells fabricating tissues.

Even the most elementary multicellular morphospace is a multidimensional nightmare, in terms of both specifying the location of each plant and computing the course of adaptive walks. Each of the three subdomains is itself a three-dimensional space, and even though some subdomains appear to share one or more axes (variables), the location of each phenotype depends on the numerical values of many variables, each of which must be calibrated against an independent axis. For example, although the cell subdomain has a "size" axis, cell size does not define the "size" of a multicellular plant (i.e., the designation of size requires at least two axes). Also, even though the cell and tessellation subdomains have an axis defining "shape" (measured by an aspect ratio), the shape of the cells used to construct a tissue need not share the same shape as the whole plant. Fortunately, the dimensionality of the entire morphospace is somewhat reduced if the three axes of the cell domain are used to define the size, geometry, and shape of the cells that fabricate a tissue. Provided this protocol is used, the simplest mathematical rendition of a multicellular morphospace requires eight axes, one for each of the following variables: cell size, geometry, and shape; type of tissue construction; tissue tessellation; and overall plant size, geometry, and shape.

Additional assumptions and simplifications are required to make this eight-dimensional morphospace computationally manageable. I freely confess that some of these assumptions are likely to inspire intellectual hemorrhages in the trained botanist, but the multicellular morphospace is computationally intractable without them. Specifically, let us assume that all the cells within a particular plant have the same size, geometry, and shape. The assumption of cellular homogeneity is vital because the number of possible permutations for these variables is on the order of 10^6. We will also assume that each plant is constructed out of a single type of tissue (i.e., tissue homogeneity) and that each plant morphology is confined to one of only four types of solid geometry (the sphere, oblate spheroid, prolate spheroid, and cylinder). Finally, we will assume that the fitness landscape through which adaptive walks proceed is defined exclusively by a plant's ability to harvest sunlight and to exchange substances with the external environment by passive diffusion.

These assumptions mock the histological, anatomical, morphological, and physiological complexity observed among many kinds of plants, yet for some algae and some bryophytes they are not outrageous. Many algae and some bryophytes are composed of cells remarkably similar in size, geometry, and shape. Also, in contrast to the extensive tissue differentiation of vascular plants, the algae and bryophytes tend to have remarkably simple tissue construction. Indeed, the multicellular alga or the bryophyte may consist of unbranched or branched filaments (e.g., *Ulothrix* and *Stigeoclonium,* respectively) or comparatively undifferentiated parenchyma (e.g., *Marchantia* and *Anthoceros*). By the same token, the overall morphology of diverse algae and bryophytes can be described in reference to comparatively simple geometries, while the more complex morphologies of other plants are composite geometries that can be deconstructed into individual elements, each described in terms of a simple solid geometry (fig. 5.11). Finally, the growth and survival of all photoautotrophs depend on access to sunlight and the ability to exchange chemicals with the ambient environment. Therefore to some degree the assumptions and simplifications underwriting the construction of the multicellular morphospace and directing the course of adaptive walks through it are largely fulfilled provided the biological frame of reference is restricted to comparatively small, nonvascular multicellular plants.

Although the same physical principles directing adaptive walks through the cell morphospace hold true in principle for multicellular plants, the equations used to model the package effect on light harvesting and to model the effects of plant morphology on passive diffusion are based on questionable assumptions for such plants, particularly large ones. For example, the package effect is modeled by assuming light does not attenuate in intensity as it passes through a cell. This assumption is undoubtedly violated by large and thick multicellular plants, whose outermost cells absorb most of the incident sunlight. Consequently an alternative method of computing light harvesting efficiency is required. Here we will use the ratio of a plant's projected surface area S_P to the plant's total surface area S_T. This ratio is size independent and varies as a function of the incident angle of light θ, which changes with the diurnal procession of the sun or when a stem bends or sways or rotates when struck by moving air or water. To compute the efficiency of light harvesting, S_P/S_T must be plotted as a function of θ over the range $0° \leq \theta \leq 180°$ ("dawn to dusk"). The area under the curve provides a crude measure of the light harvesting efficiency of a plant (Niklas and Kerchner 1984; Niklas 1992).

In terms of passive diffusion, the exchange of chemicals between the plant body and the ambient environment is also complex because Fick's law assumes that the fluid surrounding a plant moves little or not at all. For small cells or very small multicellular plants this assumption is not unreasonable, because very small organisms tend to be enveloped by a comparatively thick blanket of unmoving fluid (water or air) called the boundary layer within which Fick's law largely holds true. But larger plants live in an environment where fluids can move rapidly and strip away the boundary layer. Under these conditions, the exchange of sub-

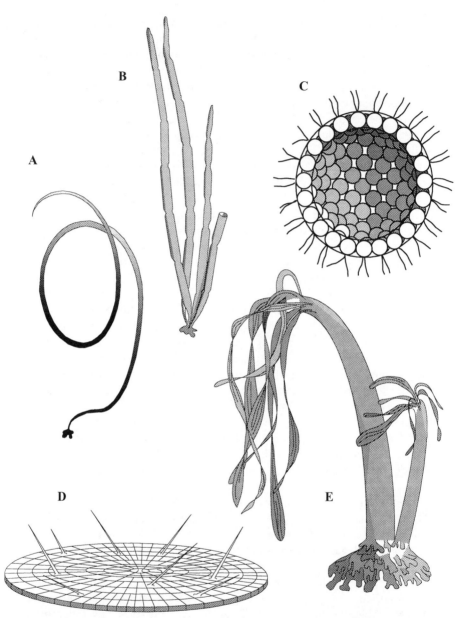

Figure 5.11. A few plant morphologies (A–E are algae, F–K are bryophytes): (A) *Chorda* (a brown alga); (B) *Scytosiphon* (a brown alga); (C) *Volvox* (a green alga); (D) *Coleochaete* (a green alga); (E) *Postelsia* (a brown alga); (F) *Riella* (a liverwort); (G) *Fossombronia* gametophyte (a "leafy" liverwort); (H) *Sphaerocarpos* gametophyte with attached sporophytes (a "thalloid" liverwort); (I) *Calobryum* gametophyte (a "leafy" liverwort); (J) *Anthoceros* thalloid gametophyte with vertical cylindrical sporophytes (a hornwort); (K) *Funaria* gametophyte with vertical cylindrical sporophytes (a moss).

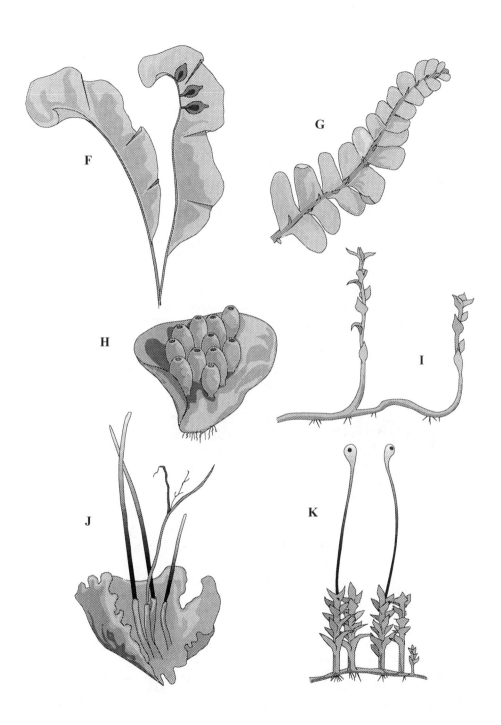

stances like carbon dioxide and oxygen is influenced by forced convection. Therefore, even though passive diffusion can dominate the physiology of very small unicellular or multicellular plants, other physical phenomena influence the rate at which chemicals are exchanged between large plants and their environments. Space does not permit a discussion of these phenomena or the mathematical tools required to estimate their influence on plant physiology (for a detailed discussion, see Nobel 1983). Here we will simply ignore this complexity and continue to use Fick's law to estimate plant fitness in terms of the ability to absorb metabolites from the ambient fluid environment. Put differently, we will tacitly assume either that our hypothetical plants are very small multicellular organisms or that they live in very slow moving water or air.

Clearly, however, when treating aquatic multicellular algae and terrestrial bryophytes, we must bear in mind the profound differences between the two media in which these organisms live—water and air. The physical properties of water and air differ significantly and have very different effects on light harvesting efficiency and passive diffusion. Water attenuates the intensity of sunlight and preferentially absorbs the red wavelengths of light. In contrast, air neither attenuates nor alters the spectral properties of light. For this reason, aquatic plants living a few centimeters below the water surface can experience significantly lower light intensities shifted in favor of blue wavelengths than plants living in air. Likewise, substances like carbon dioxide and oxygen have lower molecular diffusivities in freshwater than in air, and therefore plants living in air can acquire CO_2 and O_2 more efficiently than their aquatic counterparts. Water is roughly one thousand times denser than air, and because they have bulk densities near that of water, aquatic plants tend to be either neutrally or negatively buoyant. In contrast, terrestrial plants are roughly a thousand times denser than air and experience the compressive effects of gravity. Another important feature is that aquatic plants are continuously bathed in water, whereas terrestrial plants are exposed to the drier atmosphere. Thus land plants like the bryophytes must either conserve water or suspend metabolic activities, whereas multicellular aquatic plants are unconstrained regarding the elaboration of large external surface areas that aid passive diffusion and light harvesting. It is also apparent that aquatic plants can release waste products into their surrounding fluid medium, whereas terrestrial plants must either excrete wastes or sequester them within their tissues. In sum, even the most casual inspection of the properties of water and air and their potential effects on plant physiology cannot fail to reveal that the fitness landscapes for aquatic and terrestrial plants are very different. For these reasons the adaptive walks through the aquatic and terrestrial landscapes are governed by very different considerations in terms of the effects of size, geometry, and shape on plant physiology.

This expectation is quickly realized when adaptive walks are simulated in the aquatic fitness landscape and compared with those through the terrestrial fitness landscape (fig. 5.12). Adaptive walks in both landscapes were "seeded" with a small spherical cell morphology because the walk through the cell morphospace indicated that small spheres maximize light harvesting and passive diffusion. Recall that this walk identified other cell geometries as adaptive provided size was arbitrarily forced to increase. These alternative cell geometries were identified by the walk because the ratio of surface area to volume for a sphere disproportionately decreases with increasing cell size. Consequently, larger unicellular plants must adopt a cell geometry other than the sphere when their size increases. In contrast, multicellular plants can increase in size by adding more cells. And in terms of the most fit (optimal) plant morphology, the adaptive walk through the aquatic landscape identifies an unbranched filament composed of very small, touching spherical cells tessellated in a slender cylinder (fig. 5.12).

The walk through the aquatic landscape ends once this multicellular morphology is reached, and it can be perturbed from this adaptive peak only by arbitrarily mandating a further increase in cell size. Provided cell size is *forced* to increase, the walk identifies an unbranched filament of cylindrical cells as optimal for light harvesting and passive diffusion. Cell size can continue to increase only if cells adopt a more slender cylindrical shape. No other morphology has a higher relative fitness, and therefore the optimal morphology in the aquatic landscape is an unbranched filament composed of very slender cylindrical cells. Nonetheless, there are "minor adaptive peaks"—phenotypes that exist on the foothills of the adaptive peak. These alternative morphologies are the hollow sphere, the hollow cylinder, and a sheetlike, very flat oblate spheroid (fig. 5.12). The ability of the hollow sphere to intercept sunlight is independent of orientation because the ratio of its projected surface area to total surface area always equals 1/4 regardless of the incident solar angle. The hollow cylinder and the flattened sheetlike oblate spheroid expose every cell to the ambient environment, and therefore each cell has direct access to dissolved nutrients. The disadvantage of these two morphologies is that the ability to intercept sunlight depends on orientation, and some cells are always shaded by others. The adaptive walk through the terrestrial landscape suggests that increasing overall plant size is adaptive on land because it favors water conservation. In contrast to all previous adaptive walks, the first tentative walk onto the terrestrial landscape does not have to be "driven" by arbitrarily mandating an increase in overall plant size.

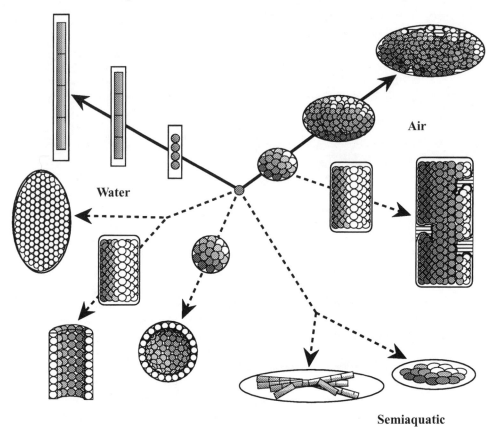

Figure 5.12. Adaptive walks through the multicellular plant morphospace. Walks begin with a small spherical cell and diverge morphologically depending on whether they progress into an aquatic (left side), terrestrial (upper right), or semiaquatic fitness landscape (lower right). Walks reaching principal fitness peaks are denoted by solid arrows; walks reaching secondary fitness peaks are indicated by dashed arrows. Plant morphologies are shown by solid lines drawn around tessellated tissues and cells.

The walk shows that a small oblate spheroid is morphologically optimal because it simultaneously reduces the external surface area through which water can be lost, provides a reasonable projected surface area for light harvesting, and has a substantial total surface area for the passive diffusion of carbon dioxide and oxygen. Larger plant sizes have higher relative fitness, and as the adaptive walk locates larger plants, it identifies the cylinder and oblate spheroid as having nearly equivalent relative fitness in terms of conserving water. To increase further in overall size, the cylinder must become more slender and the oblate spheroid must progres-

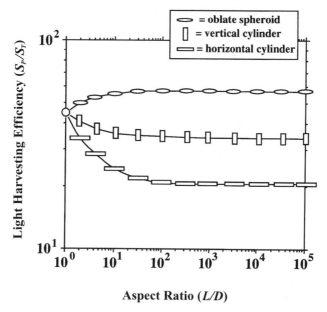

Figure 5.13. Light harvesting efficiency plotted against aspect ratios (shape) of oblate spheroid and vertical or horizontal circular cylinders. Light harvesting efficiency is calculated based on the quotient of projected surface area toward the light and the total surface area of each geometry and shape. The oblate spheroid is predicted to maximize light harvesting regardless of shape.

sively flatten; otherwise the rate at which either geometry gains carbon dioxide from the atmosphere steadily decreases. The adaptive walk also suggests that both geometries can increase in size yet maintain the same shapes if they are permitted to invaginate portions of their external surface. Both the oblate spheroid and the cylinder can maintain the same shape yet increase in overall size if portions of their external surface are internalized to produce moist spaces and chambers (see fig. 5.12). These cavities increase the surface area through which carbon dioxide and oxygen can be exchanged with the external atmosphere while simultaneously reducing that through which water vapor can be lost to the drier atmosphere.

In passing it is worth noting that, regardless of its size or shape, the orientation of a geometry with respect to the ground influences the relative fitness of hypothetical phenotypes in the multicellular morphospace. For example, a vertical cylinder that can present its curved surface to the early and late sun is more efficient at light interception than a horizontal cylinder, half of whose curved surface lies on a substrate (fig. 5.13).

The adaptive walk through the terrestrial landscape appears to identify the basic morphologies of most bryophytes. Without exception, the sporophyte generation of the nonvascular land plants has a cylindrical morphology, whereas the free-living gametophyte generation is either thalloid or cylindrical in general appearance (see fig. 5.11). As I noted, both of these morphologies optimize the functional requirements for vegetative growth. But these geometries are also compatible with the reproductive roles of the two multicellular generations of bryophytes. Consider that the free-living gametophyte must produce gametes, permit the fertilization of eggs, and subsequently nurture the developing sporophyte embryo within the archegonium. All these reproductive obligations require more or less uninterrupted access to water, which may be achieved if a plant hugs its moisture-laden substrate. The oblate spheroid and the cylinder are equally effective at gaining access to water provided they are prostrate. The gametophyte generations of some species have adopted the oblate spheroid morphology, while others have a cylindrical morphology. Thus, in contrast to the morphologically conservative sporophyte generation, the bryophyte gametophyte evinces a certain morphological schizophrenia, which may relate to the fact that the gametophyte of bryophytes is the long-lived generation and so has the primary responsibility for vegetative growth and survival from year to year. The adaptive walk through the terrestrial landscape shows that either of the two "optimal geometries" can cope with demands for reproduction as well as the vegetative growth. In contrast, the sporophyte generation is short-lived and depends on the gametophyte generation for nutrients and water. Its reproductive role is to produce, shed, and disperse spores. From a geometric perspective, this reproductive role is best accomplished by a vertical cylindrical geometry. The higher spores are released above the ground, the greater their potential for long-distance dispersal. The cylinder is an effective geometry for elevating spores because it is mechanically very stable. Perhaps the morphological conservatism of bryophyte sporophytes is a consequence of more intense selection pressure on the reproductive role of the sporophyte rather than on its vegetative role.

The Semiaquatic Landscape: The Invasion of Land

Although crude, prior adaptive walks through the multicellular morphospace highlight the important differences between the most fit morphological configurations for a multicellular aquatic plant and a multicellular terrestrial plant. Of special interest is that the phenotype with the highest relative fitness in the aquatic landscape (the unbranched filament composed of cylindrical cells) has a low relative fitness in the terrestrial landscape (fig. 5.14). But an adaptive walk through any morphospace can help to identify only phenotypes with the highest relative fitness *within a particular environmental context.* And therefore there is no necessary correlation between the fitness of a phenotype in one environment and its fitness in another. The phenotypic "reversal of fortune" shown by the adaptive walks through the aquatic and terrestrial landscapes is intriguing because it suggests two possibilities: either the transition from an aquatic to a terrestrial multicellular plant may have involved an adaptive walk through a "fitness bottleneck," or the aquatic ancestor to the land plants may have had a low relative fitness in water and so would have been less than the consummate aquatic organism. These two possibilities are not mutually exclusive, particularly if the ancestor of the first land plants lived in a habitat periodically inundated by water and exposed to the air. Because it would have to survive the physiological rigors of both physical environments, a semiaquatic organism would not occupy an adaptive peak in either the aquatic or the terrestrial landscape—it would be neither "best suited to life in air" nor "best suited to life in water."

The fitness landscape for ancient semiaquatic plants and an adaptive walk through it are difficult to model because there is no way of knowing the periodicity (if any) of water deprivation or how these plants may have coped with water stress. Dormancy is an obvious escape route. Many plants suspend their physiological and reproductive activities when deprived of water. Dormancy begs the question, however, because we are searching for the size, shape, and geometry of a truly "amphibious" plant rather than a "hibernating" plant. Toward this goal, we will make three assumptions. The first is that the semiaquatic plant must be equally well adapted to life in water and in air. The second assumption is that it is a small organism. On the average, larger organisms take disproportionately longer to reproduce than smaller organisms (see fig. 5.7). Therefore, in an ecologically unpredictable and stressful habitat small and fast-growing plants have clear advantages over large and slow-growing ones. The third assumption is that the most limiting ecological factor will dictate morphology because the limiting factor will impose the most intense selection pressure. If so, then water conservation will take priority for semiaquatic plants, although other factors will play a role in dictating morphology. Based on these assumptions, an adaptive walk in the semiaquatic landscape indicates that the optimal plant morphology is a small oblate spheroid appressed to a substrate (fig. 5.15). This combination of morphology and orientation has the highest relative fitness in this particular landscape because it best reconciles the conflicting demands of life in air and in water. Nonetheless, the semiaquatic landscape is a remarkably "rich" morphological domain—equally fit alternative morphologies exist. Among

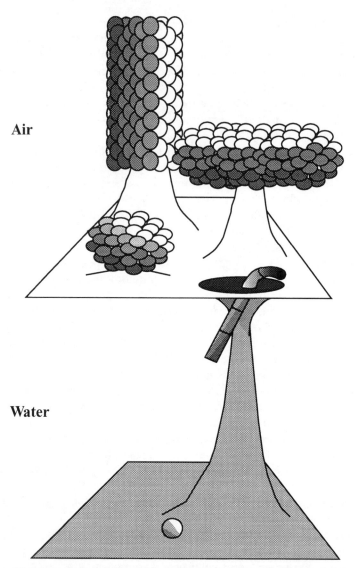

Air

Water

Figure 5.14. Hypothetical relation between fitness peaks and valleys for the interface between an aquatic and a terrestrial (aerial) fitness landscape. The morphology occupying the adaptive peak in the aquatic domain resides in a fitness valley on the terrestrial landscape.

these alternatives is a densely branched filament composed of short cylindrical cells. Significantly, the optimal morphologies predicted for a semiaquatic habitat are remarkably similar to those of the charophycean alga, particularly species of *Coleochaete* that have either a parenchymatous or a pseudoparenchymatous tissue construction tessellated into a more or less pancake-like geometry (see fig. 5.11D). Of course there is no reason to assume that the ancestor to the first land plants had a *Coleochaete*-like morphology just because this morphology occupies an adaptive peak in the semiaquatic landscape. It is just as possible that the ancestor to the land plants had a morphology poorly adapted to life in water that happened to be well suited to life on land.

Forced Moves in Design Space

The philosophy guiding this chapter and the next is that physical and chemical laws and processes limit the number of morphological or anatomical options theoretically available to plants. Only a comparatively few among the vast number of options actually work. From a functional point of view, the rest are mere junk. Thus walks through the design space are forced or "constrained" to move along certain pathways that navigate through the great bulk of junk in line with phenotypic designs that confer at least some opportunity to survive and successfully reproduce. Naturally the structure of the phenotypic space plays a critical role—a good design buried in a large pile of junk will be difficult if not impossible to reach unless phenotypic evolution can "jump" from one region to another.

Forced moves in the functional design space of plant morphology and anatomy are dictated by the genetic landscape, and the correspondence between the design space and the genetic landscape is critical. Recall that the number of genetic variants that even a modestly sized population of sexually reproductive plants theoretically can produce exceeds the number of atoms in the observable universe. Yet among this vast constellation of genotypes, comparatively few are produced by a population, and fewer still attain phenotypes that survive and eventually reproduce. Chance largely dictates the genotypes actually produced by a population, but natural selection determines which of those produced will work. Natural selection sorts among phenotypes, leaving the most fit and eliminating those that are less fit.

In much the same way, physical and chemical laws and processes afford the criteria for sorting among all conceivable organic designs and eliminating the vast number that are functional junk. That these laws and processes are invariable in time and space is a principal tenet of

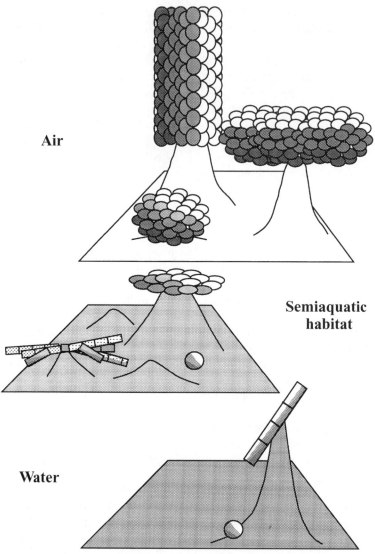

Air

Semiaquatic habitat

Water

Figure 5.15. Hypothetical relations among the fitness peaks and valleys of the interfaces between the aquatic, semiaquatic, and terrestrial (aerial) fitness landscapes. A filamentous aquatic plant is predicted to be the optimal form; a flat oblate spheroid is predicted to be the optimal form in the semiaquatic landscape; and a cylinder and large oblate spheroid reside on the dual fitness peaks of the terrestrial landscape.

physics and chemistry and engineering, and so they are constant selection pressures that act on all manner of living things at all times.

Naturally, the effects of these physical and chemical laws and processes are complex. The way these laws and processes influence organisms depends on the environment and can change as an organism grows and develops. It will also change as an organism evolves, moves into new habitats, and acquires new ways of growing and developing. This complexity is well illustrated by considering any physical law, for example, the law of buoyancy. Mathematically, this law is given by the formula $B = (\rho - \rho_f)Vg$. In this formula, ρ is the bulk density of the organism, ρ_f is the density of the fluid in which the organism is suspended, V is the volume of the organism, and g is gravitational acceleration. Barring extreme changes in temperature or solute concentration, the density of any type of fluid, like freshwater or air, is more or less constant (e.g., the difference in the density of freshwater at temperatures of 0°C and 40°C is less than 1%). Likewise, the gravitational acceleration changes little over the globe. For these reasons an organism's buoyancy is always profoundly influenced by the way the physical universe behaves. No organism can violate the law of buoyancy.

Nevertheless, the density and volume of an organism can change as it grows and develops. By changing its weight, an organism can change its buoyancy. (Notice that in the formula for buoyancy ρVg is the mass force of the organism, commonly called weight.) By virtue of its growth in size and development, an organism can change its weight. Less dense biological materials can accumulate in cells as they get bigger and older. The volume of an organism can also be increased as less dense materials accumulate in cells. The density and volume of organisms can also change as a consequence of evolutionary changes in building materials. Lighter materials can be discovered in the design space and can be incorporated as body size increases. Thus growth and development as well as evolutionary changes can alter the buoyancy of organisms.

When cast in a biological context, virtually every physical or chemical law illustrates the marvelous reciprocity between the properties of the physical universe and the properties of life. These laws and processes cannot be violated, but life has found many surprising ways of side-stepping their effects. Clearly, adaptive walks through a design space must comply with physical and chemical laws and processes. But every step in an adaptive walk is taken by a living thing that, by its very existence and unique biological properties, can modify their expression. But what of the "driving force" behind these walks? What, if anything "propels" life to climb to adaptive peaks? The answer is that no such force exists. In its place, we find that organisms are endowed with the ability to produce variants of themselves and that the environment is empowered to select among these variants. In place of an implacable evolutionary "driving force," we find that evolution is the sum of life's opportunistic responses to random and nonrandom events. In the absence of organic variation and differential sorting, evolution would cease. In this sense the words of William Shakespeare ring true: "When the sea was calm, all ships alike showed mastership in floating."

The next chapter illustrates additional examples of "forced moves" in the design space of plants.

6   The Terrestrial Landscape

> *The environment, living and non-living, of any species is actually in continual change. In terms of our diagram this means that certain of the high places are gradually being depressed and certain of the low places are becoming higher.*
>
> Sewall Wright

This chapter picks up where the previous chapter ended and is structured according to the topics discussed in chapter 4. The taxonomic framework of this chapter is the green algal lineage, particularly the charophytes, and the embryophytes. All extant bona fide land plants are embryophytes and trace their origins to some group of green algae whose closest living counterparts are found in the Coleochaetales.

Why the green algae and no other algal group successfully invaded the land may never be known with certainty. Nevertheless, part of the answer is likely buried in aspects of their physiology and development that permit the acquisition of multicellular life forms that are prerequisite for the successful colonization of the land. In this regard, the number of multicellular green algae outstrips that of all the other freshwater and soil-dwelling multicellular algae combined, although the freshwater diatoms are far more taxonomically diverse than any other group of freshwater algae. For example, among the roughly one hundred genera of multicellular freshwater algae in the United States, seventy-five are green algae (Smith 1950). The remaining twenty-five are unequally distributed among the pyrrophytes (two genera: *Dinothrix* and *Dinoclonium*), phaeophytes (five genera: *Herbibaudiella, Sphacelaria, Pseudobodanella, Lithoderma,* and *Pleurocladia*), rhodophytes (seven genera: *Compsopogon, Batrachospermum, Audouinella, Sirodotia, Thorea, Tuomeya,* and *Lemanea*), and chrysophytes (ten genera, e.g., *Tribonema* and *Bumilleria*). Likewise, among the multicellular soil-dwelling algal genera over 95% belong to the green algae. Although other algal groups have successfully invaded the habitats envisioned as the cradle of early land plant evolution, the green algae are currently the most ecologically successful and taxonomically diverse among the freshwater and soil-dwelling multicellular plants.

Multicellularity is a prerequisite for survival, growth, and reproduc-

251

tion on land. The compartmentalizing of the plant body into different types of cells and tissue systems is necessary to deal with the conflicting physiological requirements for water conservation and the exchange of gases between the plant body and the external atmosphere. Cellular differentiation is needed to regulate the passive diffusion of carbon dioxide and oxygen in relation to the diffusion of water vapor from the plant body, and also to provide a bulk flow of water to compensate for its loss by tissues exposed to the drier atmosphere. Unicellular plants can survive in moist microhabitats, but when exposed to the dry air they either die or must suspend their metabolic activities by becoming dormant. Compartmentalizing into cells is also required for successful sexual reproduction on land. The formation of sperm and egg or spores adapted to survive in the air requires a multicellular construction. Algal groups other than the green algae have evolved multicellularity, but for reasons currently not known these groups were largely unsuccessful at invading freshwater and diversifying to the degree found in the green algae. Even among the green algae, only the charophytes appear truly successful in these habitats. As such, they were poised on the threshold of a totally unoccupied landscape into which they were capable of evolving by virtue of the features reviewed in chapter 4.

Why Invade the Land?

In many ways life on land presents difficulties for plants. It means giving up unlimited access to water and coping with the compressive effects of gravity. Indeed, we may well wonder why plants colonized the land's surface at all. But two simple facts tell us they had something to gain by leaving the water. First, water attenuates the intensity of sunlight and preferentially absorbs the wavelengths of light most useful to photosynthesis. Second, carbon dioxide and oxygen are more accessible near the water's surface. The intensity of light decreases exponentially as it passes through a column of water: if 50% of the available light energy is absorbed by the first centimeter of water, the light is weakened yet another 50% as it passes through the second centimeter, and so forth. Also, the quality of sunlight changes as it penetrates water. Because wavelengths in the red end of the spectrum diminish more quickly than others, the efficiency of photosynthesis decreases with increasing water depth. The effect of water on light limits the depth plant life can sink to before the rate of photosynthesis fails to match the rate at which plants consume the foods they manufacture from sunlight and raw chemicals. The equilibrium between food production and consumption, known as the compensation point, varies among species, so different kinds of plants can grow at different depths in the oceans and deep lakes, but for most freshwater aquatic plants photosynthesis is more efficient near the water's surface. For much the same reasons, photosynthesis is more efficient on land than in water. Air is an optically transparent medium; it neither attenuates the intensity of sunlight nor preferentially absorbs its useful wavelengths. Also, the molecular diffusivities of carbon dioxide and oxygen are higher in air, so the raw materials of photosynthesis can enter or leave plant tissues more rapidly than in freshwater.

For these reasons it is not surprising that many kinds of plants live in "the shallow end of the pool"—just beneath the air-water interface, nestled in the interstices of moist soil, or wedged in the crevices of wet rocks. This tendency may account for how and why plants came to invade the land. Consider that shallow bodies of water are an ecologically unpredictable environment that can rapidly dry up. As we have seen, these environments tend to be occupied by comparatively small organisms that grow and reach sexual maturity more rapidly than their larger counterparts living in deeper water. By the same token, small organisms with high reproductive rates tend to have high mutation rates and so, as a very general rule, may evolve more rapidly than larger organisms. Therefore the shallow end of the pool provided the two major ingredients for invading the land: a habitat with intense selection pressures favoring plants that could survive exposure to the air, and small plants with high reproductive and mutation rates that had much to gain physiologically on land. Undoubtedly, ancient green algae living in shallow bodies of water or moist soil were repeatedly exposed to the air, and many of them died. But natural variation in populations of these organisms presumably left behind a few variants capable of enduring short-term water deprivation and brief exposure to the air. These plants survived and provided the source for future genetic variation and evolutionary innovation. Continued genetic "trial and error" eventually led to adaptations permitting plants to survive longer and longer periods in the air, eventually culminating in the first bona fide land plants.

A Morphospace for Vascular Plants

The embryophytes unlocked the door to the terrestrial landscape with a reproductive "double key"—a multicellular gametophyte that had to alternate with a multicellular sporophyte to complete the sexual life cycle. Once this door was opened, the embryophytes had to take one of two divergent pathways, because their life cycle does not permit concurrent elaboration of the size and complexity of both generations in the same species (see fig. 3.17D). As a group, the lineages constituting the bryophytes followed a pathway leading to longer-lived, larger, and morphologically complex gametophytes compared with their short-lived, smaller, and comparatively simple sporophytes. The vascular plant lineages took the alternative route and evolved long-lived and morphologically complex sporophytes companioned by short-lived and comparatively small gametophytes. For each pathway, the larger of the two multicellular generations in the life cycle assumed the principal responsibility for vegetative growth and survival on land. Among the bryophytes, the gametophyte generation evolved to fulfill this "vegetative obligation." In some respects this pathway may have provided the greatest morphological latitude, because previously discussed adaptive walks through the terrestrial landscape show that both the oblate spheroid and the cylinder can reconcile the manifold vegetative and reproductive obligations of the gametophyte generation, whereas the biophysical strictures on the vegetative and reproductive obligations of the sporophyte produce only one optimal phenotype, the cylinder. It is somewhat ironic that, although constrained in their morphology, the tracheophytes with their cylindrical construction came to dominate the land in terms of both standing biomass and species number. To be sure, this was not always the case—the first land plant flora undoubtedly consisted mostly of bryophytelike, nonvascular plants. But over time the nonvascular plants lost their taxonomic dominance, and the vascular sporophyte currently reigns supreme on land.

This ecological supremacy is not too difficult to understand in light of the numerous adaptations of the vascular sporophytes to life on land, or more properly life in air. Their aerial portions are enveloped by a cuticle that limits the loss of water from the plant body through stomata whose guard cells can close when water stressed. In addition to adaptations that conserve water, they contain xylem and phloem tissues, affording preferred routes of rapid transport for water and for cell sap. Armed with these adaptations, the vascular sporophytes eventually invaded every terrestrial habitat, including hot, arid deserts. And by taming the terrestrial landscape, they graciously, albeit unconsciously, created or ameliorated numerous microhabitats that have been opportunistically occupied by the smaller bryophytes as well as a host of other organisms.

The first vascular plants quickly diversified in morphological complexity. Much of this diversification was prefigured during the Devonian, a period that saw the first plants with leaves, roots, lateral meristems, and a treelike growth habit (fig. 6.1). In many ways this was the most effulgent, if not "brilliant," period of evolutionary innovation in terms of adaptive walks through the sporophyte morphospace. Nonetheless, the first vascular plants had humble morphological beginnings (fig. 6.2).

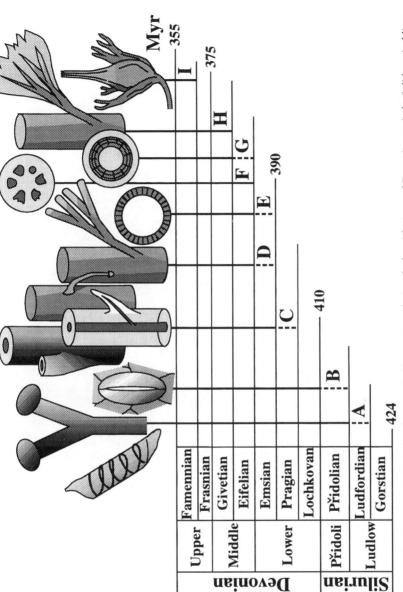

Figure 6.1. Major morphological and anatomical innovations during the late Silurian and Devonian periods. Solid vertical lines extend to unequivocal first appearance of innovation in the fossil record; dashed lines extend to equivocal earlier first occurrences: (A) tracheids and bifid sporophytes bearing terminal sporangia; (B) stomata; (C) unequal branching, vascularized small leaves ("microphylls"), and roots (from left to right); (D) sporophytes with distinct lateral branching systems; (E) periderm (corky outer layer of secondary tissue); (F) dissected vascular traces in axes; (G) secondary xylem and phloem; (H) webbed leaves with manifold vascular bundles ("megaphylls"); (I) seeds. Adapted from Chaloner and Sheerin 1979 and Niklas 1994.

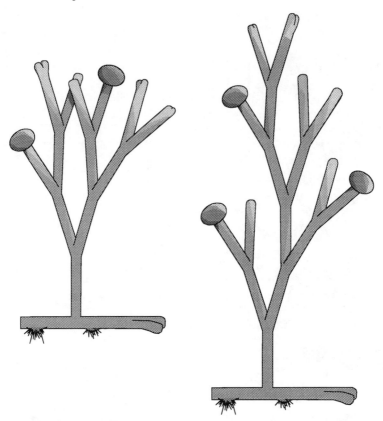

Figure 6.2. General morphology of some early vascular land plant sporophytes. Plants consist of leafless cylindrical axes, of which some are horizontal (rhizomes) and others are vertical and bear terminal sporangia. Plants lack roots but bear rhizoids (cellular extensions of the epidermis). The earliest vascular plants had sporophytes that branched more or less equally (left). Some subsequent plants fossils indicate plant axes branched unequally (right).

Their sporophytes consisted of repeatedly branched cylindrical axes, some growing vertically into aerial portions of the plant while others grew horizontally into a rhizome-like branching system. The aerial axes were either sterile organs or fertile organs bearing sporangia and in either case served simultaneously for photosynthesis and mechanical support. Unlike many modern sporophytes that have a prominent main stem produced by a dominant apical meristem, the first vascular land plants consisted of equidominant, repeatedly dichotomized axes produced by the bifurcation of apical meristems. Thus each apical meristem could produce a pair of adjoining cylindrical axes whose lengths varied depending on the frequency of apical bifurcation and the rate of subapical cellular elongation and division. As a consequence of this mode of growth, the most ancient sporophytes lacked a single main stemlike axis. Also, the sporophytes of the most ancient tracheophytes did not have lateral meristems and therefore did not produce secondary tissues like cork and wood. For this reason the axes of these plants were not much tapered along their lengths. In the absence of secondary tissues, which provide mechanical support against the effects of gravity and wind, the overall height of the most ancient tracheophytes was limited. Numerous fossil remains verify that most were short, only a few centimeters in stature.

This all too brief review of the morphology and anatomy of the most ancient vascular sporophytes suffices to show that only six variables are needed to construct a morphospace for the first tracheophytes (fig. 6.3): axis length l, axis taper t, the probability of branching p, the rotation angle between each pair of axes and the horizontal plane γ, and the bifurcation angle ϕ between paired axes (see Niklas and Kerchner 1984). High values for axial length and taper produce long, very slender axes; low values give short, squat axes. High values of p produce profusely branched sporophytes; low values result in sparse branching. Low values for the bifurcation angle give rise to vertically erect branching patterns; high values for this parameter give morphologies with broadly splayed axes. High values for the rotation angle produce bushy sporophytes; low values result in horizontally flattened (planated) branching systems. Noting that the two axes produced by a bifurcating apical meristem can take on different numerical values for each of the six variables, we see that the morphospace is complex and multidimensional. This complexity can be reduced somewhat by dividing the morphospace into two subdomains, one containing sporophytes with equal (isometric) branching (for which $p_1 = p_2$) and another containing morphologies with unequal (anisometric) branching (for which $p_1 \neq p_2$) (fig. 6.4). The fossil record verifies that the first vascular sporophytes had more or less equal branching, and so it is in the isometric subdomain that adaptive walks on the terrestrial landscape begin.

Calibrating the fitness of sporophytes within the morphospace is comparatively easy because growth, survival, and reproduction undoubtedly included the requirement to intercept sunlight, the need to mechanically sustain the weight of aerial organs, and the ability to produce and disperse spores some distance from parent sporophytes. (Because all the hypothetical plants examined here have the same basic cylindrical construction, they are morphologically indistinguishable in their ability to

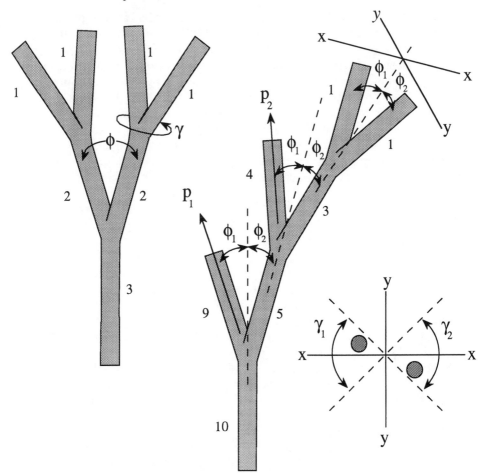

Figure 6.3. Mathematical relations among morphological features required to simulate a morphospace for early vascular land plants (see figs. 6.5 and 6.6). Three parameters are required: the bifurcation angle ϕ, the rotation angle γ, and the probability of branching p. Because sporophytes branched dichotomously (see fig. 6.2), the three parameters can vary independently for each pair of branching (e.g., ϕ_1 and ϕ_2).

conserve water and exchange chemicals with their environment. So for convenience we shall assume that all sporophytes within the morphospace are equally fit in terms of these vegetative obligations for survival and growth.) For such simple morphologies, light interception, mechanical stability, and spore dispersal are fairly easy to quantify. The same technique used to measure the light harvesting efficiency of multicellular algae and bryophytes may be used to gauge the relative fitness of even the

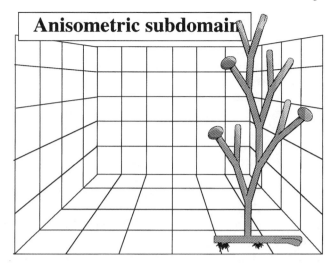

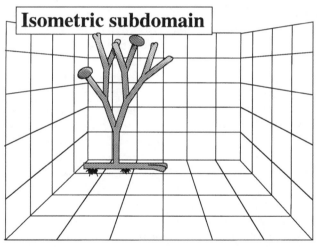

Figure 6.4. Two subdomains in the morphospace for early vascular land plants. An isometric subdomain is required to simulate sporophytes with equal branching; an anisometric domain is required to simulate sporophytes with unequal branching. Adaptive walks in the morphospace begin in the isometric subdomain and are then free to enter the anisometric subdomain.

most morphologically complex sporophyte. Regarding mechanical stability, the maximum mechanical forces that develop in a cylindrical plant axis bending under its own weight depend on its orientation and on the number and orientation of the axes it supports. The weight of each axis may vary depending on the density of its tissues and its overall size, but if we assume that all axes have the same tissue density, then the weight of each axis depends exclusively on its length and girth (i.e., volume). Thus two of the variables used to construct the morphospace (axial length and girth) define a priori the weight of each axis, assuming a uniform tissue density. By the same token, the rotation and bifurcation angles that simulate the branching pattern of a sporophyte provide the orientation of each axis. In sum, the maximum mechanical forces generated in each axis owing to the weight of tissues is automatically set by the variables used to construct the morphospace.

Turning to reproductive fitness, provided the spores produced by all sporophytes have roughly the same size and density, an elementary ballistic model suffices to describe the potential for long-distance dispersal by wind currents (Okubo and Levin 1989). According to this model, the maximum lateral distance spores can be transported is directly proportional to the square of the height at which they are produced on sporophytes. Even a small increase in height confers a selective advantage to reproduction in terms of spore dispersal. Spores can be produced anywhere along the length of a sporophyte, but the total height of each hypothetical sporophyte may be used as a rough gauge of its ability to shed spores farther than another. From a mathematical perspective, the height of each sporophyte is automatically defined by the numerical values of axial length, the probability of branching, and the rotation and bifurcation angles (see fig. 6.3). (From an engineering perspective, height is governed by stem diameter, but in these simulations stem diameter is not a variable.) Another important variable determining reproductive success is the number of spores a plant produces. Sporophytes that bear few spores are less fit than those that bear more, so the relative reproductive fitness of a particular plant depends jointly on the number of spores produced and on the height at which spores are shed. Provided sporangia are produced at the tips of branches and the number of spores per sporangium does not vary among hypothetical sporophytes, the relative reproductive fitness of each plant is proportional to both the number and the height of branch tips. Because both of these variables are automatically defined by the morphospace, the relative reproductive fitness of each sporophyte is easily computed.

The foregoing discussion implicitly assumes that the total fitness of each sporophyte is proportional to light harvesting efficiency E and relative reproductive fitness R, and inversely proportional to the maximum mechanical forces produced in aerial branching elements M. Assuming that each of these three elements contributes equally and independently to Darwinian fitness ω, the most parsimonious mathematical expression for the total fitness of a phenotype is the geometric mean of E and R divided by M; that is, $\omega = [(E)(R)]^{1/2}M^{-1}$.

The fitness landscape for each of the three functional tasks contributing to total fitness differs because each task has different phenotypic requirements. For example, phenotypes that maximize light interception will also tend to maximize the mechanical forces acting on their aerial branches. Consequently a phenotype with a high relative fitness in terms of E will have a low relative fitness in terms of M. When both of these functions are simultaneously performed and evaluated, the simplest assumption is that overall fitness increases in proportion to E/M. Here we see an unavoidable conflict in the design requirements of the same sporophyte, because E cannot be increased without increasing M, and the optimal phenotype will be the morphology that can maximize E with respect to M. Other functional obligations are more forgiving or may even reinforce the same morphological solution because their phenotypic requirements are similar or identical. For example, phenotypes that maximize long-distance spore dispersal may simultaneously reduce the mechanical forces that develop in vertical axes. A vertical unbranched and unbent morphology with sporangia at its tip will favor spore dispersal and will experience little or no bending stress along its length. Theoretically, therefore, an adaptive walk that optimizes R/M is comparatively direct and simple. When all three tasks are considered simultaneously, however, the fitness landscape becomes much more complex than the landscapes defined by only one or two tasks. Sporophytes that maximize long-distance spore dispersal (high fitness) may also minimize the stresses within their axes (high fitness), but they may minimize their ability to intercept sunlight (low fitness), since most of their branches are vertically oriented and therefore bunched together.

How far the complexity of the fitness landscape depends on the number of functional obligations a sporophyte must simultaneously perform presents an interesting opportunity. We can model adaptive walks on fitness landscapes based on one or more of these functional obligations and then compare the results of simulated adaptive walks with morphological trends observed in the fossil record of early vascular land plants. Because there are three functional obligations, each contributing to fitness, there are seven different fitness landscapes and therefore seven potentially different models for adaptive walks—one walk for each of the three individual biological tasks (light harvesting, mechanical stability, and reproduction); one for each of the three pairs of biological tasks (mechanical stability and reproduction, light harvesting and mechanical stability, and light harvesting and reproduction); and one adaptive walk through a fitness landscape defined by performing all three tasks simultaneously. By comparing the adaptive walks in the seven fitness landscapes with trends in the fossil record, we can begin to answer what happens to adaptive walks when the factors influencing fitness become more complex.

The answer to this question is not intuitively obvious, because fitness landscapes defined by only one biological task contain only a few adaptive peaks, whereas fitness landscapes defined by the simultaneous performance of two or more functional obligations have more (figs 6.5 and 6.6). Taken at face value, the adaptive walks through the seven fitness landscapes indicate that the number of equally fit morphologies *increases* as the fitness landscape becomes more complex. Additionally, as the number of adaptive peaks increases, their height decreases (fig. 6.7). That is, the maximum relative fitness in a landscape is inversely proportional to the number of adaptive peaks. These results are in perfect agreement with engineering optimization theory, which treats complex systems performing manifold tasks (Brent 1973; Gill, Murray, and Wright 1981). As I noted earlier, when designing a machine to perform a single task, an engineer typically finds only one configuration that can do this task superbly well. Alternative configurations may exist, to be sure, but they are less efficient and are typically ignored, for obvious reasons. When designing a machine that must perform numerous tasks at the same time, however, an engineer often finds that many possible configurations serve the same purpose equally well. Although the number of options increases, the efficiency with which each task is performed by a multitask machine is much less because tasks often impose conflicting design requirements that must be reconciled. Clearly, vascular sporophytes are not machines, and therefore only a crude analogy can be drawn between their behavior and the behavior of engineered devices. Nevertheless plants are "natural" machines, "engineered" by natural selection, so it is not surprising that the properties of adaptive walls on the fitness landscapes of the sporophyte morphospace are compatible with the relations optimization theory predicts for machines.

If these adaptive walks serve any useful function, it is to show that organic complexity may not impose the severe limits on evolution that are sometimes envisioned. Indeed, the reverse may be true: the evolution of more complex organisms may lessen the burden of climbing the adaptive peaks of a fitness landscape by converting a landscape with a single adaptive peak into a "gently rolling plain" whose hummocks afford comparatively easy passage. Curiously, however, one of the hallmarks of a gently rolling plain is the general absence of topological cues to the locations of valleys and hills. By crude analogy, the fitness landscape for a "complex" multitask organism may result in selection pressures that have little or no direction because fitness varies little from one place to another. In this

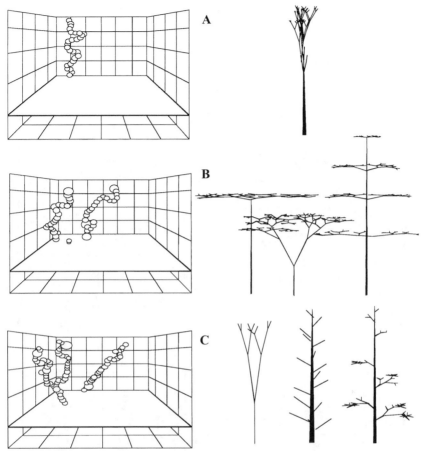

Figure 6.5. Single-task-driven adaptive walks through the morphospace of early vascular land plants. The morphospace (shown to the left) is depicted as a box consisting of two sub-domains, a lower isometric subdomain and an upper anisometric subdomain (see fig. 6.4). All walks begin in the isometric subdomain with a *Cooksonia*-like morphology (not shown). The walks through the morphospace are depicted as a series of "bubbles," each representing a step in a walk. The diameter of each bubble reflects the volume of the morphospace that had to be searched to locate the next most fit plant morphology. Sketches to the right of each adaptive walk depict the morphologies residing on adaptive peaks for each walk: (A) walk maximizing spore production and dispersal (reproductive success); (B) walk maximizing light interception; (C) walk maximizing mechanical stability. Adapted from Niklas 1995.

sense an adaptive walk over a gently rolling landscape can wander for a long time, picking up little or no gain or loss in relative fitness over much of its course. In this way much of the fitness landscape can be explored and much of the morphospace occupied with impunity.

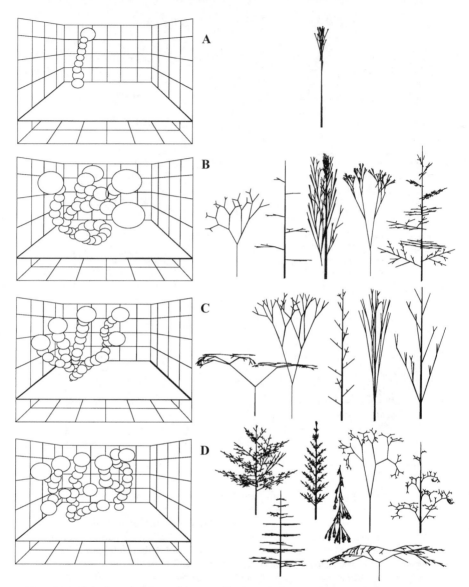

Figure 6.6. Multiple-task-driven adaptive walks through the morphospace of early vascular land plants (see fig. 6.5). The morphospace (shown to the left) is depicted as a box consisting of two subdomains, a lower isometric subdomain and an upper anisometric subdomain (see fig. 6.4). All walks begin in the isometric subdomain with a *Cooksonia*-like morphology (not shown). The walks through the morphospace are depicted as a series of "bubbles," each representing a step in a walk. The diameter of each bubble reflects the volume of the morphospace that had to be searched to locate the next most fit plant morphology. Sketches to the right of each adaptive walk depict the morphologies residing on adaptive peaks for each walk: (A) walk optimizing mechanical stability and reproductive success; (B) walk optimizing light interception and mechanical stability; (C) walk optimizing light interception and reproductive success; (D) walk optimizing light interception, mechanical stability and reproductive success. Adapted from Niklas 1995.

However entertaining the metaphor of "craggy mountainsides" and "gently rolling hills" may be, we must bear in mind that the adaptive walks treated here have four obvious weaknesses. First, "fitness" is measured in terms of comparatively few biological tasks that are further assumed to contribute to fitness in an independent manner. The obvious epistatic relation between photosynthesis and reproduction (Gates 1965; Nobel 1983) therefore was entirely neglected, as was the possibility that some tasks are more important to fitness than others (Franklin and Lewontin 1970; Lewontin 1974; Ewens 1979). Second, walks are simulated as continuous processions among more fit phenotypes; alternatively plausible types of walks were not considered (Gillespie 1983, 1984; Kauffman 1993), and genetic revolutions could result in rapid transitions between adaptive peaks. Third, the fitness landscape was assumed to be spatially stable, in pointed neglect of evident changes in the environment, which are predicted to shift the location of fitness peaks (Wright 1932). In this sense the fitness landscape is a rippling body of water whose contours may change as the environment changes. And fourth, all walks are assumed to be unfettered by genetic or developmental constraints. Even for plants, which arguably may be more phenotypically "plastic" than animals, this is a naive expectation (Maynard Smith et al. 1985).

On the other hand, the approach illustrated here has some obvious strengths. First, adaptive walks are simulated in a dimensionally complex morphospace containing phenotypes representative of the entire spectrum of vascular land plant morphology. Second, although only three tasks were considered, the functional obligations used to define and quantify the various fitness landscapes are nonetheless biologically realistic, at least for most past and present terrestrial plants. Third, although the environment, living and nonliving, is always changing, it is nonetheless true that the early evolution of vascular land plants occurred in a biologically uninhabited environment and was dominated by unchanging physical laws and processes. Here the fitness landscape for the first occupants of the terrestrial landscape was painted in the primary colors of biophysics rather than the subtle hues of complex biotic interactions characterizing subsequent plant history. This crude portrait appears to be legitimate in terms of what is known about the fossil record and the paleoenvionments of the first tracheophytes. And fourth, the phenotypes occupying adaptive peaks and the morphological transformations predicted by simulated walks are, in very broad terms, compatible with those seen in the fossil record of early tracheophytes.

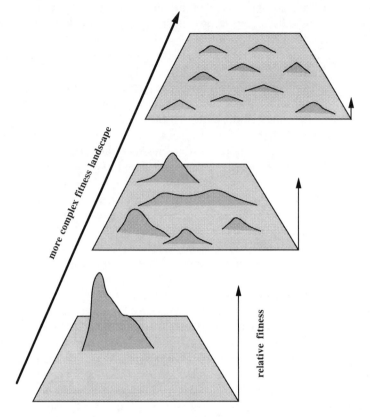

Figure 6.7. Relation posited between the complexity of a fitness landscape and the number of biological obligations an organism must perform simultaneously to survive and reproduce. The fitness landscape defined for an organism that must perform only one biological task is rugged and contains one or only a few adaptive peaks (bottom landscape). As the number of biological tasks that must be performed simultaneously increases, the number of adaptive peaks in the landscape also increases, but the relative height of the peaks (maximum relative fitness) decreases (middle and upper fitness landscapes).

This last point is important because the compliance of predicted and observed morphological transformations provides a reasonable criterion for evaluating the credibility of these adaptive walks. The walks through the morphospace for early vascular plants indicate that taller plants have a selective advantage over their smaller counterparts in terms of photosynthesis and reproduction. Taller plants can avoid the shadows cast by neighboring plants and can use the faster wind speeds above the ground to transport their spores. Also, the taller plants on adaptive peaks are mechanically stable and photosynthetically efficient by virtue of a treelike morphology. The main stem or "trunk" of these phenotypes can support lateral, planated branching systems that are functionally analogous to the foliage leaves of modern plants. With very few exceptions, the lateral branching systems on these phenotypes are held aloft in ways that avoid eccentric loading forces that can induce mechanical instability of the main vertical stem. It must be noted, however, that the fossil record shows large, treelike plant species are more prone to extinction than their herbaceous counterparts when the enironment changes rapidly (see chapter 8). Large size appears to confer advantages in stable habitats, but not in rapidly changing ones.

There can be little doubt that the phenotypes occupying adaptive peaks, particularly those in the most complex "tripartite" fitness landscape, are complex and morphologically similar to some upper Devonian vascular plants (see fig. 6.6D). Without undue imagination, the paleobotanist can find the large treelike lycopods (e.g., *Lepidodendron*), arborescent horsetails (*Calamites*), progymnosperms (e.g., *Archaeopteris*), seed ferns (e.g., *Medullosa*), and the more ancient trimerophytes (e.g., *Psilophyton*) among the morphologies identified by adaptive walks that can optimize light interception, mechanical support, and spore dispersal and production. Perhaps more important, in addition to reaching phenotypes that look very like actual fossil plants, the adaptive walks predict sequences of morphological transformations that some paleobotanists believe characterize the early history of vascular plants. Every adaptive walk began with a *Cooksonia*-like morphology, a small more or less equally branched phenotype. And with very few exceptions, each walk proceeded from this starting point by finding phenotypes that were progressively more "overtopped" and that bore progressively more "planated" and "reduced" lateral branching systems. Overtopping occurs when one axis in each pair "overgrows" the other; planation results when axes shift from a three-dimensional branching pattern to one in which the truss of axial branches aligns in a single plane; and reduction occurs when the growth of some axes is suppressed relative to the growth of others. Naturally the hypothetical phenotypes in a morphospace cannot grow. They are simple mathematical constructs. But as an adaptive walk progresses from one phenotype to its neighbor, it serves to identify mathematical differences among neighboring phenotypes that reflect morphological transformations corresponding to differences in the growth patterns among hypothetical plants. In this sense "overtopping," "planation," and "reduction" describe evolutionary changes in growth patterns. Although these three terms may be unfamiliar to some biologists, they figure prominently in the telome theory that attempts to describe the early evolution of complex sporophyte morphologies from comparatively simple plants like *Cooksonia* and *Rhynia* (Zimmermann 1952; see Stewart and Rothwell 1993). The term "telome" refers to any terminal naked axis on an ancient tracheophyte. The overtopping of telomes is thought to account for the evolution of plants with a single main vertical stem, and planation and reduction are believed to be the evolutionary processes giving rise to lateral branching systems that prefigured the leaves of plants like ferns.

One of the deficiencies of the telome theory is that the theory never explains why some lineages of ancient vascular plants overtopped or planated or reduced their axes while others did not. What, if any, are the advantages of becoming more overtopped or planated or reduced? The telome theory also offers no predictions about the sequence of transformations separating an ancestor from its more derived descendants. Do axes become overtopped and then become planated and reduced, or do planation and reduction evolutionarily precede overtopping? In this sense the telome theory is less a theory and more a lexicon of descriptive terms that can be applied almost willy-nilly to fit any set of circumstances. In large measure, the adaptive walks through the morphospace of early vascular land plants correct some of these deficiencies. They precisely identify both the magnitude and the direction of phenotypic transformations required to get from a *Cooksonia*-like morphology to that of an arborescent lycopod or horsetail. They also explicitly predict how differences in overtopping, planation, and reduction contribute to the relative fitness of a phenotype. In this sense adaptive walks contribute the evolutionary syntax for the terminology of the telome theory.

Leaves and Patterns of Arrangement

Leaves broadly defined have evolved independently in at least six plant lineages (liverworts, mosses, lycopods, horsetails, ferns, and seed plants). Botanists and mathematicians alike have long been intrigued by the arrangement of leaves on stems (phyllotaxy), which is under tight genetic control and varies little as a consequence of changes in the external environment (with the exception of day length). The developmental precision with which leaves are produced by the apical meristems of some plants is impressive. The angular divergence between successively formed leaves on the apical meristem of *Epilobium hirsutum* has a standard error of 0.26°, which in a meristem measuring 100 μm in diameter, equals a distance of 0.2 μm, or far less than the diameter of an average individual cell. Traditionally, biological features that are under tight genetic control and that have very narrow norms of reaction are believed to be adaptive. Thus it is not surprising that the study of phyllotaxy figures prominently in the literature dealing with plant modifications aiding photosynthesis as well as in that treating the evolution of plant morphology and development. Here we will construct a morphospace for the arrangement of leaves on stems and examine adaptive walks governed exclusively by the ability of phenotypes to harvest sunlight.

The first task is to find a convenient, yet biologically realistic way to describe phyllotaxy. Fortunately, for reasons unknown the different phyllotactic patterns observed among real plants may be represented by a procession of fractions called the Fibonacci series—1/1, 1/2, 1/3, 2/5, 3/8, 5/13, 8/21, 13/34, 21/55, and so forth. Mathematicians have long been fascinated with this series because it converges on the decimal fraction 0.38197, called the "golden mean" or *sectio aurea*.* Our interest is less esoteric and purely utilitarian because the numerator of each fraction in this Fibonacci series equals the number of circuits around a stem separating superimposed leaves, while the denominator of each fraction equals the number of leaves separating superimposed leaves (fig. 6.8). One of the simplest phyllotactic patterns, the distichous pattern, results when one circuit around the stem reaches two superimposed leaves. The distichous phyllotaxy is represented by the second fraction in the Fibonacci series, 1/2. Another phyllotactic pattern observed among plants, the tristichous leaf arrangement, occurs when one circuit around a stem produces the superimposition of every third leaf. This condition is represented by the third fraction in the Fibonacci series, 1/3. Yet another spiral phyllotactic pattern is produced when two circuits and five leaves obtain two superimposed leaves, or 2/5. Notice that with the aid of the Fibonacci series the number of leaves a stem can produce before it bears two or more overlapping leaves is easily computed. More important, each step in the Fibonacci series expresses the fraction of the circumference of the stem axis spanned between two successive leaves in a spiral phyllotactic pattern. When multiplied by 360°, each fraction in the series obtains the divergence angle of a particular phyllotactic pattern (see fig. 6.8). The

*The *sectio aurea* refers to the distance from the end of a line at which, if the line is cut, the smaller fraction of the line is to the larger as the larger is to the whole; for example, 0.38197:0.61803 = 0.61803:1.0. Although there are many other series of fractions that converge on 0.38197, the Fibonacci series has played a prominent role in discussions of classical architecture and biological symmetry.

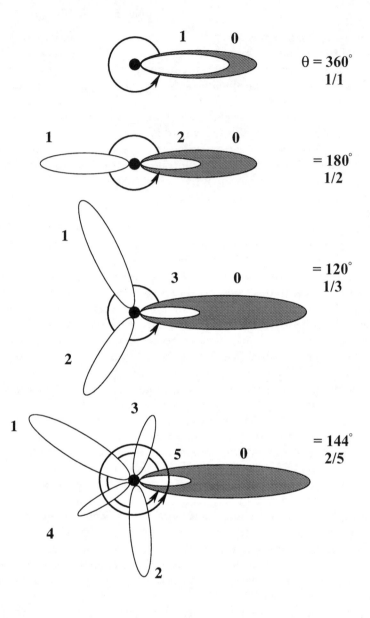

distichous leaf arrangement has a leaf divergence angle equal to 180°—the product of 1/2 and 360°. The tristichous phyllotactic pattern has a leaf divergence angle of 120°—1/3 times 360°; and so on. Because the Fibonacci series converges on the "golden mean" decimal fraction 0.38197, phyllotactic patterns conforming to the higher fractions in the series will have divergence angles that also converge on 137.5092°.

Much has been made of this fact because the golden mean decimal fraction is an irrational number—that is, just like π, its numerical value is never brought to closure. As a consequence, whenever successive leaves are formed at just this angular distance around the axis of a stem, no leaf will ever lie precisely above any other, even for a stem bearing an infinite number of leaves (Wright 1873). Any adaptive walk through a phyllotactic morphospace will gravitate toward leaf arrangements with divergence angles approaching 137.5° because these minimize leaf overlap and therefore maximize a plant's ability to harvest sunlight. Therefore the real question is not whether phyllotactic patterns produced by divergence angles near the golden mean angle are efficient—they will invariably be found to be so—but rather whether other phenotypic features can compensate for the effects of "inefficient divergence angles" on light interception.

We begin constructing the phyllotactic morphospace by making a few simplifications to keep the number of variables manageable. The first of these is that each phenotype consists of a single vertical stem and that all the leaves on the stem have an elliptical outline (fig. 6.9). Because of this assumption, we will not consider the effects of leaf geometry on light interception, although this is clearly an important feature. The morphospace is further simplified when every plant bears the same number of leaves with the same total leaf area. With these simplifications, only two variables are required to define the overall morphology of each phenotype: the distance of the first leaf from the base of the stem h_0 and the distance between successive leaves I, the internodal distance. The total height H of each stem equals h_0 plus the product of the internodal distance and the number of leaves n; that is, $h_0 + nI$. Since every phenotype bears the

Figure 6.8. Geometry of leaf arrangement (phyllotaxy). Leaf arrangements are viewed along the lengths of vertical stems. Leaves are depicted as elliptical outlines and are consecutively numbered from the base to the tip of each stem (e.g., 0 = first, basal leaf and 5 = last, distal leaf in bottom diagram). The leaf divergence angle θ is the angle spanned between two successive leaves on a stem. The leaf divergence angle can also be rendered as a fraction whose numerator indicates the number of gyrations around the axis of the stem required to find two superimposed leaves and whose denominator indicates the number of leaves found along the number of gyrations. A divergence angle of 360° is rendered by the fraction 1/1 (upper diagram); a divergence angle of 144° is rendered by the fraction 2/5 (bottom diagram). The sequence of fractions 1/1, 1/2, 1/3, 2/5, and so forth is called the Fibonacci series.

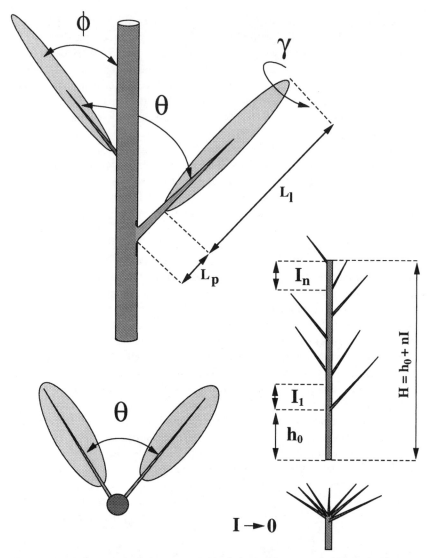

Figure 6.9. Mathematical relations among morphological features required to simulate a morphospace for a phyllotactic morphospace. In addition to the leaf divergence angle θ (lower left; see fig. 6.8), seven additional parameters are required: the leaf rotation angle γ, the leaf deflection angle ϕ, leaf-blade length L_l, petiole length L_p, the height of the first stem internode h_o, the height of successive stem internodes I, and the number of stem internodes n. Total stem height H equals h_o plus nI. When I equals zero, a rosette morphology is produced (lower right).

same number of leaves, n is a constant and only h_o and I vary among phenotypes. Note that as the internodal distance approaches zero, phenotypes progressively acquire a rosette growth habit, while long internodal distances obtain tall phenotypes with equally spaced leaves.

Because all leaves in the model are elliptical in outline, leaf size may be measured by the length of the ellipse's major axis. Leaf shape is conveniently defined as the quotient of leaf blade width and length (the quotient of the ellipse's minor and major axes). A little biological reality can be introduced into the morphospace by allowing the blade of each leaf to be attached to the stem by a cylindrical petiole of length L_p. Leaves flush to the stem occur when $L_p = 0$ (see fig. 6.9). Nonetheless, only three variables are required to define the orientation of each leaf with respect to the direction of ambient sunlight: the angle swept between the longitudinal axes of the leaf and the cylindrical stem ϕ; the rotation angle of the leaf blade with respect to the horizontal plane γ; and "the all important" divergence angle swept between successively attached leaves on the stem θ (see fig. 6.9). Thus, in addition to orienting a leaf with respect to the direction of sunlight, the divergence angle also defines the phyllotactic pattern. Low values for ϕ produce leaves that are nearly vertical in orientation and whose upper surfaces are appressed to the surface of the cylindrical stem; horizontally cantilevered leaves result when $\phi = 90°$. The leaf blade parallels the Earth's surface when $\gamma = 0°$ and lies perpendicular to it when $\gamma = 90°$.

Even though every adaptive walk invariably shows that divergence angles converging on 137.5° maximize light interception, particularly for phenotypes with a rosette growth habit, these walks also show that the divergence angle per se does not determine the light harvesting efficiency of a phenotype (figs. 6.10–6.12). Numerous other phenotypic features, like leaf shape, orientation, and internodal distance, equally affect the capacity of phenotypes to harvest sunlight and can compensate for "poor" phyllotactic divergence angles. For example, leaf shape profoundly influences the light harvesting efficiencies of phenotypes with a rosette growth habit; long and very slender grasslike leaves are more efficient than leaves with a circular or near circular outline because they minimize overlapping leaf blades (fig. 6.10). Alterations in the angle between leaves and the vertical stem all virtually eliminate the influence of leaf arrangement on light harvesting because variously oriented leaf blades avoid excessive shading of one another (fig. 6.11). And large internodal distances maximize the ability to intercept sunlight regardless of the leaf divergence angle because leaves are more widely spaced and, once again, avoid shading one another (fig. 6.12).

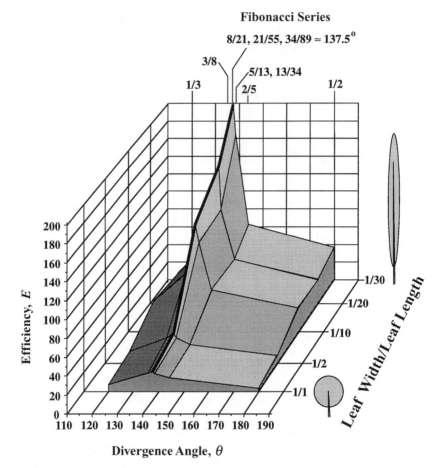

Figure 6.10. Adaptive walk maximizing light interception efficiency through a phyllotactic morphospace for which leaf aspect ratio varies (leaf width/leaf length). The optimal divergence angle invariably equals 137.5°, the angle converged upon by the Fibonacci series of fractions. Morphologies with circular leaves (leaf width/leaf length = 1/1) are the least efficient at harvesting light regardless of the leaf divergence angle (foreground); very long leaves (leaf width/leaf length = 1/30) are the most efficient (background).

The adaptive walks through the phyllotactic morphospace serve as an object lesson regarding the role of "developmental constraints" in evolution, which has received much attention, although insufficiently documented in plants (see Gould 1980; Alberch 1980, 1981; Arthur 1984). The notion of a developmental constraint is that development, which must be highly organized to achieve the final size and form of an organism, necessarily imposes limits on how far morphology or anatomy can be

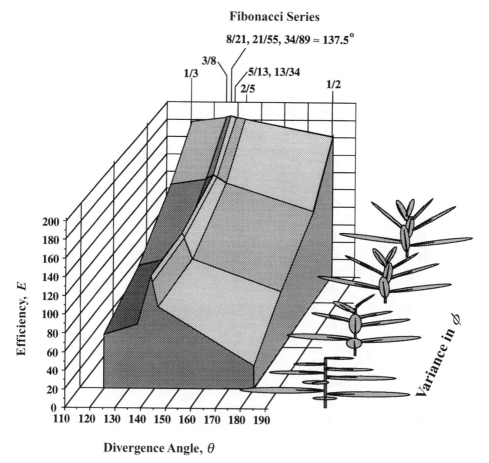

Figure 6.11. Adaptive walk maximizing light interception efficiency through a phyllotactic morphospace for which the leaf deflection angle ϕ is varied among the leaves on a stem. The optimal divergence angle invariably equals 137.5°, the angle converged upon by the Fibonacci series of fractions. Morphologies with horizontally oriented leaves are the least efficient at harvesting light regardless of the leaf divergence angle (foreground); morphologies for which the leaf deflection angle varies from zero to 60° (from base to tip of stem) are the most efficient (background).

varied in an adaptive manner. But a subtle distinction must be made when dealing with "constraints," because a developmental system can limit adaptive solutions in one of two ways. It can set strict boundary conditions on the expression of morphology and anatomy that the organism cannot transgress in any circumstances, or it can necessitate compensatory changes in other features not directly under the control of the

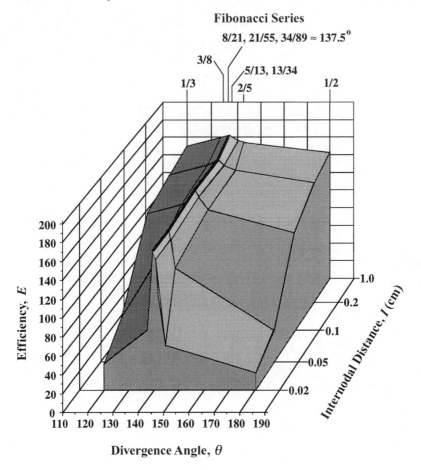

Fibonacci Series

8/21, 21/55, 34/89 ≈ 137.5°

Figure 6.12. Adaptive walk maximizing light interception efficiency through a phyllotactic morphospace for which internodal distance *I* varies. The optimal divergence angle invariably equals 137.5°, the angle converged upon by the Fibonacci series of fractions. Morphologies with very short internodes are the least efficient at harvesting light regardless of the leaf divergence angle (foreground); morphologies with long internodes are the most efficient (background).

developmental system, giving an organism some adaptive latitude. In the first case, morphology or anatomy is limited by the development of the organism. This is what is traditionally meant by developmental constraint. But in the second case, even though the developmental system by its nature imposes limits, it may permit and even require changes in other facets of the developmental repertoire. In this sense it is not an unavoidable obstacle but may even inspire evolutionary changes.

The developmental repertoire of an organism should be thought of not as a single monolithic edifice, but rather as a complex network of spatially and temporally interconnecting systems whose collective and coordinated efforts ultimately achieve the mature phenotype. Modifications or changes in one component of this complex network, in either space or time, may require modifications or changes in other components. Examples of developmental systems that provoke or at least require compensatory changes in other morphological features are rare in the literature. But as early as 1895 Roux drew attention to two phases in the process of development: an early self-differentiation phase dominated by genetically determined features largely insensitive to environmental stimuli, and a later interactive phase during which morphological changes are possible in response to external environmental factors. The phenomenon of phyllotaxy may be a case illustrating these two phases. As I noted, the precision with which leaves are organized and produced on the apical meristems of stems is remarkable and fits in with Roux's self-differentiation phase: it is tightly controlled by genetics and little affected by external stimuli. However, the unyielding precision of phyllotaxy does not operate in a vacuum. The pattern of leaf arrangement is only part of a broader developmental repertoire involving an interactive phase during which a number of phenotypic features can be modified in a fashion that may compensate for ineffective leaf arrangements. Numerous experiments verify that leaf size and shape, petiole length, and so forth can be varied in accordance with ambient light conditions to achieve extremely efficient light harvesting morphologies. This "plasticity" does not imply a lack of interdependence or a lack of organization and control among the constituents of the development repertoire. Experiments confirm that exchanges of assimilates among plant organs are highly correlated and physiologically sophisticated. But plants are developmentally versatile organisms and have evolved a variety of ways to accommodate diurnal, seasonal, and long-term changes in their light environments. Indeed, some adaptations must give us pause when we speak too glibly of the "sedentary" nature of plants, some of which have evolved solar tracking leaves that move and reorient their blades from hour to hour with respect to the direction of sunlight.

Emphasis on adaptive walks in a phyllotactic landscape can give the erroneous impression that all plants rely exclusively on photosynthesis for their nutrition and that leaves serve only to capture sunlight. A great variety of terrestrial and aquatic plants bear leaves adapted to capturing animals, which are digested for their protein and other nutrients (fig. 6.13). Over the long course of plant evolution, foliage leaves have been converted into pitchers containing digestive juices rich in hydrochloric acid (the pitcher plants, *Sarracenia* spp. and *Nepenthes* spp.), cagelike traps that are sprung when plant hairs are touched not once but twice (the Venus flytrap, *Dionaea muscipulata*), tonguelike extensions covered with sticky droplets of deadly bait (*Drosera* spp.), or tiny transparent capsules whose internal vacuum cleaner action pulls in microscopic aquatic animals through a trapdoor when they brush against surrounding delicate bristles (the bladderworts, *Utricularia* spp.). Leaves serve a multitude of biological functions besides capturing sunlight.

An Adaptive Walk for Anatomy

Up to now, adaptive walks through a morphospace for vascular land plants have involved the general morphology of sporophytes and some of the particulars regarding the arrangement of their leaves. These adaptive walks indicate that increasing plant size is adaptive on land and that treelike morphologies bearing lateral planated branching systems or foliage leaves occupy adaptive peaks. Here the objective is to construct a simple adaptive walk for the anatomical modifications that attended the evolution of these morphologies. Logically this walk must be consistent with the starting conditions, predicted trends, and end points of the morphological adaptive walks previously reviewed. In other words, the anatomical

Figure 6.13. Leaves of some insectivorous plants. (A–B) The erect leaves of trumpet plants, *Sarracenia flava* (A), and the pendulous leaves of pitcher vines, *Nepenthes* spp. (B), contain a basin of digestive juices rich in hydrochloric acid. When insects lose their footing on the rim of the leaf, they may fall into the basin, drown, and be digested by the inner surface of these leaves. (C) The hinged, cagelike leaf of the Venus flytrap, *Dionaea muscipulata*. The rim of each half of the leaf blade is lined with spikes, beneath which a band of nectar glands attracts insects. The open face of the each half of the leaf carries a few isolated plant hairs that trigger the cagelike leaf to close when touched twice, not once. When the trap is sprung (bottom left), the inner surfaces of the leaf begin to secret digestive juices and later absorb the nutrients released from the insect's body. After the insect is fully digested, the two halves of the leaf blade separate once again. (D) The leaves of sundews, *Drosera* spp., are covered with glandular plant hairs that secrete a sweet nectar mixed with digestive juices. An insect touching these hairs becomes enmeshed, and the neighboring hairs begin to bend toward the victim. Gradually the entire leaf enfolds the insect, and the hairs further digest and eventually absorb the substance of the victim's body. (E) The tiny transparent, capsulelike leaves of aquatic bladderworts, *Utricularia* spp., absorb the water trapped within them to create a vacuum. When a small aquatic creature touches the bristles near the tip of the leaf a small trapdoor opens, water rushes in, and the creature is vacuumed into the leaf, closing the trapdoor behind it. Glands on the inner surface of the leaf release digestive acids, and the captive is killed, dissolved, and consumed. Within two hours, the vacuum within the bladderlike leaf is recreated and the trap is reset.

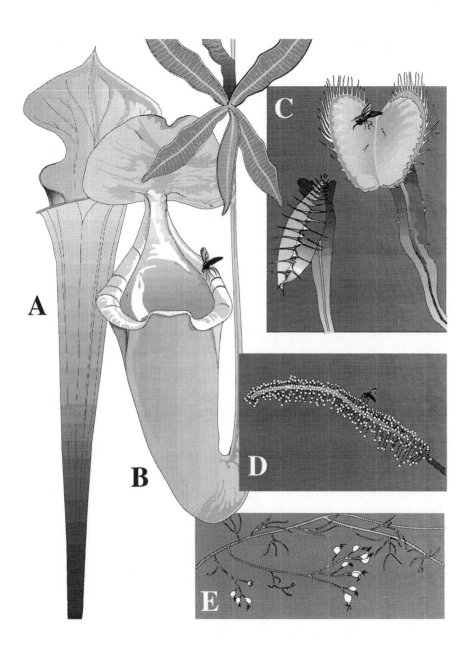

modifications predicted to increase fitness on land must begin with the simple morphologies of the earliest land plants, produce increasingly taller plants, concord with the "overtopping" and "planation," and ultimately come to rest on treelike phenotypes bearing lateral leaflike appendages.

The first step in this adaptive walk involves reconciling the potentially conflicting requirements for photosynthesis and mechanical support. The earliest land plants had cylindrical stemlike axes that functioned simultaneously as photosynthetic and mechanically supportive organs. The optimal location for photosynthetic tissues in a cylindrical plant stem is just below the surface of the stem where cells have direct access to light and atmospheric gases. Experiments with modern plants show that as much as 90% of the photosynthetically useful light is absorbed by the first few layers of cells in green stems and leaves. And because carbon dioxide has a lower molecular diffusivity than oxygen, the rate at which carbon dioxide can be supplied to cells decreases from the surface to the center of a cylindrical stem. Thus it is not surprising that the bulk of the green tissues in the photosynthetic stems and leaves of modern plants tends to be just beneath the surface of these organs. But engineering theory also shows that the optimal location for a stiffening agent in the stem is the same. Herein lies a design problem. The best place to put the stiffest material in a vertical stem is just under the surface, where mechanical forces resulting from bending and torsion reach their maximum intensities. Anatomical comparisons among the stems of living vascular plants show that stiff plant tissues like sclerenchyma are typically just beneath the epidermis.

Naturally there is no design problem in the placement of photosynthetic and mechanical tissues if one kind of tissue can do both things simultaneously. But the best mechanical tissues are elastic, strong, and lightweight so that they can restore a bent stem to its original orientation, resist breakage, and contribute as little weight as possible to the stem. These properties tend to preclude photosynthesis. Elastic and strong tissues require cells with very thick walls that tend to exclude carbon dioxide and oxygen. Additionally, a thick-walled and elastic mechanical tissue surrounding a photosynthetic tissue would likely filter and attenuate sunlight. Lightweight tissues tend to be dead when mature, consisting solely of empty cell walls, and a dead tissue is an impossible design for photosynthesis. Because two different kinds of tissues cannot occupy the same place at the same time, a spatial conflict arises when a stem has to contain photosynthetic tissues and simultaneously provide mechanical support for very large loads.

There is one mitigating factor, however. Specialized mechanical tissues are required only for tall, heavy stems. Short, light stems can be supported by thin-walled living tissues, whose cells function hydrostatically. When inflated with water, the living protoplast of a hydrostatic cell exerts pressure on its cell wall, making it more rigid and stiff. Because water is incompressible and because plant cell walls are remarkably strong in tension, a tissue composed of thin-walled cells can resist the compressive force of gravity. There are two limitations to hydrostatically supported stems. First, their mechanical stability depends on a continuous supply of water. Second, they are composed of a heavy material—water—and cannot grow too tall before they bend under their own weight. Plants with hydrostatic stems are ecologically restricted to moist habitats and are limited to comparatively little height. The advantage of a mechanically hydrostatic stem is that it can be a photosynthetic stem (provided the water lost to the atmosphere during photosynthesis is replenished). The best location of a water-conducting tissue is at the center of such a stem, where water can diffuse equally in the radial direction to the perimeter where water is lost at the stem's surface. The center is also a good location for a conducting tissue composed of hollow tubular cells, because the center of a bent or twisted cylindrical stem experiences the lowest levels of mechanical force and so offers tubular conducting cell walls a refuge from mechanical failure. The ecology, size, and anatomy of the first vascular land plants are consistent with those predicted for plants with hydrostatic stems. These organisms lived in moist habitats and had comparatively short stems with very little tissue differentiation except a slender vascular strand running through the center of their mostly green stems.

That the vascular land plants eventually evolved to occupy every corner of the terrestrial landscape and achieved the stature of modern trees indicates that plant evolution explored options other than the hydrostatic anatomical design. The precise steps taken during this adaptive walk are not easily simulated by computers, but the trail of plant evolution can be followed with other tools. One of these was provided by the great mathematician and engineer Leonhard Euler, who found a formula interrelating plant height, stem diameter, and the mechanical and physical properties of plant tissues:

$$H_{max} = C\left(\frac{E}{\rho}\right)^{1/3} D^{2/3},$$

where H_{max} is the maximum height to which a vertical cylindrical stem can grow before it elastically buckles under its own weight, C is a proportionality constant, E is Young's elastic modulus, ρ is the bulk density of the all tissue used to construct the stem, and D is stem diameter. Euler's formula states that for any stem diameter, the maximum height to which the stem can grow depends on the stiffness and density of its tissues. Pliable and waterlogged hydrostatic tissues like parenchyma are less desirable building materials than stiff and lighter-weight tissues like wood—lumber is better than limber.

This prediction is easily verified by comparing the actual heights of plants with the maximum heights of hypothetical stems built exclusively of one kind of tissue (fig. 6.14). Euler's formula shows that the stems of most land plants reach heights that cannot remotely be mechanically supported by hydrostatic stems. Beyond the sizes of the stemlike axes of moss sporophytes, which are so short that a hydrostatic tissue is more than sufficient to keep them vertical, plant stems must contain stiffer and lighter tissues.

The fossil record of early land plants complies well with the predictions of Euler's formula (fig. 6.15). The oldest vascular land plants had very slender stems, comparable in diameter to those of modern mosses. Although their anatomical preservation is extremely poor, it appears that these Silurian plant stems were composed predominantly of undifferentiated parenchyma, a tissue very much like potato or apple flesh. Parenchyma is a hydrostatic tissue; it is very pliable and has a density very near that of water. This tissue must have been abandoned as the principal supporting tissue in progressively larger and taller plants during the Devonian period. The maximum stem diameter reported for any time interval in the Devonian continuously increases and reaches the size of modern trees by the end of the Upper Devonian. As expected, tree-sized Upper Devonian stems were composed predominantly of wood or comparably stiff and light tissues. Unfortunately the heights of these plants when alive are not known with certainty because Silurian and Devonian plant stems are rarely found in one piece. Nevertheless, plant height can be estimated fairly accurately from stem diameter based on the statistical relation between the height and stem diameter of living plants. Using this relation, the maximum stem diameters reported for fossil plants reveal that plant height steadily increased from the Silurian to the end of the Devonian. Smaller plants with slender hydrostatic stems persisted throughout this time, perhaps living in the understory sheltered by their larger treelike companions, much as many mosses do today. But the evolutionary trend of increasing plant size throughout the Devonian is unmistakable, as is the fact that this trend required the evolutionary innovation of stiffer and lighter tissues.

In very broad terms, the adaptive walk for plant anatomy is comparatively easy to sketch (fig. 6.16). It begins with a short stem consisting of a photosynthetic "rind" of tissue just beneath the surface and a central water-conducting strand running through a hydrostatically inflatable

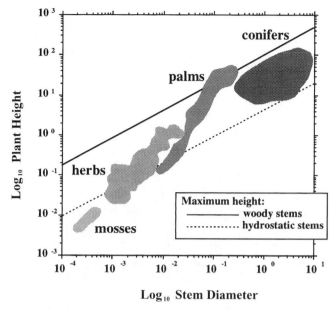

Figure 6.14. Plant height plotted against basal stem diameter for living plants. The theoretical buckling heights for stems composed of wood and a hydrostatic tissue like parenchyma are shown by diagonal lines. These lines are computed from Euler's formula (see text). The data for mosses fall below the theoretical buckling heights of hydrostatic stems. With the exception of some tree palms, the data for other kinds of plants fall between the theoretical buckling heights of woody and hydrostatic stems. Adapted from Niklas 1994.

inner "core" of nonphotosynthetic tissue (fig. 6.16A). The stems of plants taller than mosses require some type of specialized mechanical tissue whose optimal location conflicts with that of the photosynthetic rind. The second-best location for a mechanical tissue is between the photosynthetic rind and the inflatable hydrostatic core (fig. 6.16B). The thickness of this mechanical tissue can be increased to support taller stems (fig. 6.16C). But the amplification of mechanical tissues between the rind and the core would eventually present a physical barrier to the exchange of metabolites between the water-conducting and photosynthetic tissues. Beyond a certain size, the solution is to compartmentalize the functions of the plant body into organs devoted exclusively to photosynthesis and organs dedicated to mechanical support and transport of nutrients. The mechanical and water-conducting obligations of a stem could be simultaneously met by a tissue containing thick-walled cells devoted exclusively to mechanical support (e.g., xylem fibers) and hollow tubular cells devoted exclusively to water transport (e.g., tracheids). Taller stems could evolve by increasing the number of mechanically supportive cells in this

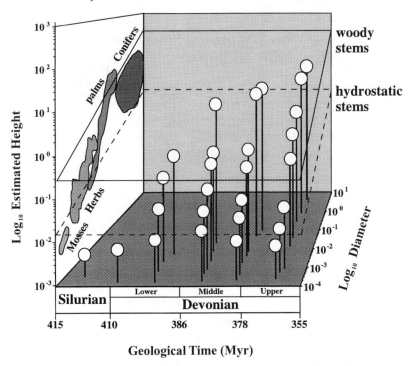

Figure 6.15. Estimated plant height and stem diameter plotted against geological time for late Silurian and Devonian plant fossils. The relation between the height and stem diameter of living plants is shown to the left (taken from fig. 6.14). Estimated heights of fossil plants are depicted by spiked spheres. The three-dimensional surfaces of theoretical buckling heights for woody and hydrostatic stems are shown to posit when fossil stems had to evolve tissues stiffer than hydrostatic tissues.

tissue, and the efficiency of water transport could be concurrently improved by increasing the diameter of water-conducting cells (fig. 6.16D). Anatomical modifications in the appearance of the stem's mechanical and conducting tissue would likely attend the evolutionary amplification in the size and specialization of photosynthetic organs (fig. 6.16E). The final step in the evolution of taller plants was the appearance of stems with an internal core of wood and an external layer of porous bark (fig. 6.16F). The evolution of bark was essential because the living layer of cells that produces wood, called the vascular cambium, requires oxygen for respiration. These cells must also be protected from damage. The bark serves both purposes. The great beauty in the design of woody stems lies in its ability to deal with Fick's law for passive diffusion as well as the

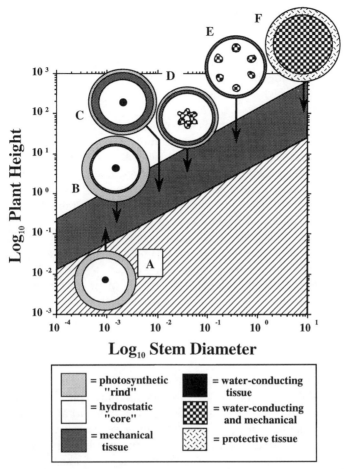

Figure 6.16. Adaptationist hypothesis for the evolution of vascular plant stems. The hypothesis is rendered in terms of the graphic relation between plant height and stem diameter for woody and hydrostatic stems (see fig. 6.14). Plant heights falling in the domain indicated by diagonal lines on the graph can theoretically rely on hydrostatic tissues for mechanical support of stems. Plants heights falling in the darkly shaded domain of the graph must possess tissues stiffer than hydrostatic tissues like parenchyma. Hypothesized anatomical transformations of stems are as follows: (A) ancient stem with a photosynthetic rind surrounding a hydrostatic core through which there runs a slender water-conducting tissue; (B) same as (A) but with a thin layer of mechanically supportive tissue sandwiched between the photosynthetic rind and the hydrostatic core; (C) same as (B) but with a thicker mechanical layer of tissue; (D) amplification of vascular tissues to supply water and nutrients to specialized photosynthetic organs; (E) further modification of conducting tissues and total elimination of photosynthetic rind; (F) evolution of secondary xylem and periderm (protective outer tissue).

mechanical requirements for vertical growth. As new layers of wood are deposited toward the center of the stem, the vascular cambium increases in circumference to contain the greater volume of wood. The bark grows to accommodate this increase, but the vascular cambium maintains the same distance from the external atmosphere supplying oxygen. This is essential because Fick's law shows that distance of transport and time of transport are proportional.

The evolution of woody stems also conferred a tremendous geometric advantage to plants. The vascular cambium produces a new layer of wood every year. Therefore over the years growth layers of wood accumulate and woody stems become progressively thicker. In contrast to nonwoody plant stems, which cannot increase in girth as they age and grow in length, the girth of a woody stem with a vascular cambium "evolves" every year. As indicated by Euler's formula, this is an excellent growth pattern because thicker stems can grow proportionately taller, regardless of the tissues used to construct them. It is no coincidence that the largest terrestrial organisms are trees (fig. 6.17).

Space does not permit a detailed discussion of the evolution of wood and its constituent cell types, but there is a superb book devoted to this subject (Carlquist 1975).

Steles and Bower's Hypothesis

The adaptive walk for stem anatomy cannot ignore the evolution and anatomical elaboration of the stele, the portion of the plant body consisting of the primary vascular tissues (xylem and phloem) and, when present, the conjunctive tissues (the pericycle and the pith tissue). The stele is the "internal plumbing" of stems that supplies water to photosynthetic organs and conducts cell sap to and from living cells. The steles of living and fossil plants differ in their size and general appearance (fig. 6.18), but it is now generally agreed that the most ancient kind of stele consisted of a central strand of xylem enveloped by a hollow cylinder of phloem (Beck, Schmid, and Rothwell 1982), a configuration called a haplostele (fig. 6.18A). In some ancient and modern plants the xylem strand is fluted and, in transverse section, appears star shaped. This type of stele is called an actinostele (fig. 6.18B). The stele of other kinds of plants consists of interconnecting longitudinal strands of xylem that in transection appear as separate plates of xylem embedded in a field of phloem. This type is called a plectostele (fig. 6.18C). Among fossil and living ferns, the xylem may consist of a tube that surrounds a central core of parenchyma, called the pith. Strands of xylem and phloem interconnect the stele, called a siphonostele, with lateral leaves borne by the stem (fig. 6.18D).

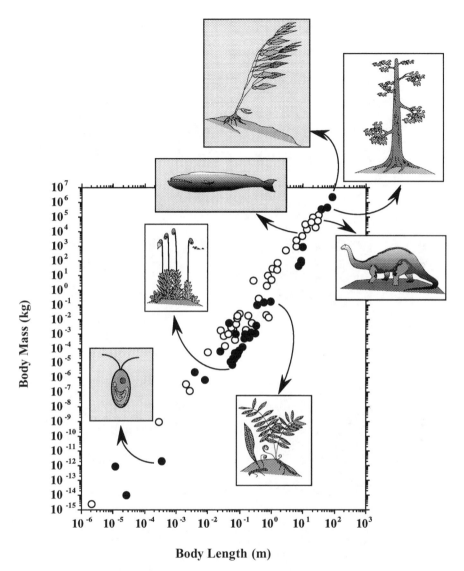

Figure 6.17. Relation between body mass and body length. Data from plants are shown as solid circles, data from animals as open circles. Some representative plant morphologies are indicated in boxes. The largest organisms in this data set are the great Pacific kelp (upper left) and *Sequoia sempervirens* (upper right). The smallest plant in the data set is a unicellular freshwater alga. Adapted from Niklas 1994.

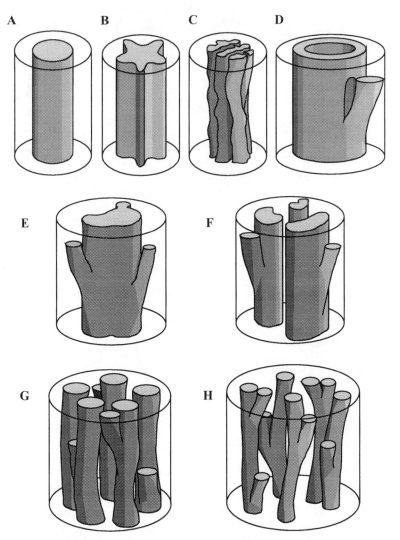

Figure 6.18. Stelar anatomy. Diagrams render the basic anatomical appearance of the primary xylem (shaded areas) in the plant stem (open cylinders): (A) haplostele; (B) actinostele; (C) plectostele; (D) siphonostele with leaf trace and leaf "gap"; (E–H) anatomical transformations posited for the evolution of the eustele (H) from a lobed haplostele (E). Adapted from Gifford and Foster 1989.

And last, among gymnosperms and dicotyledonous flowering plants, the stele consists of more or less discrete vascular bundles running the length of the stem, separated by areas of parenchyma. In transection the vascular bundles are arranged more or less in a ring, surrounding a pith, and vascular traces interconnect the vasculature of the stem with lateral leaves. This kind of stele is called a eustele (fig. 6.18H); it is believed to have evolved through the modifications of simpler types of steles (fig. 6.18E–G).

The size and appearance of the stele have received much attention because it provides taxonomic characters that may be used to distinguish among species, genera, and sometimes entire lineages. How far the size and appearance of the stele influence biological functions like the transport of nutrients and mechanical support is poorly understood. Early plant anatomists noted that the stele differs in size, shape, and geometry as a function of the absolute size of plant stems. A juvenile fern plant may have a haplostele yet acquire a siphonostele when its stems have reached full size. Indeed, the extent to which the steles of other plants are lobed or dissected into interconnecting strands varies along the length of stems and often increases during stem ontogeny (fig. 6.19). These ontogenetic modifications in the size and appearance of the stele are mirrored, in very broad terms, by phylogenetic differences among plants. Small and ancient plants had small and simple steles, whereas more recent plant lineages have larger and more complex ones. The ontogenetic and phylogenetic increase in the size and complexity of the stele appears to have obvious mechanical consequences. Regardless of its shape or location, the size of the stele must increase in proportion to overall stem size whenever the vascular tissues are used to provide mechanical support. The phylogenetic lobing and dissection of the stele place vascular tissues closer to the perimeter of vertical stems. As we have seen, mechanical theory indicates that tissues toward the perimeter of a stem are more effective than those placed near or at the center of the stem. Therefore the ontogenetic and phylogenetic changes in the size and appearance of plant steles are entirely consistent with adaptive evolution favoring increased mechanical stability (Speck and Vogellehner 1988).

Likewise, ontogenetic and phylogenetic modifications in size, shape, and geometry are said to be adaptive in terms of the physiological role of the stele, which, in addition to transporting food and water longitudinally through the stem, must supply water and nutrients laterally to its surrounding tissues. Long ago Galileo pointed out that as any object increases in size, its volume enlarges as the cube of body length but its surface area increases only as the square. As I noted earlier, Galileo's "2/3-power rule" holds true only for a series of objects differing in size

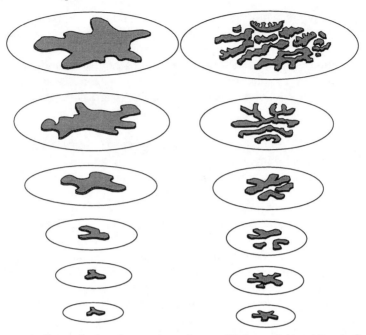

Figure 6.19. Changes in the xylem anatomy of stems of *Psilotum* (left) and *Lycopodium* (right). Xylem shown as shaded region(s); other tissues occupy the space bounded by elliptical outlines of stems. Adapted from Bower 1935 and Bierhorst 1971.

but sharing the same geometry and shape. If so, the ratio of the surface area to the volume decreases as a 2/3 power of increasing size. Frans O. Bower (1935) was the first botanist to apply Galileo's rule to plants by suggesting that ontogenetic and phylogenetic modifications in the geometry and shape of the stele maintain the same ratio of exchange surface area to vascular tissue volume as plant size increases during growth or over evolutionary history. Bower thought that the ratio is adaptively kept constant to obtain "physiological similitude" among plants differing in absolute size. The decrease in the surface area relative to volume can be avoided by changing geometry or shape.

Ontogenetic changes in the stele's ratio of surface area to volume are consistent with Bower's adaptive hypothesis. David Bierhorst (1971) has shown that differences in the perimeter of the stele running through the length of *Psilotum nudum* plants are directly proportional to changes in the cross-sectional area of the vascular tissue. Because the perimeter is proportional to the lateral exchange surface of the stele and because the cross-sectional area of the vascular tissue is proportional to the volume of the stele, the ratio of the surface area to the volume of the stele in

P. nudum remains more or less constant regardless of the size of the plant, as first emphasized by Bower, who showed the same relation for the vascular plant *Lycopodium scariosum*. The constancy of the ratio of surface area to volume can occur only if the geometry and the shape of the stele change ontogenetically as a function of the absolute size of plant stems (see fig. 6.19).

The phyletic corollary to Bower's hypothesis appears incorrect, however. The ratio of stele surface area to volume for progressively larger and taller late Silurian and Devonian plants steadily declines with increasing stem diameter (fig. 6.20). Ontogenetic changes in the steles of plants like *P. nudum* and *L. scariosum* thus do not recapitulate phylogenetic changes in the surface area relative to the volume of the stele. In fact it is naive to believe that the ratio of the stele's surface area to volume has any bearing on the rate of exchange of metabolites between the stele and other stem tissues. The exchange of food among cells occurs through plasmodesmata and involves active physiological transport rather than passive diffusion, so the surface area of the stele does not intrinsically determine the rate of exchange of food. By the same token, water initially passes from the stele into the rest of the stem through pores within cell walls and then subsequently through the cell walls and protoplasts of living cells. The size of these pores, the thickness and chemistry of cell walls, the physiological rates of living cells in the stem, and the efficiency of stomata collectively dictate the rate at which water passes from the stele into the living cells of the stem. None of these factors are known to correlate with the surface area of the stele. Ontogenetic changes in the size and appearance of the stele more likely reflect the requirement to interconnect stems and leaves (if present) with vascular tissues that must be variously dissected or lobed in such a way that the surface area of the stele increases relative to the volume of vascular tissues in stems. The phylogenetic trend of decreasing surface area relative to volume likely reflects the fact that the vascular tissues became increasingly important for mechanical support and therefore increase in volume relative to the rest of the stem. The changes in the appearance of the stele envisioned to precede the appearance of the eustele are consistent with this hypothesis (see fig. 6.18E–H). The ratio of the surface area to the volume of the steles of plants implicated in this evolutionary sequence of anatomical modification actually declines over geological time. And the phylogenetic pattern of lobing and dissection correlates with the evolutionary appearance and progressive specialization of lateral branching systems and foliage leaves that had to be connected to larger stems by means of separate vascular strands supplying nutrients to lateral organs. In summary, Bower's hypothesis, though elegant, is wrong.

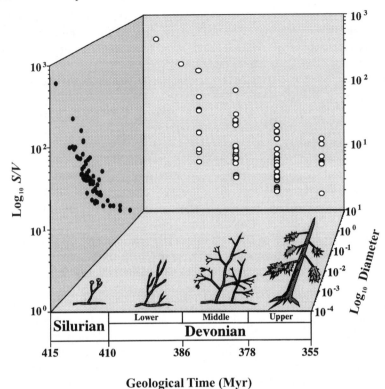

Geological Time (Myr)

Figure 6.20. Relations between the ratio of xylem surface area to volume and stem diameter for fossil plants differing in geological age. The ratio of surface area to volume decreases with increasing stem diameter (left panel) and decreases from the late Silurian through the Devonian (back panel). Crude renderings of representative plant morphologies are provided on the "floor" of the three-dimensional graph.

The Adaptive Walk of Early Seeds

Here we turn our full attention to plant reproduction. The topic is remarkably complex and cannot be treated in the detail it richly deserves. Also, it is not possible to simulate adaptive walks throughout the entire domain of plant reproductive organs. Flowers, for example, are very diverse in size, shape, number and position of parts, hue, and scent. This floral diversity in large part reflects adaptive modifications in response to animal pollinators, whose behavior cannot be easily reduced or predicted by mathematical equations. Yet many plants have evolved wind pollination, which is amenable to rigorous mathematical analysis and simulation in wind tunnels. Some wind pollinated plants are extremely ancient, like the pines; others are comparatively new organisms, like the grasses. But regardless of their antiquity or recency, wind pollinated plants have many features in common. They all tend to produce light, small pollen grains, often in vast quantities. Presumably these features aid pollen transport by wind currents and increase the probability that some pollen will hit a reproductive mark. Wind pollinated plants also have adaptations to filter pollen grains from the air. The stigmata of grass flowers are featherlike and so provide a large surface area for capturing pollen. They also extend well above the rest of the flower, whose parts may inadvertently capture pollen to no reproductive advantage. Likewise, the flowers of grasses are exerted well above the foliage leaves of their stems, which may also capture pollen. The ovules of pine and spruce are aggregated into cones, increasing the probability that some airborne pollen may eventually make contact with them. Ovulate cones also tend to be produced on the tips of branches of trees, where pollen is less likely to be trapped by leaves and stems. These and other modifications strongly suggest that wind pollinated species are aerodynamically adapted, as the ecological success of the grasses and the great antiquity of pines attest. Fortunately the science of aerodynamics is robust and provides a foundation for exploring adaptive walks in the landscape created by plants growing in the wind. I shall employ this science to discuss the early evolution of seed plants, which most biologists believe were wind pollinated organisms.

The early seed plants evolved during the Devonian and diversified in species number throughout the Carboniferous. The seeds of these plants were diverse in morphology, ranging from those with a deeply lobed integument lacking a micropyle to those having a well-defined integument and micropyle. This diversity has inspired theories for the evolution of the seed and the selection pressures leading to the integumented seed (Long 1977; Niklas 1981; Rothwell and Scheckler 1988; Rowe 1992). One theory relies on a sequence of morphologies showing a progressive reduction and fusion of what is believed to have originally been a sterile truss of lobes surrounding the integument (fig. 6.21). The most primitive morphology in the hypothetical sequence is represented by the fossil called *Genomosperma kidstonii* (fig. 6.21A). In place of an integument, the megasporangium of *G. kidstonii* was surrounded by a whorl of free telomelike lobes attached at the base. The fossil called *Salpingostoma dasu* is believed to be more evolutionarily derived (fig. 6.21B). The telomelike lobes of this seed clasped the megasporangium but were free along much of their length. A yet more derived condition is posited for the fossil called *G. latens*, in which the shorter lobes are partially fused at their base and appressed around the apex of megasporangium to form the integument (fig. 6.21C). Fusion of the lobes is almost complete in the

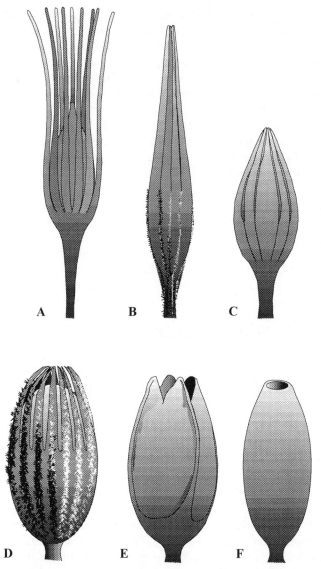

Figure 6.21. Morphologies of some early Paleozoic seeds arranged in a sequence purporting to show the evolution of the integumented ovule: (A) *Genomosperma kidstonii;* (B) *Salpingostoma dasu;* (C) *G. latens;* (D) *Physostoma elegans;* (E) *Eurystoma angulare;* (F) *Stamnostoma huttonense.* Adapted from Long 1977 and Niklas 1981, 1992.

seeds called *Physostoma elegans* and *Eurystoma angulare*, whose integuments and micropyles are better defined (fig. 6.21D–E). The hypothetical sequence culminates with the fossil seed called *Stamnostoma huttonense*, which had a morphology much like that of modern seeds (fig. 6.21F).

The sequence shown in figure 6.21 does not do justice to the range of morphological variation reported for Devonian and early Carboniferous seeds. Flattened seeds and seeds with winglike integuments are known from the late Devonian, in addition to radially symmetrical seeds lacking wings. Consequently the hypothetical sequence for the evolution of the seed at best offers only a general outline for the kinds of morphological modifications that may have led from a nonintegumented megasporangium, like *G. kidstonii*, to a seed with a true integument and micropyle, like *S. huttonense*.

Protection, optimization of wind pollination, and seed dispersal have all been proposed as selection pressures favoring the morphology of *S. huttonense* over that of *G. kidstonii*. The progressive fusion and reduction of telomelike lobes would have provided the megasporangium protection against herbivores as well as desiccation. Invertebrates with piercing or sucking mouth parts make their appearance in the Devonian well before the evolution of seed plants, and so coexisted with the earliest seed plants (see Taylor and Taylor 1993). Wounded and pierced Devonian plant fossils occur with sufficient frequency to lead us to believe that the reproductive organs of early seed plants required protection against herbivorous insects. Superficially, reproductive organs like *Genomosperma* seem ill equipped to protect the megasporangium from insect attack, in contrast to fully integumented organs like *Stamnosperma*.

Aerodynamic studies also show that the sequence of morphological modifications envisioned as leading to the seed may have increased the efficiency of wind pollination. The available evidence indicates that most if not all of the most ancient seed plants were wind pollinated, although some may also have been pollinated by insects (Taylor and Millay 1979). The ability of fossil seeds to capture pollen or spores from the air was tested empirically. Scale models of Devonian and Carboniferous seeds are easily constructed, and their aerodynamic properties can be measured and quantified in a wind tunnel. These studies show that the lobed "integument" of *G. kidstonii* offered a large surface area to capture airborne pollen or spores from the air before they could reach the tip of the megasporangium and effect pollination. In contrast, morphologies with more fused and reduced lobes could deflect and channel windborne pollen toward the apex of the megasporangium, increasing the probability of pollination. Aerodynamically, *S. huttonense* was the most efficient morphology for deflecting airborne pollen into the micropyle (see Niklas

1992). Finally, the integuments of ancient seeds may also have functioned to disperse mature seeds (fig. 6.22A). Like the autogyroscopic seeds and fruits of modern species, the winged integuments of these fossil seeds may have slowed the rate of descent through the air and so have increased the distance of lateral transport from parent plants (fig. 6.22B–D).

Two of the three adaptive hypotheses (protection and wind pollination) predict very similar if not identical adaptive morphological trends. The progressive reduction and fusion of sterile lobes simultaneously afford the megasporangium increasing protection against herbivory and increasing aerodynamic streamlining favoring wind pollination. Therefore a complement of selection pressures may have favored seeds with a well-defined integument and micropyle. In contrast, the suggestion that winged integuments served as a mechanism for wind dispersal is not an adaptive hypothesis per se because it begs the question of how the integument evolved. In this sense it simply suggests how the integument, once formed, may have subsequently become modified for wind dispersal (see Rowe 1992). Essentially, the adaptive hypothesis for winged seeds begins where the other two hypotheses leave off.

Leaving these adaptive scenarios aside for the moment, it is important to note that the fossils sketched in figure 6.21 are found in rocks that are very much the same age, suggesting that the early seed plants morphologically diverged in many directions in a comparatively short time. Seen in this light, the hypothetical sequence purporting to show a *linear* evolutionary trend is probably misleading because it suggests an ancestor-descendant relationship that cannot be verified and may indeed not exist. It is preferable to envision an "adaptive radiation" of seed morphology whose variants were culled by natural selection to leave only a few survivors (fig. 6.23). Plants with exposed megasporangia may have lost more of their seeds to insect predators than those with partially or fully integumented seeds. Likewise, plants with seeds that were poorly designed aerodynamically may have been wind pollinated less frequently than plants with streamlined seeds and so left behind fewer viable mature seeds to establish the next generation. The loss of some seeds each year would have established a vicious cycle, because the efficiency of wind pollination is dictated in part by the number of pollen grains released in the air by conspecific plants, which in turn depends on the number of plants in a given community. If the number of plants in the community gradually declined because fewer viable seeds were produced each year, then fewer pollen grains would have been released each year and fewer ovules would have been pollinated. In other words, natural selection would have taken its toll, and species producing vulnerable and less frequently pollinated seeds would have been winnowed away while

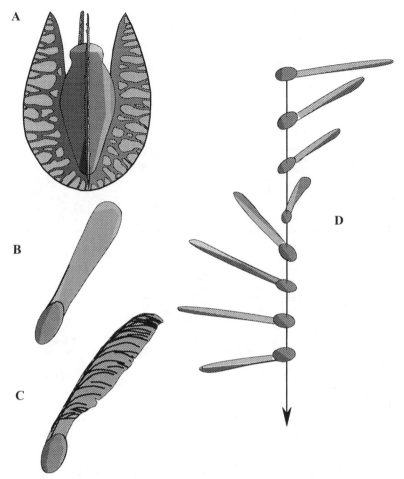

Figure 6.22. Autogyroscopic behavior of winged seeds and fruits: (A) fossil winged seed from late Devonian, described by Rowe 1992; (B) winged seed of pine; (C) winged fruit of maple; (D) autogyroscopic behavior of descending pine seed or maple fruit.

species producing integumented and more efficient seeds would have flourished.

The early evolution of seed plants was a period of great morphological and reproductive innovation that may have been inspired by some kind of genetic revolution. The diversity of early seed morphology must have been underwritten by great genetic and developmental variation within populations and among species. Regardless of how ovule variation was genetically achieved, it is reasonable to assume that the early seed plants

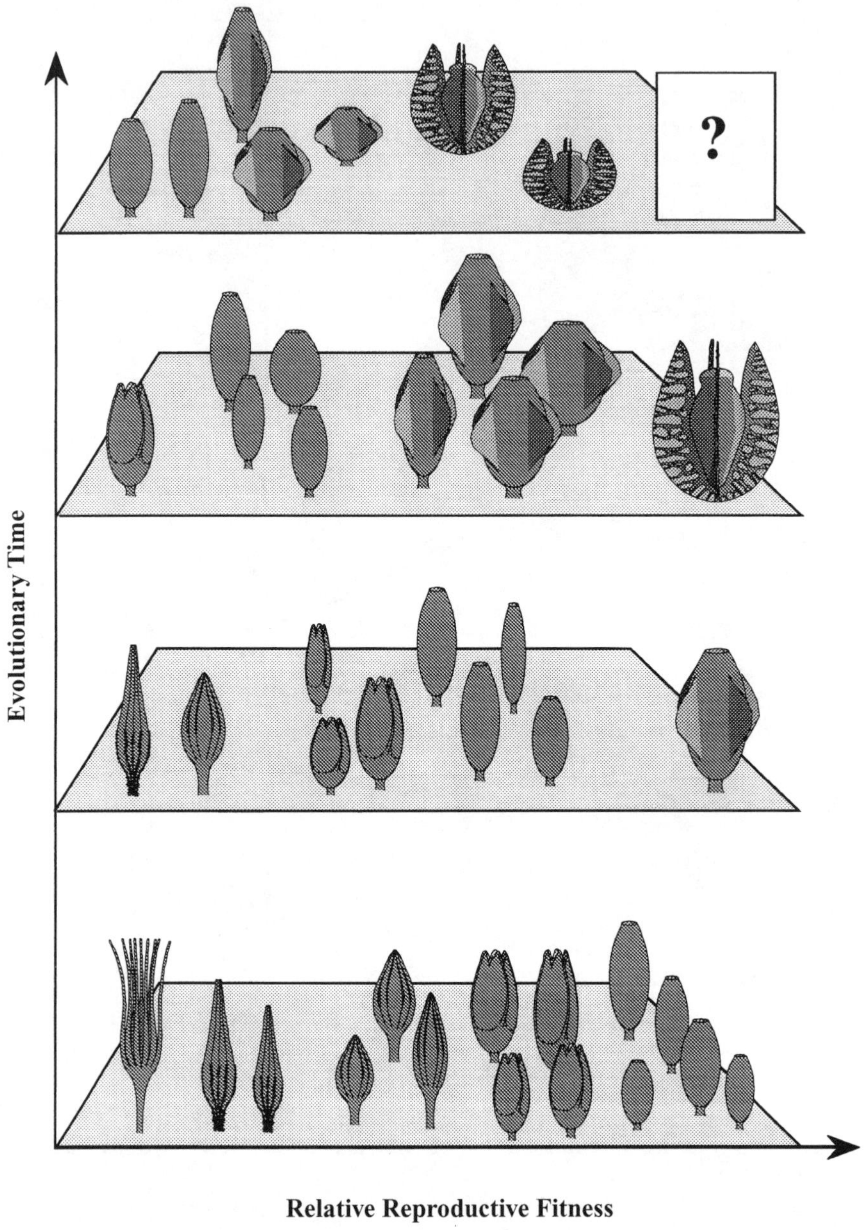

Relative Reproductive Fitness

Figure 6.23. Alternative view to the evolution of the seed. In contrast to a series of morphological transformations (see fig. 6.21), a radiation of early seed plants results in a plethora of seed morphologies, some more fit than others in terms of protecting megagametophytes and embryos (bottom fitness landscape). Over time, these early morphologies are culled by natural selection until only a few fit morphologies remain. Adaptive fine-tuning results in some seeds with winged integuments (upper panel) and may have engendered additional modifications that have not yet been discovered.

were poised at the edge of a relatively unexplored reproductive environment and that selection on early variants was relaxed compared with later times. Over time, selection pressures acting on the raw material of this variation removed plants (and ultimately entire species) that were less reproductively fit than their contemporaries. The appearance of winged and variously elaborated seeds may reflect adaptive fine-tuning, some of which may have resulted from selection pressures favoring long-distance dispersal, which reduces intraspecific competition among seedlings and increases the probability of colonizing new sites. The invasion of new habitats by airborne seeds may have favored allopatric speciation in geographically splintered populations. As reviewed in chapter 2, geographic isolation favors genetic and phenotypic divergence among subpopulations owing to the different selection pressures exerted by different habitats. But because the aerodynamic principles and laws governing wind pollination hold true in each habitat, splintered subpopulations and even different species would have converged on very similar seed morphologies in the absence of adaptations to insect pollination or some other mechanism. If so, the early history of seed plants would have been characterized by an initial burst of morphological variation followed by convergence on a few seed morphologies concurrently adapted to wind pollination, protection against herbivory, and seed dispersal.

Part 4 Long-Term Trends

Plate 4. Convergent evolution between representative species of the New World plant family Cactaceae (shown on the left) and the Old World plant family Euphorbiaceae (shown on the right). Arborescent, broad-leafed species found in tropical regions, like *Pereskia sacharosa* and *Euphorbia melifera* (upper left and right, respectively), are believed to reflect the ancestral condition. Small species with succulent, short, and sparsely branched stems, like *Ferocactus* sp. and *E. valida* (lower left and right), are highly derived and adapted to arid desert conditions. Numerous species, such as *Pilocereus lanuginosus* and *E. canariensus* (middle left and right), can be arranged into a morphological transformation series purported to show the reduction in leaf size, woody stem tissues, branching, and overall plant size attending the radiation of cacti and euphorbs into arid environments. Detailed phyletic and comparative morphological analyses are required to determine the precise ancestor-descendant relations among these species and to determine whether this is an example of convergent or parallel evolution. (The photographs of *Pereskia sacharosa* and *Pilocereus lanuginosus* were printed from color slides generously provided by James D. Mauseth, University of Texas, Austin; photographs of *Euphorbia melifera* and *E. canariensus* were printed from color slides generously provided by Grady L. Webster, University of California, Davis.)

7 Divergence and Convergence

*The saying "Many roads lead to Rome" is as true in
evolution as in daily affairs.*

Ernst Mayr

It's déjà vu all over again.

Yogi Berra

Three long-term phenotypic trends, called divergence,
parallelism, and convergence, are seen in evolution-
ary history. These trends are important for two reasons. First, each of
them can be understood in terms of long-term stabilizing, directional,
or diversifying selection (see fig. 1.13). That is, each trend reflects on a
grand scale the way organisms have phenotypically responded to the di-
rectionality (or lack thereof) of environmental pressures over long periods. Thus the three phenotypic trends are noteworthy not because they
are frequently observed (only three general possibilities exist), but because they give insights into long-term adaptive responses to sustained or
changing selection pressures.

The second reason these trends are important is that they are integral
to our perception of whether phenotypic traits are homologous or analogous characters. Homologous features are equivalent parts of two or more
organisms with a last common ancestor, regardless of whether these parts
differ in form, size, or function. Analogous traits are different parts of unrelated organisms that nevertheless perform the same function. There are
three components to the concepts of homology and analogy: whether the
parts being compared are developmentally equivalent; whether the organisms have a last common ancestor; and whether the parts are functionally equivalent. Notice, however, that the concepts of homology and analogy place only two of these components in direct opposition (equivalent
versus different parts, common versus divergent ancestry). The third component is not a diametric proposition: homologues *may or may not*
be functional equivalents, whereas analogues are *invariably* functionally
equivalent parts. Thus, strictly speaking, homology and analogy are not
precisely opposite concepts because they posit opposing conditions for
two but not all three of their salient ingredients. Notice further that any

judgment of whether two phenotypic characters are homologues or analogues requires information from developmental studies to resolve the issue of equivalent versus nonequivalent body parts and a precise phyletic hypothesis for whether the organisms being compared share or do not share a last common ancestor.

I shall return to "the problem of homology" later. Here I discuss the nature of parallelism, convergence, and divergence in terms of the three modes of natural selection. Although these phenotypic trends are easily diagrammed by means of two lines drawn in parallel or converging on or diverging from one another, the evolutionary patterns are better cast in terms of a cladistic hypothesis whose ramifying branches more realistically render phyletic relations among species or higher taxa (fig. 7.1). From such a hypothetical cladogram, we can see that parallelism occurs whenever the transformations in phenotypic characters obtaining contemporaneous descendants correspond so that the phenotypic states of the character being compared between the descendants are as similar to one another in appearance as their forerunners were. This is illustrated in the cladogram by the two identical transformation series $\square \rightarrow \triangle \rightarrow \diamond$ and $\square \rightarrow \triangle \rightarrow \diamond$. Parallel phenotypic transformations can result when stabilizing selection occurs, that is, when descendants experience similar or steadfast environmental conditions over a long period such that corresponding phenotypic transformations occur among related species. In contrast, divergence occurs when closely related species diversify in appearance. This is illustrated by the transformation series $\square \rightarrow \triangle$ and $\bigcirc$, and $\triangle \rightarrow \diamond$ and $\bullet$. In both cases a single taxon diverges to produce two very different phenotypes. Divergence in one or more traits can occur as a consequence of geographic isolation or when a species colonizes an environmentally heterogeneous habitat, that is, whenever descendants experience some form of disruptive selection. On a large scale, divergent evolution is called adaptive radiation. In contrast to divergent and parallel evolution, both of which involve closely related species, convergent evolution occurs when different species diverge in appearance from their respective ancestors and evolve phenotypic correspondences as a result of identical selection pressures. This is illustrated by the transformation series $\square \rightarrow \triangle \rightarrow \bullet$ and $\square \rightarrow \bigcirc \rightarrow \bigcirc$. Directional selection can lead to convergent phenotypes among unrelated organisms.

Of the three phenotypic trends, divergence and convergence are the more easily recognized. Two closely related species with different phenotypes are sufficient to adduce that divergent evolution has occurred. Likewise, two species with corresponding phenotypes but descended from different ancestors are sufficient to verify convergent evolution. In contrast, recognizing parallel evolution is far more difficult because three

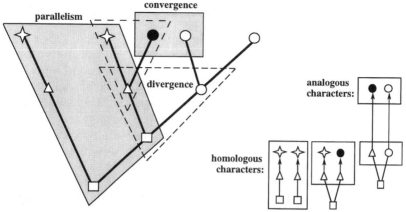

Figure 7.1. Evolutionary parallelism, divergence, and convergence diagrammed in terms of a hypothetical cladogram for eleven phyletically related taxa. Symbols denote different states of a single phenotypic trait. Arrows connecting symbols show character transformations among the states of the trait. Homologous character states are those that are derived directly from the same previous character states, are comparable parts of the same character transformation series, or are identical with that of the last common ancestor. Analogous characters appear to be similar because they serve the same function, but they are derived from different character states.

pairs of contemporaneous descendant species are required, with the two members of each pair belonging to different lineages (see fig. 7.1). Because a minimum of six species is needed to prove the existence of parallelism and because the phyletic relations among these species may be difficult to resolve owing to a poor fossil record, unequivocal examples of parallelism are extremely difficult to find. Evolutionary parallelism is also difficult to deal with conceptually because phenotypic dissimilarities among the various parts of related species can always be found if one looks hard enough. In plane geometry, parallelism is easily proved and understood with the aid of an intersecting line—biology has no such simple and elegant Euclidean proof. Although most evolutionists recognize that parallelism can occur, many give the concept only passing mention when discussing long-term phenotypic trends.

Divergent, parallel, and convergent evolution may all contribute to the evolutionary history of a clade with a long and rich history (see fig. 7.1). Historically complex phenotypic patterns, however, present a serious challenge to the evolutionist in terms of the taxonomic level that warrants the use of these three terms. For example, all embryophytes are believed to share the same last common ancestor; that is, the embryophytes are traditionally viewed as a monophyletic group of plants. Consequently the phenotypic divergence of the embryophytes during the early Paleozoic, which shortly followed the invasion of land, may be seen as one of the more striking examples of divergent evolution, justly qualifying as a "grand adaptive radiation." This radiation culminated in the appearance of numerous embryophytic lineages, each with its own historical pattern and fate. Modern lineages thus do not share an immediate common ancestor, even though they all trace their phyletic legacy to a shared ancestor. Over the course of their long history, some embryophytic lineages have continued to diverge phenotypically, while others have converged in appearance or structure. As formally defined, each of the three trends is distinguishable based on two important features: first, the direction taken by phenotypic modifications during the evolution of a group (whether species diverge phenotypically, evolve in a parallel manner, or converge in appearance or structure), and second, how far species or higher taxa diverged from the same last common ancestor (propinquity of descent). The first of these two features, the direction of phenotypic transformations, may be clear-cut and unambiguous, but only with the aid of a phyletic hypothesis that proposes clearly defined taxonomic relationships among species and the nearness of descent of higher taxa. In the absence of such a hypothesis, the use (and meaning) of the terms divergence, parallelism, and convergence becomes highly problematic.

Analogy or Homology?

From the hypothetical cladogram shown in figure 7.1 we see that the time species or higher taxa diverge from one another can profoundly influence whether phenotypic features are analogous or homologous. Analogous structures are those that have superficial similarities but are not the products of the same genetic and developmental background. Homologous features may be widely divergent in appearance or function, or they may be phenotypically and functionally similar. The salient feature of homologues is that they share the same basic genetic background or developmental pathway because they are equivalent parts of closely related species. It is apparent, therefore, that propinquity of descent dictates whether phenotypic correspondences among species owing to similar selection pressures are analogous or homologous. Species sharing an immediate common ancestor are expected to share very similar genetic backgrounds, and so they should have very similar biological dispositions to respond to similar selection pressures. Conversely, distantly related species are likely to have divergent genetic or developmental capabilities and so are likely to take alternative adaptive routes even when confronted with the same selection pressures. Divergent phenotypic evolution therefore is more likely to involve homologous structures than analogous ones. Conversely, because convergent evolution involves species that do not share a last ancestor, convergent phenotypic correspondences are more likely to reflect the outcomes of the path of least genetic or developmental resistance, which will differ among unrelated species.

The notion that adaptive modification typically takes the path of least resistance was first suggested by the botanist W. F. Ganong, who wrote, "When, through a change in some condition of the environment, the necessity arises for the performance of a new function, it will be assumed by the part which happens at the moment to be most available for that purpose, regardless of its morphological nature, either because that part happens to already have a structure most nearly answering to the demands of the new function, or because it happens to be set free from its former function by change of habit, or for some other non-morphological reason" (1901). Ganong's reasoning implies that, for unrelated species, convergent phenotypic correspondences reflect mutations influencing different developmental pathways that result in functionally corresponding structures under similar selection pressures. Conversely, among closely related species, divergent phenotypes likely reflect mutations affecting similar developmental pathways with disparate results under dissimilar selection pressures (see Schluter 1996).

The concept of homology is clearly context driven in terms of the times organisms phyletically diverge from their last common ancestor, which in turn is influenced by the taxonomic level of analysis, and in terms of whether the character transformations attending these divergence times differ, which depends on the level of developmental detail available. Nevertheless, it is generally accepted that two or more phenotypic character states are homologues if they are derived directly from the same previous character state, if they are both part of the same character-state transformation series, or if the most recent common ancestor of the two organisms also had the character state (Simpson 1961; Hennig 1966, Mayr 1969; Wiley 1981). Casting these three possibilities in terms of figure 7.1, we see that ◇ and ●, on the one hand, and △ and ○, on the other, are homologous character states because in each case they share a common ancestor (with the character state △ and □, respectively). Likewise, □, △, and ◇ are homologous character states because they are both part of the same transformation series. And finally, ○ and ○ are homologues because they descend from the common ancestor ○. In contrast, ● and ○ are analogous character states, despite sharing □ as a distant ancestor state, because they immediately descend from two different last common ancestors (with character states △ and ○, respectively).

Sometimes it is very easy to determine whether parts of two or more kinds of organisms reflect the same series of character transformations.

In other cases the judgment on whether parts are homologues or analogues is very difficult. Consider the wings of insects, birds, bats, and pterodactyls. The fossil record shows that insects and vertebrates phyletically diverged from one another long ago. This is reflected by the fact that the basic body plans of insects and vertebrates are profoundly different. The wings of insects are extensions of the cuticle flapped by flight muscles bending the thorax cuticle. Insect wings are presumably derived from the gill covers of ancient aquatic ancestors, perhaps originally extended to aid in thermoregulation or in skimming the water surface to avoid predators or capture prey (Kingsolver and Koehl 1985; Wooton and Ellington 1991). In contrast, the wings of birds, bats, and pteridactyls are all developmentally derived from the vertebrate forelimb. Despite their functional correspondence, therefore, insect and vertebrate wings are not homologues—they serve the same function, but they involved widely different character transformations in two groups of animals whose last common ancestor existed in the extremely remote past.

But now consider the issue of homology for vertebrate wings. All these wings are derived from character transformations involving the vertebrate forelimb. Nevertheless, the fossil record shows that birds, bats, and pterodactyls are members of very different branches of the large and complex clade of amniotic animals. Each of these branches can be traced back to a last common ancestor, but they undoubtedly diverged from this ancestor at different times in the past. Detailed comparative analyses of vertebrate wings also shows subtle but significant differences. Consequently the character transformations that produced the wings of bats, birds, and pterodactyls were not equivalent. Therefore it would be a flight of fancy to argue that all vertebrate wings are strict homologues—it is much more reasonable to view these structures as analogues that serve as good examples of convergent evolution within the clade of amniotic animals.

A comparable example from the plant kingdom is seen when we consider vessels. Among living tracheophytes, vessels are found principally in angiosperms. But these water-conducting cells have also been found in some horsetails (*Equisetum*), lycopods (*Selaginella*), filicalean ferns (*Pteridium aquilinum*), gymnosperms (*Ephedra*), and an enigmatic fossil plant group (Gigantopteridales) (see Carlquist 1975; Li, Taylor, and Taylor 1996). Even though all of these vessel-bearing plants ultimately descend from a common tracheophyte ancestor, the vastly different divergent times of the horsetails, lycopods, ferns, gymnosperms, and angiosperms as well as the different ways their vessels develop and mature leave little doubt that these conducting stuctures have evolved independently many times, presumably as an adaptation to rapid water conduction. There is now even good reason to believe that vessels may have had multiple rather than few origins among the dicots (Schneider and Carlquist 1995). Thus vessels are an excellent example of analogous structures.

In the absence of a robust fossil record or a well-reasoned phyletic hypothesis based on comparisons among living organisms, three criteria or "tests" have been used to determine whether phenotypic correspondences are analogues or homologues. These tests ask if the structures have equivalent positions in the general organization of the organisms being compared, if they have equivalent special properties, and if they are morphologically or anatomically bridged by a series of structures with intermediate forms (Remane 1952; Hagemann 1975; Kaplan 1977). When applied to the parts of the flower (sepals, petals, stamens, and carpels) and the foliar leaves of vegetative stems, these three tests verify that these organs are homologous rather than analogous. Sepals, petals, and so forth initially develop from similar small "blebs" of cells called primordia, borne at the growing tip of the floral bud. Because floral primordia are organized on modified reproductive stems in much the same way that foliar leaf primordia are organized at the tips of vegetative stems, the organs that develop from them have equivalent positions in the general organization of the flowering plant body. The parts of the flower also are morphologically and anatomically similar to foliar leaves. These similarities may be obscured by the various ways sepals, petals, and so forth have been functionally modified, but all these organs share many equivalent "special qualities." Finally, intermediate forms between foliar leaves and sepals, sepals and petals, petals and stamens, and stamens and carpels are often found even on the same mature flower, leaving no doubt that these organs are developmental variants of the same basic organ type. The terms "homonome" and "serial homologues" are sometimes used to describe the situation seen for structures like leaves. Both refer to structures that share the same general developmental pathways but are found on different parts of related organisms (Riedl 1978).

In theory, morphological or anatomical correspondences between two or more organisms can encompass the entire phenotype. Complete phenotypic correspondence is exceptionally rare, however, first, because selection pressures tend to have differential effects on the various phenotypic features making up the whole organism, and second, because the phenotypic target of selection may shift with the passage of evolutionary time as different selection pressures affect the same organism. As a consequence, most species may be thought of as phenotypic mosaics, with each component reflecting a different facet of history. Mosaic evolution—different phenotypic features evolving at different tempos and in different directions—has happened frequently in plants (e.g., Knoll et al.

1984). One of the consequences of mosaic evolution is that homologous and analogous features can appear in the same organism.

Consider the phenotypic correspondences between the desert species of New World Cactaceae and Old World Euphorbiaceae. Members of both these dicot families adaptively radiated into arid environments and were subjected to similar selection pressures that evoked strikingly similar features. Some of these features are homologues, while others are best considered analogues. For example, the stems of some desert cacti and euphorb species are succulent, storing appreciable amounts of water in fleshy pith and cortex cells. Because the same tissues are used for water storage, the succulent stems of cacti and euphorbs are homologues that have been amplified by convergent evolution, as they are in many other desert-dwelling species (fig. 7.2). Likewise, the stems of cacti and euphorbs are often short and broad and little branched, which reduces the ratio of surface area to volume. Perhaps because they require less mechanical support, the stems of many species produce comparatively small amounts of wood from a vascular cambium that takes the form of narrow strips of tissue. As a consequence, the wood in their stems often has the appearance of vertical "rods" or interweaving "networks" (Mauseth 1988). Other phenotypic correspondences between cacti and euphorbs are analogues rather than homologues. For example, desert cacti often bear spines that are modified leaves, whereas desert euphorbs bear thorns that are modified branches. Both spines and thorns serve the same function—they convectively dissipate heat, and they deter herbivores from stems. But spines and thorns are developmentally derived from very different organs (leaves and branches, respectively), and so they are functionally analogous, not homologous.

The analogous and homologous phenotypic correspondences seen between some desert cacti and euphorbs raise the more general issue of whether it is proper to speak of divergence, parallelism, or convergence in terms of phyletic trends seen in individual characters or whether these terms should be reserved for evolutionary trends at the level of the entire phenotype. If the biologist elects "parts" over the "entire" organism, then the three phenotypic trends can obviously occur simultaneously in the same lineage. Recurrent examples of plant mosaic evolution show that the individual characters of some species can diverge from, converge on, or parallel those of other species. From a purely practical perspective, if we apply terms like convergence only to the entire phenotype, these terms could hardly ever be legitimately used, because different parts of the same kind of organism have undoubtedly evolved in different directions or at different rates for most species. But the trends seen in different organs borne by the same organism are often correlated, because natural

Figure 7.2. Succulent stems of desert-adapted species in three very different flowering plant families. From left to right: *Euphorbia* sp. (Euphorbiaceae), *Senecio stapeliformis* (Asteraceae), and *Stapelia* sp. (Asclepiadaceae).

selection acts on the entire phenotype and because the developmental pathways engendering different phenotypic features tend to be coordinated (and so correlated with one another). Consequently many, albeit not all, of the phenotypic features of an organism are similarly affected by the same selection pressures. In light of this a not too strict perspective is desirable, one that admits the freedom of different parts of the same plant or animal to change at different rates and in different directions depending on the intensity and number and kind of selection pressures experienced, yet that necessarily preserves the conviction that the entire organism is the true vehicle of evolution.

Divergence and Adaptive Radiation

Opportunities for phenotypic divergence and adaptive radiation often occur when organisms gain access to geographic areas that are wholly unoccupied or poorly colonized by other species (Carlquist 1980). Isolated from the home range of their main populations and unfettered by competition with members of their own or other species, a few isolated organisms may quickly establish splinter populations that rapidly diversify and evolve into new species through the combined effects of natural selection and the random processes of mutation and genetic drift. Some of the best examples of phenotypic divergence are those occurring on islands geographically well distanced from mainlands, like the Hawaiian archipelago (Carlquist 1965, 1980). As I noted in chapter 2, the oldest of the eight major Hawaiian islands, Kauai, is from 3.8 to 5.6 million years old. The youngest island, Hawaii, is roughly 1 million years old (fig. 7.3). These dates, however, do not necessarily mean that allopatric speciation has occurred in the window of 1 to 7 million years, because the history of the Hawaiian island chain extends back at least 28 million years and because some species may have evolved in the distant past only to successively hopscotch from older to younger islands. Some species are endemic only to comparatively young islands like Maui, however, and so must be less than 7 million years old. Another line of evidence for the recency of some Hawaiian speciation events is the ability of many Hawaiian species to hybridize without obvious interspecific sterility (see Carlquist 1995). Thus it is fair to say that phenotypic divergence and speciation (perhaps even transspecific evolution) can occur in what amounts to a geologically brief period.

Numerous examples of phenotypic divergence and adaptive radiation within and among closely related genera abound on the Hawaiian islands. Two examples, the Hawaiian alanis (*Pelea,* in the Rutaceae) and the silversword complex (*Dubautia-Argyroxiphium-Wilkesia,* in the Asteraceae), suffice to illustrate the diversity in growth habit and size that can be achieved in a comparatively short time within an initially small taxon (Carlquist 1980). *Pelea* species occupy the dry forest to high, wet forests and bogs (*P. waialealae*) and range in size from diminutive shrubs (*P. orbicularis*) to trees (*P. barbigera*), with one species achieving a vine growth habit (*P. anisata*). But regardless of how impressive it may be, the ecological and morphological diversification of *Pelea* species is nothing

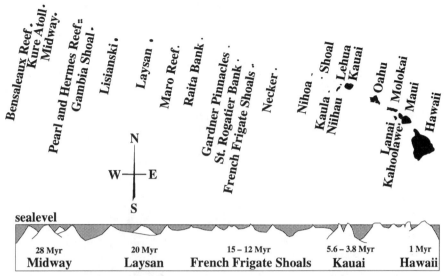

Figure 7.3. Geography and age of the Hawaiian archipelago. The Hawaiian island chain resulted from volcanic eruptions and is a continuous oceanic range of mountains extending from the southeast to the northwest (upper panel). Remnants of the oldest islands are mere atolls and reefs that lie to the northwest of the most recently formed island, Hawaii. Potassium-argon ratios (lower panel) indicate that Hawaii is approximately 1 million years (Myr) old; the oldest of the eight major Hawaiian islands are between 3.8 and 5.6 Myr old, but the history of the emergent chain of atolls and reefs extends back well over 28 Myr. Adapted from Carlquist 1980.

compared with what occurred in the silversword complex (fig. 7.4), which outclasses even Darwin's finches as a classic example of adaptive radiation (Carlquist 1965, 1980). The silversword complex undoubtedly traces its ancestry to some kind of tarweed, originally from the northern coast ranges of California, that colonized Kauai about 4 million years ago (Baldwin and Robichaux 1995). The mainland tarweeds, such as *Madia* and *Raillardiopsis,* tend to be sprawling, narrow-leaved perennial plants that are woody at the base of their stems and bear clusters of yellow flowers. Many of the Hawaiian *Dubautia* species share these features (*D. linearis*), and so we have a faint inkling of what the *Dubautia* ancestor may have looked like roughly 4 million years ago. However, other *Dubautia* species have evolved the tree growth habit (*D. reticulata*).

The silverswords and greenswords (placed in the genus *Argyroxiphium*) and the iliau (*Wilkesia gymnoxiphium*) are closely related to *Dubautia,* as witnessed by a naturally occurring hybrid between a single pair of *Dubautia* and *Argyroxiphium* species. Some silverswords grow in bogs (*A. caliginii*), while others grow at high elevations on volcanic cinder cones

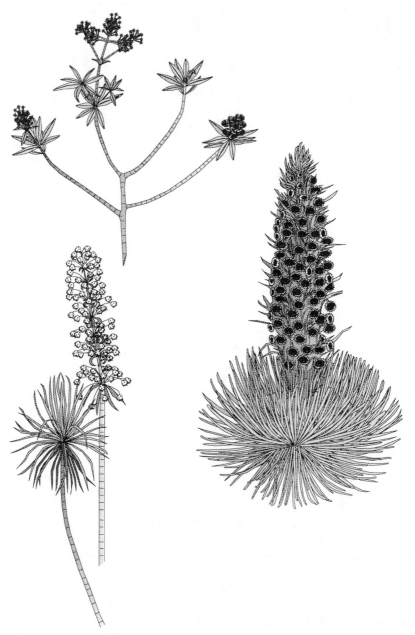

Figure 7.4. Divergent evolution in the *Dubautia-Argyroxiphium-Wilkesia* complex of Hawaiian plants. From top to bottom: *Dubautia linearis*, *Argyroxiphium sandwichense*, and *Wilkesia gymnoxiphium*.

(*A. sandwichense*). Many species are beautifully adapted in different ways to these divergent habitats. For example, juvenile *A. sandwichense* plants grow as deep-seated rosettes whose gently recurving spearlike leaves are densely covered with plant hairs called trichomes. This growth habit and leaf covering cope well with the freezing night temperatures, high wind speeds, water stress, and intense ultraviolet radiation that occur at high elevations. Additionally, spaces within leaves are filled with a gelatinous substance that can store large amounts of water when it occasionally rains. This water is used during dry periods for vegetative growth, and it is recruited from leaves when the stems of mature plants bolt to form massive inflorescences covered with numerous maroon flowers. The plants die after the flowers release their seeds.

The greenswords mimic the silverswords in growth form, but their leaves lack the felty covering of plant hairs. One species, *A. grayanum*, grows in bogs. Another species is found growing on cinder cones (*A. virescens*), a habitat shared by their silverword compatriots.

Unlike the silverswords and greenswords, *Wilkesia gymnoxiphium* grows in moderately dry scrubby forests and has grasslike leaves arranged in whorls (rather than in spirals as in *Argyroxiphium*; see fig. 7.4). The leaves are clustered at the top of a gangling polelike stem that can be woody at the base and reach 3 m or more in height (Carlquist 1980). Like the silverswords, *Wilkesia* plants die after flowering; that is, they are monocarpic.

Although the term "adaptive" radiation is often used to describe the rapid divergence of closely related species, Sherwin Carlquist (1980) points out that many of the morphological and anatomical features of some island plants have no immediately obvious adaptive value and may simply be expressions of diversification unfettered by selection pressure ("diffusive evolution"; see chapter 5). This is a subtle but important distinction because the process of adaptation can be understood (and legitimately said to occur) only in the context of environmental sorting of variants within populations. If no sorting occurs, then "adaptive" radiation sensu stricto cannot be said to have occurred. For example, the tree growth habit is often achieved by plants descended from originally herbaceous species isolated on islands. But it is naive to believe a priori that the tree growth habit is invariably an expression of adaptive convergence. In the absence of proof, it is just as likely to be the by-product of unrestricted plant growth in a habitat favored with abundant water, equitable temperatures, and the absence of large herbivores. A wet forest habitat can foster unfettered amplification in overall plant size, but this affords no evidence that this habitat environmentally sorted against small plant size (as evinced in some habitats by the abundance of cohabiting herbaceous and shrubby plant species). With characteristic clear logic, Carlquist notes that the term "adaptive" radiation should be reserved to describe phenotypic divergence resulting from clearly identifiable selection pressures sorting for particular variants. Because disruptive selection is more likely to occur when closely related species invade different habitats, we may wish to distinguish two types of radiations—those that occur within one habitat and those that occur across two or more habitats. Within a single, presumably ecologically fine-grained habitat, diffusive evolution may obtain phenotypic divergence among closely related species as a result of weak selection pressures. In contrast, a radiation across many habitats affords an opportunity for disruptive selection and can result in phenotypically divergent species, warranting the epithet "adaptive" radiation.

Convergence on the Tree Growth Habit

Phenotypic correspondence among unrelated species provides strong circumstantial evidence for adaptive evolution because it shows that organisms differing in genetic and developmental capabilities can converge on comparable solutions to life's exigencies when confronted with the same or very similar selection pressures. Here we shall consider a much touted example of convergent evolution—arborescence, the tree growth habit.

Strictly speaking, a tree is any perennial plant with a permanent, woody, self-supporting main stem or trunk, ordinarily growing to a considerable height, and usually developing branches at some distance above the ground. The tree growth habit therefore involves many rather than a few phenotypic features and so appears to serve as an example of the molding of the entire phenotype by selection pressures. But what are these selection pressures? As I noted in the previous section, habitats favoring luxuriant plant growth may permit the tree growth habit, but they do not invariably select for arborescent growth forms. As we shall see, however, the tree growth habit has been achieved by numerous plant lineages and is associated with biomechanical traits favoring the elevation and display of photosynthetic and reproductive organs. Arguably, convergence among so many different plant lineages for many biomechanical traits rather than only one or a few important ones strongly suggests that these traits are the result of adaptive evolution and not fortuitous expressions of diffusive evolution.

Arborescent species are found in many families of flowering and non-flowering plants. Tall and branched dicot tree species like oak and maple and gymnosperm trees like pine and spruce are familiar sights in northern latitudes. The trunks of all these plants are invested with an outer corky layer of tissue (phellem), produced by a cork cambium (phellogen), surrounding a vascular cambium that produces secondary xylem or "wood" to the inside of the stem and secondary phloem to the outside (fig. 7.5). The successive layers of secondary xylem and secondary phloem produced by the vascular cambium are laterally spanned by vascular rays that provide avenues for the radial transport of water and other materials. Regardless of their differences in size and shape and despite their significant reproductive differences, the trunks of flowering plants and gymnosperms all share the same basic developmental pathways and mature anatomical construction.

In contrast, in many significant respects the stems of other types of trees differ anatomically and developmentally from those of oaks and pines. For example, the bulk of cycad stems consists of a massive inner pith and outer cortex through which vascular leaf traces pass and ultimately depart on the opposite side of the stem to which each leaf is attached (fig. 7.6). Instead of a single vascular cambium, successive vascular cambia are produced from the inside toward the outside of the cycad stem. These cambia are concentrically arranged, much as they are in a beet stem. Regardless of the number of vascular cambia produced, the trunks of even very old cycads contain comparatively small amounts of mechanically soft wood and significant amounts of pith and cortex. In lieu of stiff wood, the principal stiffening agent of the cycad stem is an external layer of persistent leaf bases that are mechanically very strong yet lightweight. Despite the paucity of stiff wood and the presence of a heavy, water-laden pith and cortex, the stems of some cycads can reach considerable heights, rivaling those of many dicot trees. The Cuban cycad *Microcycas calocoma* can reach 9 m in height. *Macrozamia hopei*, a cycad native to Queensland, Australia, can grow 18 m high.

Because of their anatomical and morphological differences, cycads and oak trees appear to be adaptively convergent. But appearances can be deceiving. Some evolutionary biologists believe that cycads and all other seed plant lineages ultimately trace their evolutionary origin to the same last common ancestor, an extremely ancient, now extinct group of Devonian plants called the Progymnospermophyta (see fig. 4.15). Some members of this ancient plant group had the ability to produce copious amounts of wood from a vascular cambium. If, as some believe, all seed plants evolved from the progymnosperms and none lost the ability to produce a vascular cambium, then the different expressions of tree growth habit seen among seed plants are the consequence of adaptive divergence

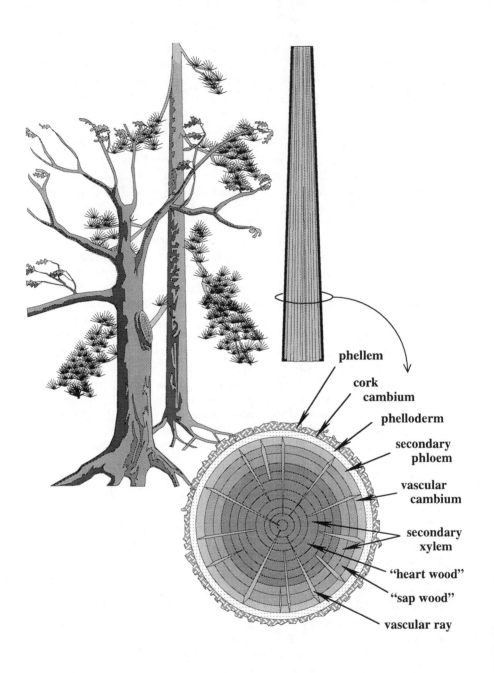

phellem

cork
cambium

phelloderm

secondary
phloem

vascular
cambium

secondary
xylem

"heart wood"

"sap wood"

vascular ray

rather than true evolutionary convergence. Alternatively, some evolutionists believe that the angiosperms reflect a seed plant lineage that lost the capacity to produce a vascular cambium, only to reattain this feature later in history. If this evolutionary scenario proves correct, then the tree growth habit of dicots is homologous neither with that of the progymnosperms nor with that of other living seed plants.

These opposing evolutionary scenarios pose little difficulty for our present purposes because the fossil record clearly shows that the tree growth habit has evolved independently in the lycopod lineage and in the horsetail lineage as well as in the seed plants. The lycopods are an extremely ancient group of plants, descending from a Devonian group called the Zosterophyllophytina (see fig. 4.15). Evidence for this descent comes from the anatomy of stems and reproductive structures. When anatomical details are sufficiently preserved, the fossil stems of ancient lycopods are seen to possess a central strand of primary xylem. The first xylem cells to differentiate and mature (protoxylem cells) occur on the outside of the lycopod vascular strand. The innermost cells of the strand differentiated and matured into primary xylem cells later in ontogeny (metaxylem cells). This "outside to inside" (or exarch) pattern of xylem maturation is also seen in well-preserved zosterophyllophyte stems. Exarch xylem maturation provides one line of evidence supporting the conviction that the lycopods ultimately trace their ancestry to some type of zosterophyllophyte (fig. 7.7). Another line of evidence is the shape of sporangia. Zosterophyllophytes and lycopods alike have kidney-shaped (reniform) sporangia that split apart laterally (much as the two valves of a clam separate) to release spores. Reniform sporangia differ in shape and mode of rupture from the sporangia produced by other plant lineages.

The most ancient lycopods were herbaceous, comparatively short plants lacking secondary xylem. By Carboniferous times some lycopods, collectively called lepidodendrids, grew to heights achieved by modern dicot and gymnosperm tree species. Lepidodendrids were a conspicuous floristic element of the Carboniferous landscape, and their fossil remains are often beautifully preserved because these plants grew in swamps especially

Figure 7.5. Rendering of the morphology and woody stem anatomy of angiosperms and conifers (e.g., oak and pine). The trunks of trees consist of concentric growth layers of secondary xylem (wood) beneath the living vascular cambium. The innermost and oldest layers of wood serve as mechanical supportive tissue (the heart wood). The outermost and youngest layers of wood conduct water (the sap wood). Materials can be transported laterally from one layer of wood to another by means of vascular rays. Secondary phloem is produced by the vascular cambium to the outside of the trunk. The external surface of the trunk is called the phellem ("corky outer bark"). The phellem is produced from a living layer of meristematic cells (the cork cambium; technically called the phellogen).

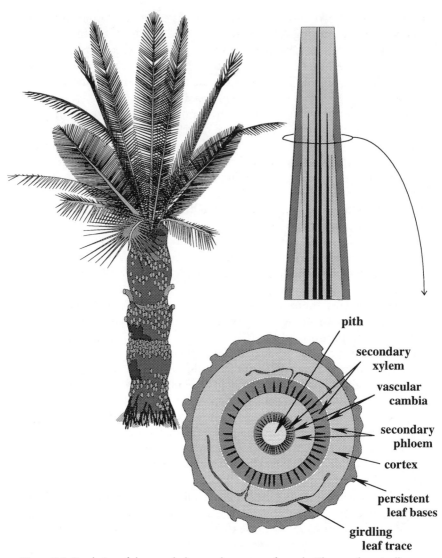

pith

secondary
xylem

vascular
cambia

secondary
phloem

cortex

persistent
leaf bases

girdling
leaf trace

Figure 7.6. Rendering of the morphology and anatomy of cycads. The cycads typically bear pinnately compound leaves. The bulk of their vertical stems consists of a massive cortex and pith. Vascular cambia develop in the stem and increase in number from the tip to the base of old stems. Each cambium produces a limited amount of comparatively soft wood toward the inside of the stem and a small amount of secondary phloem toward the outside. The stem is mechanically supported principally by an external layer of persistent leaf bases.

conducive to preserving plant tissues. The best-known lepidodendrid is the genus *Lepidodendron* (fig. 7.8). The largest intact trunk of this plant measures approximately 35 m long and has a basal diameter of 2 m. Because the specimen is incomplete at both ends, we know it is only a portion of a larger organism. From fossil remains, it is also known that the massive trunks of *Lepidodendron* branched dichotomously and produced a crown of numerous smaller branches supporting leaves and well-organized clusters of sporangia called strobili. The leaves, fossils of which may measure as much as 1 m long, were deciduous, dropped from older portions of branches, and left behind conspicuous leaf cushions on the external stem surfaces. The basal portions of *Lepidodendron* consisted of dichotomizing axes that bore helically arranged, laterally attached leaf-like appendages that probably functioned as roots. Like foliage leaves, these roots abscised from older portions of the plant.

Anatomical evidence indicates that, unlike most present-day tree species, some *Lepidodendron* species were determinate in their growth in height. Plants of these species reached a developmentally predetermined height, produced reproductive organs, and subsequently died (just like modern silversword plants). Also, it is believed that the *Lepidodendron* tree had a unifacial vascular cambium. Unlike the bifacial vascular cambium of most living tree species, which produces secondary phloem and secondary xylem, the cambium of *Lepidodendron* produced only secondary xylem toward the inside of the stem (Eggert 1961). As *Lepidodendron* grew and reached its final height, the amount of wood produced by the unifacial cambium progressively decreased toward the top of the plant, so that the anatomy of terminal twigs maintained the appearance of juvenile stems. In this sense *Lepidodendron* "grew itself out" as it ultimately reached a developmentally predetermined maximum height. Also unlike the trunks of modern dicot and gymnosperm trees, that of *Lepidodendron* contained comparatively little wood. The bulk of the mature stem was composed of a massive cortex, whose outermost layer had longitudinally oriented, anastomosing bands of fibrous cells. The outer cortex of the oldest portion of stems developed into a barklike tissue called lycopod periderm. Despite the superb preservation of many *Lepidodendron* stems, it is not possible to determine the way lycopod periderm formed, but this tissue likely served as an important stiffening agent for the mechanical support of vertical trunks.

Thus the tree growth habit of lepidodendrids was achieved very differently from that of seed plants. Determinate growth in height, a unifacial vascular cambium, a paucity of wood, and a massive mechanically supportive periderm are among the features that distinguish lycopod "trees"

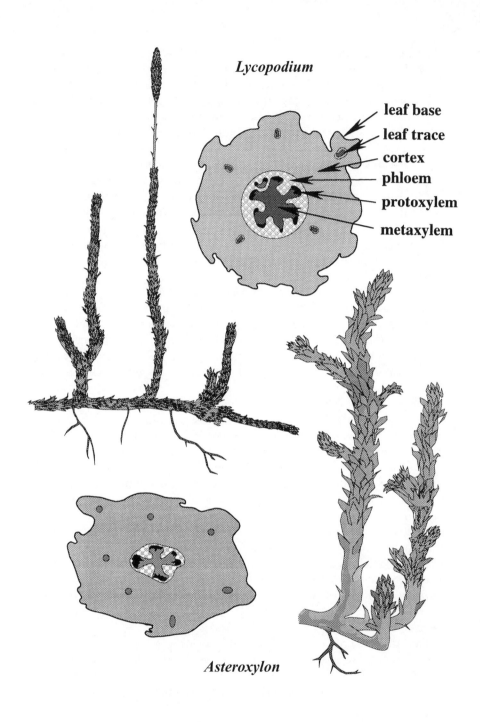

Lycopodium

leaf base
leaf trace
cortex
phloem
protoxylem
metaxylem

Asteroxylon

from other plants. Because they do not share an immediate ancestor, lepidodendrid and seed plant trees provide an outstanding example of convergent evolution, even to the extent of independently evolving leaves and roots.

The history of the horsetails, Sphenophyta (see fig. 4.15) parallels that of the lycopods. Both groups of plants first appear in the Devonian period, both attained their maximum species diversity during the Carboniferous, both evolved a tree growth habit, and both experienced a gradual decline in species diversity from the Carboniferous to the present, during which time arborescent species became extinct. The only living sphenophyte genus is *Equisetum,* whose twenty species are characterized by stems with distinct jointlike nodes bearing leaves or branches separated by intervening stem internodes (fig. 7.9). Anatomically, the stems of horsetails are unique among plants and serve to link living species with their ancient ancestors, whose remains bear striking phenotypic correspondences. Among these are the various air-filled canals running the lengths of stem internodes. The most prominent is a large central "pith canal." Vallecular canals in the cortex and carinal canals associated with vascular bundles complete the picture of the strangely beautiful anatomical organization that, among other features, distinguishes horsetails from lycopods, ferns, and seed plants.

The vertical stems of *Equisetum* are determinate in growth both in girth and in height, but their stature varies greatly among species. The shortest species is *E. scirpoides,* which reaches a maximum of 12 cm. The aerial stems of taller species, like *E. giganteum,* can reach heights over 4 m but are mutually supported by the interweaving of whorls of lateral branches. Vertical stems are interconnected by horizontally growing subterranean stems called rhizomes, which can continue to grow in length and to branch throughout the lifetime of plants. Fragmentation and transport of rhizomes by streams and rivers or soil removal may be the principal form of reproduction. Although vertical stems bear clusters of sporangia arranged in conelike strobili, the spores of *Equisetum* are short-lived, and the gametophytes they develop into are prone to desiccation in dry air.

Figure 7.7. Generalized morphological and anatomical comparisons between the living genus *Lycopodium* and the Devonian fossil plant *Asteroxylon.* Both of these plants are herbaceous and have a rhizomatous growth habit with horizontally and vertically growing stems. The stems bear small leaves through which runs a single vascular strand ("microphylls"). The vascular anatomy of the stems of both plants is characterized by exarch primary xylem differentiation (i.e., the first-formed protoxylem develops external to the second-formed metaxylem; see drawings of stem cross sections). Adapted from Bierhorst 1971, Gifford and Foster 1989, Stewart and Rothwell 1993, and Taylor and Taylor 1993.

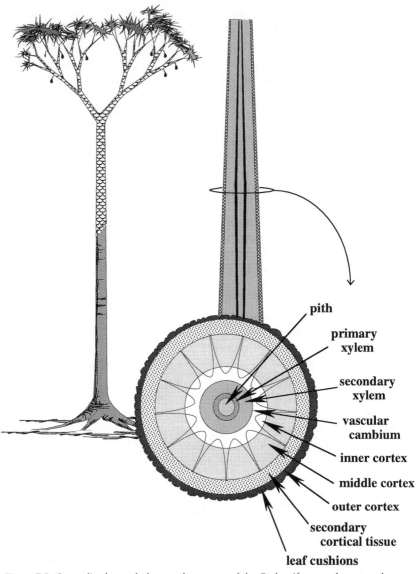

pith

primary
xylem

secondary
xylem

vascular
cambium

inner cortex

middle cortex

outer cortex

secondary
cortical tissue

leaf cushions

Figure 7.8. Generalized morphology and anatomy of the Carboniferous arborescent lycopod genus *Lepidodendron*. The canopy of the tree consisted of dichotomously branched
stems bearing large strap-shaped "microphylls" and strobili. The tree was supported at its
base by a system of dichotomously branched stems bearing rootlike modified leaves. The
unifacial vascular cambium of the trunk surrounded a comparatively small pith and produced secondary xylem but no secondary phloem. The bulk of the trunk consisted of an
inner and outer cortex beneath an external layer of persistent leaf cushions. Adapted from
Stewart and Rothwell 1993 and Taylor and Taylor 1993.

The ancestral group of plants from which the horsetails evolved is controversial. Some paleobotanists believe the sphenophytes trace their origins to the Hyeniales, a group of Devonian herbaceous vascular plants of which *Protohyenia* is an example (see fig. 7.9); others believe the Hyeniales are not closely aligned with the sphenophytes. But this controversy cannot obscure the fact that every reasonable candidate for the horsetail ancestor belongs to some kind of small herbaceous Devonian plant group that bears no immediate phyletic allegiance to the lycopod or seed plant lineages, and that the most ancient horsetails evolved from some kind of nonarborescent ancestor lacking a vascular cambium and the ability to manufacture wood.

The best-known arborescent horsetail is the fossil genus *Calamites* (fig. 7.10). The trunks of this plant could grow 20 m tall and were attached to large subterranean rhizomes that produced roots at their nodes. In some respects *Calamites* was very similar to modern *Equisetum*. Lateral branches and leaves were produced in whorls at nodes, although older branches bearing leaves may have been shed as *Calamites* trunks continued to mature and grow in height. *Calamites* also shares anatomical and developmental features with *Equisetum*, leaving no doubt that these two organisms are phyletically related. For example, the internodes of horizontal rhizomes and vertical trunks had a large centrally located pith cavity surrounded by a ring of longitudinally aligned strands of primary xylem with carinal canals. Like modern *Equisetum,* the aerial stems of *Calamites* were determinate in growth in height (Eggert 1962). The number of primary vascular strands decreased in each successive order of branching and was attended by a progressive decrease in the size of the associated pith cavity. As a consequence, the terminal twigs of *Calamites* had a small number of vascular strands and lacked a central pith cavity. Although we do not know for sure, *Calamites* may have continued to grow indefinitely in overall size by means of its subterranean rhizomes. Using modern horsetails as a scale for gauging overall size, we can estimate that the sporophytes of *Calamites* may have been the largest organisms that ever lived.

In other respects *Calamites* stems differed from those of living *Equisetum*. Wedge-shaped sectors of secondary xylem were produced from a vascular cambium external to the primary vascular strands. In some fossil specimens, sectors of calamite wood measure 12 cm thick (Andrews 1952). Much as in the arborescent lycopods, the vascular cambium in *Calamites* was unifacial, producing only secondary xylem toward the inside of the stem (Eggert 1962). The adaptive significance, if any, of this convergence is obscure. In primary growth, *Calamites* and *Lepidodendron* both may have produced amounts of primary phloem sufficient for

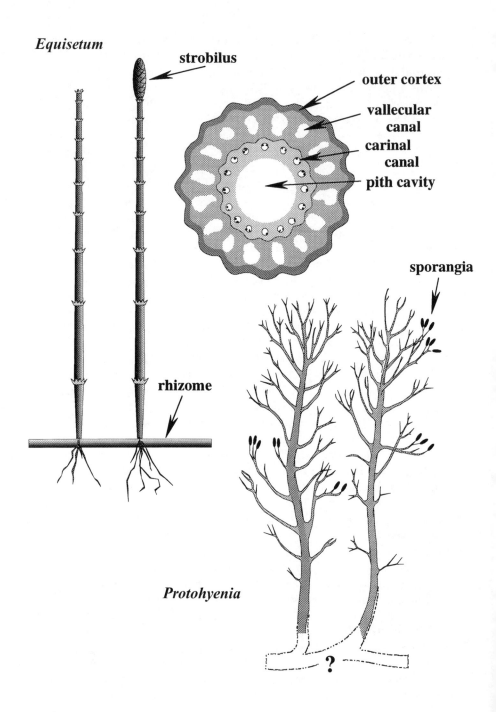

Equisetum

strobilus

outer cortex

vallecular canal

carinal canal

pith cavity

rhizome

sporangia

Protohyenia

?

the future physiological needs of the mature shoots that were determinate in growth. Regardless, it is clear that the trunks of *Calamites* and *Lepidodendron* were mechanically supported by different types of tissues (secondary xylem in the case of *Calamites;* periderm in the case of *Lepidodendron*) and differed dramatically from those of modern trees.

Other Kinds of Trees

The mature stem of the Andean wax palm *Ceroxylon* can be 50 m tall; the stems of the tree fern *Cyathea* are reported to grow to 24 m. Yet tree palms and ferns grow and develop in ways very different from *Lepidodendron, Calamites,* or modern dicots and gymnosperms. The ways arborescent palms and ferns achieve their great heights are worth reviewing because they provide excellent examples of convergent evolution between phyletically very dissimilar plant lineages.

Tree palms grow in height by means of a single apical meristem at the center of a crown of leaves whose bases typically wrap around the shoot apex (fig. 7.11). Each new leaf is initiated within the ensheathing leaf base of the next older leaf, expands and develops, and is eventually left behind by the subsequent growth of the apical meristem as it produces more new leaves. The old leaves of some palms cleanly abscise, leaving a leaf scar on the outside of the stem. In other species the remnants of older leaf bases persist for a time, enveloping the upper portions of the palm "trunk" in a comparatively stiff girdle of fibrous material.

The bulk of a palm stem is composed of parenchymatous ground tissue through which there interweave numerous primary vascular bundles associated with leaves (see fig. 7.11). These bundles persist throughout the lifetime of the palm tree and are the sole means of transporting water and nutrients. (The living phloem cells at the base of the stems of long-lived palms, like the royal palm, *Roystonea,* may be over a hundred years old.) The vascular bundles can be roughly divided into two classes that tend to take different courses as they run the length of the stem. One class of bundles has a peripheral course, hugging the perimeter; bundles belonging

Figure 7.9. Generalized morphology and anatomy of the living horsetail genus *Equisetum* and the morphology of a purported horsetail ancestor *Protohyenia*. The stem internodes of the rhizomes and vertical axes of *Equisetum* are perforated by longitudinally aligned chambers or canals and possess very little primary xylem (in the carinal canals). The anatomy of *Protohyenia*, a Devonian plant fossil, is not known. The arrangement of lateral branches, some bearing sporangia, is reported to be verticellate (arranged in whorls), which is reminiscent of the whorled leaves and lateral branches produced on the aerial stems of *Equisetum*. Adapted from Stewart and Rothwell 1993 and Taylor and Taylor 1993.

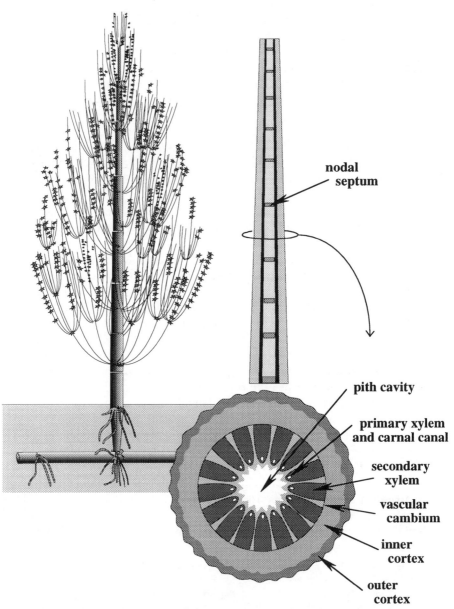

nodal septum

pith cavity

primary xylem and carnal canal

secondary xylem

vascular cambium

inner cortex

outer cortex

Figure 7.10. Generalized morphology and anatomy of the Carboniferous arborescent horsetail genus *Calamites*. The vertical treelike portions of this plant were attached to large horizontally growing axes. The stem internode anatomy is similar to that of *Equisetum* with the exception of the presence of secondary xylem (wood) in older portions of calamitean stems. Adapted from Stewart and Rothwell 1993 and Taylor and Taylor 1993.

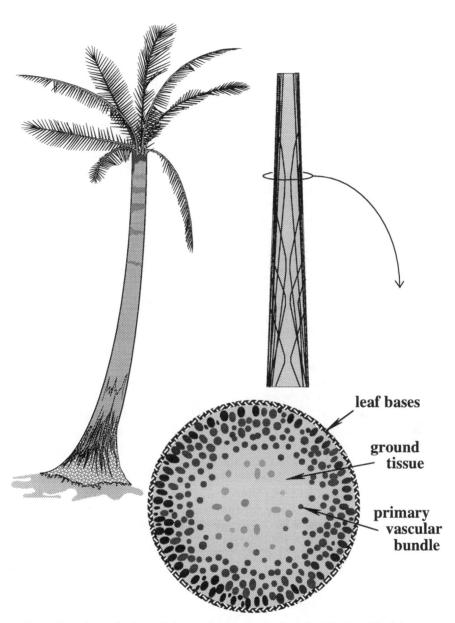

leaf bases

ground
tissue

primary
vascular
bundle

Figure 7.11. Generalized morphology and anatomy of a tree palm. The "trunk" of the palm is supported by a complex system of large and small primary vascular bundles whose number increases from the center to the perimeter of the stem. The bulk of the stem consists of ground tissue. The external surface of upper portions of the stem is enveloped in leaf bases.

to the second class diverge from the stem perimeter in their lower course, approach the center of the stem in midcourse, and approach the perimeter once again before they exit the stem into leaf bases (Zimmermann and Tomlinson 1965, 1967, 1974). Thus, meandering vascular bundles tend to be more crowded near the stem perimeter and more widely scattered and less numerous toward the center of the stem.

The texture and hardness of palm "trunks" tend to be correlated with the number of vascular bundles. Because the number of bundles is greater toward the perimeter than at the center of the stem, the outermost portion of the stem is stiffer and harder than the inner core. But the physical properties of stems are also related to the cell wall thickness of vascular and ground tissues and to the size and number of intercellular spaces within these tissues. Tall palm species tend to have stems with numerous vascular bundles, thick cell walls, and a densely packed ground tissue. In many of these species, the walls of cells increase in thickness with age (Schoute 1912; Tomlinson 1961), so stem stiffness and hardness increase toward the base of the "trunk." Indeed, the basal parts of very old palm stems have very much the same mechanical properties as wood (Rich 1986).

Palm "trunks" are often narrower at the top than at the base, but the degree of stem taper varies widely among groups of palms (Schoute 1912; Rich 1986). Sustained cell expansion is the underlying mechanism for increasing palm stem diameter, but taper can be augmented by the proliferation of adventitious roots, which form a "root tuft." Stem taper is mechanically desirable, especially for very tall plants. Its appearance in palms, conifers, dicots, and cycads is likely the result of adaptive convergence. Engineering theory shows that a tapered stem requires less tissue to construct and is more mechanically stable than a cylindrical stem of equivalent height and basal diameter. Also, theory shows that a tapered stem can have a homogeneous composition, whereas very tall untapered vertical stems require stiffer materials at their base to remain mechanically stable. Sustained cellular expansion was the adaptive route taken by palms. In contrast, the taper of woody stems arising from the growth of the vascular cambium is achieved by the telescoping of very young internodes well in advance of the appearance of the cambium in older portions of the stem. Because the vascular cambium produces layers of wood annually, successively older portions of the stem contain larger amounts of wood than younger portions. For these reasons the trunks of dicot and conifer trees are tapered in girth toward their shoot apical meristems. Palm stems cannot develop a vascular cambium but are tapered nonetheless by a very different developmental mechanism.

The trunks of some tree ferns are also tapered, but the mechanism responsible is very different from that of palm trees or dicot and conifer trees, as may be seen from examining the magnificent Upper Paleozoic tree fern genus *Psaronius* (fig. 7.12). In general appearance this fern looked much like that of a palm tree or an arborescent cycad. Fossils show that *Psaronius* grew to 8 m tall and produced a crown of palmlike leaves measuring as much as 1.5 m long. As older leaves died and dropped off, they left leaf scars on the uppermost portions of the stem. Secondary tissues are absent in the *Psaronius* stem, whose primary vascular architecture was extremely complex and strangely beautiful (fig. 7.12). In the absence of a vascular cambium and secondary stem tissues, some other mechanism for mechanical support must have existed. In *Psaronius* the "trunk" was draped with a mantle of branching adventitious roots whose overall thickness increased toward the buttressed base of the plant, giving it a tapered appearance. The outermost roots in the mantle, which had thick-walled and undoubtedly very stiff cells, were bonded to form a mechanical support system.

Another way to build a tree is shown by the North American fossil genus *Tempskya,* which grew during Cretaceous times (fig. 7.13). This plant is known only from the petrified trunklike fragments of its stems, some of which exceed 40 cm in diameter and may have grown to a height of 4 or 5 m. The "trunk" of *Tempskya* had an usual structure; it was composed of many small interweaving and branched stems bonded by adventitious roots. The "false trunks" of *Tempskya* are an adaptive convergence on the tree growth habit.

Living tree ferns also reach considerable heights, but their growth and development differ from those of ancient ferns like *Psaronius* or *Tempskya*. The stems of young plants are obconical, tapering to a point at both ends. Younger portions of stems are mechanically supported by a stiff outer cortex and a persistent armor of external leaf bases at their top and by adventitious roots developing within the stem and emerging from the cortex downward, where they eventually exit the base of the plant to form a stiff, densely packed columnar base of interweaving roots. Because roots rarely emerge from the perimeter of the stem, the "trunks" of most living tree ferns look untapered. With age, older portions of the tree fern stem die, and their tissues eventually decompose. As the stem continues to grow from its apical meristem, its living portion progressively grows upward and the height of the columnar base of roots steadily increases. If felled, the "trunk" of an extremely tall tree fern may consist almost entirely of roots.

Engineers have long known that the best location for mechanically supportive materials is just beneath the surface of a vertical column, where mechanical bending and torsional forces reach their maximum intensities. This strategy is particularly important when the quantity of the

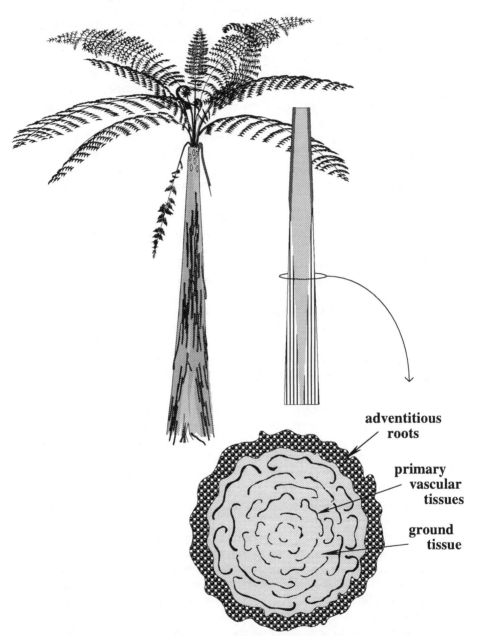

Figure 7.12. Generalized morphology and anatomy of the fossil arborescent tree fern genus *Psaronius*. The "trunk" of this plant lacked secondary tissues and was mechanically supported by a mantle of adventitious roots surrounding the primary stem. Adapted from Stewart and Rothwell 1993 and Taylor and Taylor 1993.

stiffest building material is limited, perhaps for economy in design. It is no coincidence that the stiffest tissues in a variety of plants tend to develop just beneath or very near the external surface of vertical stems. The persistent armor of leaf bases on cycad stems, the stiff lycopod periderm produced by the lepidodendrids, the internal girdle of tough vascular bundles in palm stems, and the pertinacious mantle of adventitious roots around the stems of *Psaronius* are but a few examples of this mechanical strategy. In each case the stiffest building material available is ideally located to provide stems with mechanical support (fig. 7.14). Clearly, the materials used to support stems differ among plant lineages—stiff leaf bases, periderm, girdling primary vascular bundles, draping roots, or a stiff outer cortex—and virtually every type of plant tissue and organ has been employed at one time or another. Presumably, in each case the material used was the best available option for adaptive convergence, illustrating Ganong's principle of "the path of developmental least resistance."

Reproductive Convergence

The seed plants, horsetails, lycopods, and ferns have converged reproductively as well as vegetatively. In each of these lineages, heterosporous species have evolved. Recall that heterosporous plants produce two kinds of spores, microspores and megaspores (see chapter 4). Even though the prefixes "micro" and "mega" imply that the principal difference between the two kinds is size, the sine qua non of heterospory is the evolution of unisexual gametophytes. Among modern vascular plants, experiments verify that the sex of each kind of gametophyte is determined by the metabolic microenvironment created within the sporangium as spores develop (Sussex 1966; Bell 1979). This microenvironment is undoubtedly determined by a number of factors, but the proximity of a sporangium to substances supplied by the sporophyte appears to be important. For example, when sporangia producing microspores and megaspores occur in the same strobilus, those containing microspores are typically located toward the tip of the strobilus, farthest from the substances supplied by the sporophyte, while sporangia producing megaspores develop toward the base of the strobilus, closer to the source of nutrients and hormones. Position effects like this provide evidence that the sexuality of spores and the gametophytes that develop from them is in part under the control of the sporophyte. In contrast, sexual dimorphism in bryophytes seems to be a consequence of sex chromosomes (Sussex 1966; Bell 1979).

In almost all known cases, heterospory among vascular plants is associated with micro- and megagametophytes that develop and mature almost entirely within the confines of their surrounding spore wall, a

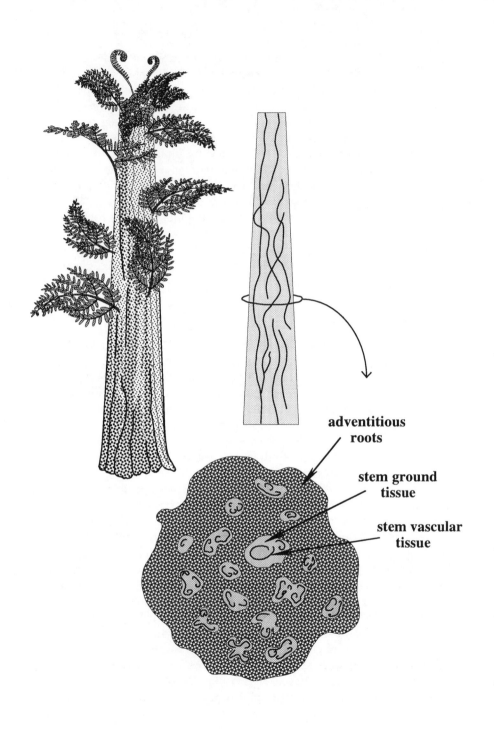

adventitious
roots

stem ground
tissue

stem vascular
tissue

condition called endospory, literally "within the spore." With few exceptions, endosporic gametophytes are nonphotosynthetic plants that fix little or no carbon, so an endosporic gametophyte depends almost exclusively on the nutrients initially supplied to the spore from its parent sporophyte. The difference in the size of microspores and megaspores is likely attributable to the metabolic self-sufficiency of gametophytes, on the one hand, and the different sexual obligations of endosporic micro- and megagametophytes on the other. Recall that the sexual task of the endosporic microgametophyte is completed once its antheridia mature and release sperm cells. Because the microgametophyte need not be long-lived, the spore it develops from does not require a large initial metabolic investment, hence its comparatively smaller size. In contrast, the sexual obligation of the megagametophyte is to produce archegonia and to nurture the developing sporophyte embryo after fertilization. Because the endosporic megagametophyte is a comparatively long-lived nonphotosynthetic organism, it likely requires a larger spore to contain a sufficient amount of nutrients to ensure its growth and survival and that of the embryo that will develop within it.

The adaptive benefits conferred by heterospory are controversial. One view is that unisexual gametophytes ensure outcrossing because sperm and egg cannot come from the same gametophyte. Thus the evolution of unisexual gametophytes ensures a high degree of heterozygosity in the next generation of sporophytes produced by sexual reproduction. This argument is somewhat countered by the fact that a temporal lag typically exists between the development of antheridia and archegonia on the bisexual gametophytes of homosporous species. This lag can have much the same genetic effect on heterozygosity because it ensures some outcrossing, especially in a closely spaced population of bisexual gametophytes differing in age and therefore in sexual maturation.

Another view is that the differential allocation of nutrients to microspores and megaspores is adaptively important because it enhances the probability that "female" gametophytes will survive and grow to sexual maturity. It also favors the subsequent development of the sporophyte embryo, and concurrently economizes on the nutrients invested in "male" gametophytes, which are dispensable once they release their sperm. Yet another adaptive benefit conferred by heterospory may be the rapid com-

Figure 7.13. Generalized morphology and anatomy of the fossil arborescent tree fern genus *Tempskya*. The "false trunk" of this plant consisted of adventitious roots surrounding vertically growing branched primary stems. Adapted from Stewart and Rothwell 1993 and Taylor and Taylor 1993.

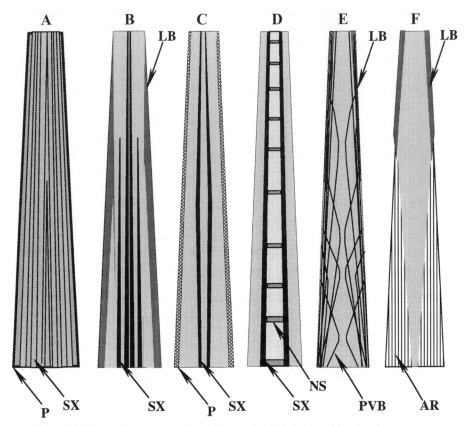

Figure 7.14. Comparisons among the various mechanical designs of fossil and extant arborescent plants. (A) woody dicots and conifers (see fig. 7.5); (B) cycads (see fig. 7.5); (C) Lepidodendrids (see fig. 7.8); (D) *Calamites* (see fig. 7.10); (E) tree palms (see fig. 7.11); (F) tree ferns (see figs. 7.12–13). Key to anatomical or morphological features: P = periderm, SX = secondary xylem, LB = leaf bases, NS = nodal septum, PVB = primary vascular bundles; AR = adventitious roots.

pletion of the sexual phase in the life cycle of the species (DiMichele, Davis, and Olmstead 1989; Niklas 1994). On the average, endosporic gametophytes take less time to develop and reach sexual maturity than the bisexual gametophytes of homosporous species that develop outside the confines of their spore walls (exosporic gametophytes). Recall that the gametophytes of free-sporing plants have a separate existence from the sporophyte generation. Thus gametophytes, whether unisexual or bisexual, are subject to desiccation and herbivory. Under these conditions, rapid closure of the sexual phase in the life cycle is very beneficial. In this regard it has been suggested that the evolutionary transition from homosporous plants with exosporic bisexual gametophytes to heterosporous plants with endosporic unisexual gametophytes involved the precocious onset of sexual maturity relative to vegetative maturity of the gametophyte generation (DiMichele, Davis, and Olmstead 1989). If so, then the evolution of heterospory involved progenesis, the acceleration of sexual maturity relative to somatic maturation (Gould 1977; Niklas 1994).

The rapid sexual development of endosporic unisexual gametophytes may influence the ecological requirements of heterosporous species, limiting them to aquatic or semiaquatic habitats as opposed to drier, more mesic conditions. The shedding of spores that rapidly grow and mature into unisexual gametophytes is ecologically equivalent to releasing sperm and egg cells into the environment (DiMichele, Davis, and Olmstead 1989). Because a time lag exists between the sporophyte's "reading" the environment for the availability of water and its forming and releasing megaspores and microspores into the environment, heterospory may be more successful in habitats where water is always reliably available. The authors of this hypothesis point out that, among modern heterosporous species adapted to drier habitats, most rely on asexual rather than sexual reproduction. However, a shift of emphasis from sexual to asexual reproduction attending a shift of habitat preference from wet to drier conditions is expected for homosporous as well as heterosporous species. All free-living gametophytes, regardless whether they develop endosporically or exosporically, depend on water for survival, growth, and successful fertilization. And among homosporous species adapted to drier conditions, most rely more heavily on asexual than on sexual reproduction.

The fossil record of sphenophytes and lycopods provides a far more convincing line of evidence that the evolution of heterospory is governed by the ecological preferences of species. For both lineages, the zenith of heterospory was achieved by the ecological dominants of the great Pennsylvanian coal swamp floras, the calamites and lepidodendrids. Among the sphenophytes, the fossil calamite cone *Calamocarpon* is the best example of heterospory (fig. 7.15). *Calamocarpon* cones, which have a wide geographic distribution in North American Pennsylvanian sediments, range from 4 to 12 mm in diameter and 8 cm in length. Two types of cones are known, monosporangiate and bisporangiate. Monosporangiate cones contain only one kind of sporangium, either micro- or megasporangia; bisporangiate cones contain both micro- and megasporangia. Despite these differences, all *Calamocarpon* cones have the same general morphology. Each consists of numerous whorls of bractlike leaves alternating with whorls of recurved sporangia-bearing branches called sporangiophores. Four elongate sporangia were produced per sporangiophore. The recurvature of the branched tips of sporangiophores oriented

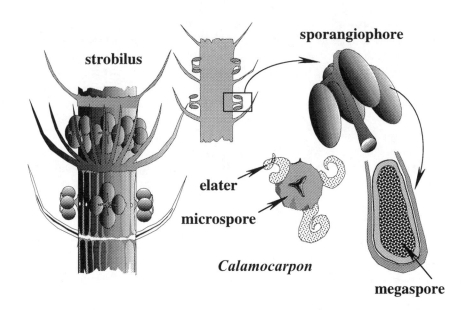

strobilus

sporangiophore

elater

microspore

Calamocarpon

megaspore

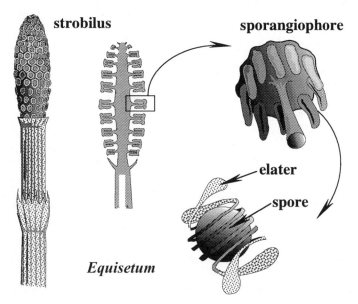

strobilus

sporangiophore

elater

spore

Equisetum

sporangia perpendicular to the longitudinal axis of the cone. Megasporangia are rectangular and measure up to 3 mm in length. A single functional megaspore was produced within each megasporangium. Well-preserved endosporic megagametophytes have been discovered within detached megasporangia, suggesting that the megasporangium and its one functional megaspore were shed as a single unit as cones matured. In contrast to the massive megaspores, the microspores of *Calamocarpon* are very small, 30–40 μm in diameter. The outer spore wall of microspores developed into three coiled winglike structures called elaters.

Although they differ in significant ways, *Calamocarpon* cones share two features with the strobili of modern *Equisetum* that leave no doubt these two genera belong to the same lineage. Both cone and strobilus have sporangia attached to highly reduced branches, forming sporangiophores arranged in whorls, and both produce spores whose outer walls are modified to form elaters (see fig. 7.15). The absence of whorls of bract-like leaves in the strobili of *Equisetum* may reflect a reductive evolutionary trend in sterile appendages associated with reproductive units. Elaters are hygroscopic structures, coiling and uncoiling in response to changes in relative humidity, much as a thermostat responds to changes in air temperature. Sporangia dehiscence exposes elaters to the external atmosphere, and by expanding and contracting, they literally push the closely packed spores within the sporangium into the air. Elaters derived from the spore wall are found nowhere in the plant kingdom except among sphenophytes. Unlike the elaters of *Equisetum* spores, which remain attached throughout the lifetime of the spore, those of *Calamocarpon* microspores appear to break off during or shortly after spore dispersal. Large numbers of calamite sporangia have been found containing detached elaters but no spores. Clearly the most significant departure of *Calamocarpon* from modern *Equisetum* species is that *Calamocarpon* was heterosporous whereas all *Equisetum* species are homosporous. But even among living sphenophytes a form of heterospory exists. Although all the spores of *Equisetum* have the same size and shape (morphological

Figure 7.15. Generalized comparisons between the reproductive organs of the fossil horsetail organ genus *Calamocarpon* and the living genus *Equisetum*. Both genera produce sporangia on modified branching systems (sporangiophores). The sporangiophores are whorled and organized into a conelike strobilus. Each sporangium is reflected backward toward the axes of stems. Spores possess structures called elaters that are derived from the outer spore wall. *Calamocarpon* was heterosporous; one large megaspore occupied the volume of each megasporangium, and the megaspore and megasporangium were often released as a single unit from the sporangiophore. Adapted from Stewart and Rothwell 1993 and Taylor and Taylor 1993.

homospory), some develop into antheridia-bearing gametophytes, which remain "male" throughout their lifetime, while other spores develop into archegonia-bearing gametophytes that can develop antheridia later if eggs are not fertilized (Duckett 1970a, b, 1977). In this sense modern *Equisetum* spores are functionally heterosporous, even though all spores have the same shape and size.

As I noted earlier, the *Calamocarpon* cones were mono- or bisporangiate, contained one functional megaspore per megasporangium, and when mature, released the megasporangium with its single functional megaspore as a unit. These features are also seen among the swamp-dwelling arborescent lycopods, the lepidodendrids. These lycopods produced their sporangia on leaves, called sporophylls, that were organized into well-defined conelike strobili (Phillips 1979). Like the heterosporous sphenophytes, some lepidodendrids produced monosporangiate "cones" containing either micro- or megasporophylls, but not both (e.g., *Lepidostrobus* and *Lepidocarpon*), whereas others bore bisporangiate strobili, typically with distal microsporophylls and more basal megasporophylls (e.g., *Flemingites*). Species producing bisporangiate strobili tend to be more ancient (e.g., *Cyclostigma*), suggesting that the monosporangiate strobilus is an evolutionarily derived condition in the lycopod lineage. Reduction in the number of functional megaspores per megasporangium is another evolutionary trend seen the lepidodendrids. For example, the Upper Devonian genus *Cyclostigma* produced approximately twenty-four megaspores per megasporangium. In contrast, the megasporangia of the Pennsylvanian *Lepidocarpon* contained a single functional megaspore (fig. 7.16). *Lepidocarpon* represents the zenith of lycopod heterospory. The megasporophyll was shed with its attached megasporangium, containing a single functional megaspore. This reproductive unit appears to have been adapted for water dispersal (Phillips 1979). The lamina of the megasporophyll folded around the megasporangium to form a longitudinally oriented slit. The rest of the lamina was strap shaped and held at a vertical angle to the rest of the sporophyll. A keel-like outgrowth was produced at the junction (see fig. 7.16). When scale models of *Lepidocarpon* megasporophylls are dropped in water, the bulky megasporangium surrounded by the folded leaf floats beneath the water surface. The unfolded leaf lamina with its keel-like protuberance extends above the water. Much like the sail of a ship, the emergent portion of the megasporophyll could have functioned to convey the entire reproductive unit away from parent plants (Phillips 1979). It may also have aided in dispersal as an autogyroscopic "wing" much like a maple fruit samara (see fig. 6.22). The endosporic megasporangium could have been fertilized by sperm released in the water by floating endosporic microgametophytes.

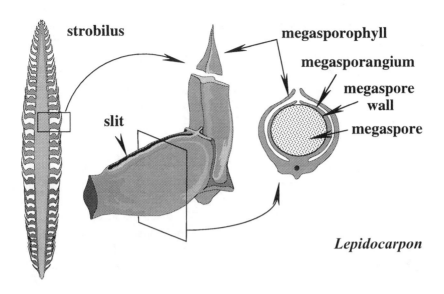

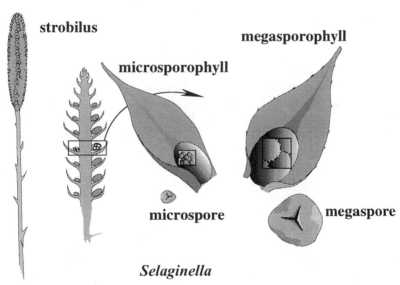

Figure 7.16. Generalized comparisons between the reproductive organs of the fossil organ genus *Lepidocarpon* and the living genus *Selaginella*. Both genera are heterosporous and bear micro- and megasporangia in conelike strobili. A single functional megaspore was produced in the megasporangia of *Lepidocarpon*. The lamina of *Lepidocarpon* was folded over the megasporangium to form a slit; the distal portion of the megasporophyll formed a sail-like structure that may have aided dispersal in water. Adapted from Phillips 1979 and Gifford and Foster 1989

Sperm could gain access to the eggs of a megagametophyte by swimming through the slit formed by the enfolding portions of the leaf lamina. Beneath this slit, the megasporangium opened to expose the gametophyte and its egg within. Beautifully preserved multicellular embryos have been found within the megagametophytes of *Lepidocarpon* (Phillips, Avcin, and Schopf 1975). The surfaces of these fossil embryos are highly convoluted, presumably an adaptation for absorbing nutrients from the endosporic megagametophyte.

Lepidocarpon has no counterpart among living lycopods, but modern reproductive analogues exist for some of the more ancient lepidodendrids that produced bisporangiate cones and shed their megaspores and microspores. Perhaps the best example is the largely tropical genus *Selaginella,* which contains approximately seven hundred species, all small herbaceous plants. Much like the more ancient lepidodendrids, *Selaginella* species produce bisporangiate strobili (see fig. 7.16). Megasporophylls tend to be arranged along the lower side of horizontally oriented strobili, whereas microsporophylls tend to be along the upper side. There is some evidence that position influences the sexuality of the spores that develop within sporangia. When oriented vertically, strobili tend to produce megasporophylls toward their base, while microsporophylls occur toward the tip. The location of mega- and microsporangia within a strobilus therefore may be influenced by gravity as well as by proximity to the source of nutrients or hormones. Like some ancient lycopods, the number of megaspores per megasporangium is not fixed in *Selaginella,* and it can vary between forty-two and four, even among the strobili produced by an individual plant (Bold 1967). In all cases, mature megaspores and microspores are shed from their sporangia, just as they were among the more ancient lepidodendrids. Among *Selaginella* species, some mechanically eject spores from sporangia; in other species microspores and megaspores are wafted by air currents.

Convergence on the Seed?

All known seed plants are heterosporous organisms that retain the megagametophyte within the megasporangium. As a consequence, fertilization of the megagametophyte and development of the resulting embryo occur within sporophytic tissue, the megasporangium. An additional feature shared among all known seed plants is the formation of a single functional megagametophyte per megasporangium. Many of these features are seen in *Calamocarpon* and *Lepidocarpon*. *Calamocarpon* produced one functional megaspore per megasporangium, which developed into a single megagametophyte that was retained within the megasporangium when shed from mature cones. Although fossils have not been found, it is possible that the endosporic megagametophyte of *Calamocarpon* initially developed within the confines of the detached megasporangium. Unfortunately it is not known whether *Calamocarpon* embryos developed within the megasporangium. In the case of *Lepidocarpon*, convergence on the seed is clearer. A single functional megaspore developed within each megasporangium, and well-developed embryos have been found within megagametophytes retained in megasporangia. That the megagametophytes of *Calamocarpon* and *Lepidocarpon* were fertilized after megasporangia detached from the sporophytes does not exclude these reproductive structures from being considered convergent on seeds. In certain climates the seeds of the maidenhair tree, *Ginkgo biloba*, are shed before fertilization occurs, yet no botanist would dismiss the maidenhair tree as a bona fide seed plant.

Still another feature shared among all known seeds is the investment of the megasporangium with an integument, a sterile layer of sporophytic tissue that develops around it. The integument is not a complete investiture; its final form must provide sperm access to the eggs produced by megagametophytes. In the distant past, the integuments of some seed plants were morphologically ill-defined and the functional role of the micropyle was performed by the modified apex of the megasporangium (e.g., *Genomosperma kidstonii*; see fig. 6.21A). This morphology was abandoned, however, and among modern seed plants the integument forms a small pore, the micropyle, through which sperm cells may enter the seed either directly or indirectly. The integument and its micropyle have no analogue among fossil or living sphenophytes. The developing megasporangia of *Calamocarpon* were likely protected within the cone by whorls of sterile bractlike leaves. Among modern sphenophytes, sporangia are protected by peltate sporangiophores until spores are ready to be shed. In contrast, the megasporangia of some lycopods have an analogue to the integument. The megasporangium of *Lepidocarpon* was surrounded by the basal portion of the leaf lamina, which formed a slit providing sperm with access to the eggs of the megagametophyte. In this sense the case for convergent evolution on the seed is stronger for *Lepidocarpon* than for *Calamocarpon*.

In other respects the case for reproductive convergence among horsetails, lycopods, and the seed plants is unconvincing. The "seed concept" extends well beyond the mere appearance of reproductive organs and encapsulates both the modus vivendi and the modus operandi of sexual reproduction. The seed is an adaptive solution to life on land functionally analogous to the way amniotic animals reproduce. In both cases the maternal parent can provide water if sperm need it to swim to egg cells. The developing sporophytic tissues of the seed are modified to retain the megagametophyte and the subsequently formed sporophyte embryo, just as the womb retains the egg and embryo of mammals. By the same token, the egg and embryo are nurtured and protected from desiccation and predation within parental tissues until the juvenile reaches the requisite size and form to survive and grow outside the environment created within the parent. The evolution of the seed unquestionably released vascular plant reproduction from the ecological requirement for an external source of water for the growth and survival of the megagametophyte, the fertilization of its eggs, and the early development of the sporophyte embryo. Thus the seed opened the door for the adaptive radiation of vascular plants into habitats drier than those favoring the survival and reproductive success of free-sporing plants. It must be admitted that the reproductive organs of heterosporous calamites and lepidodendrids, which reached their zenith in *Calamocarpon* and *Lepidocarpon,* lack the adaptive piquancy intrinsic to the meaning of the seed concept. The view taken here is that these organs evolved as adaptations to sexual reproduction in predictably wet habitats rather than as a consequence of selection pressures favoring survival and reproductive success in mesic habitats. The magnificently preserved fossils of calamites and lepidodendrids indicate that these plants were highly adapted to living in the Pennsylvanian swamp environment and show no evidence of adaptive innovations for life in drier environments. Therefore, although they are superficially similar to the seed, the sophisticated morphology of the *Lepidocarpon* megasporophyll and its megasporangium is perhaps best viewed as adaptively crafted for the dispersal of megagametophytes, the fertilization of its eggs, and the growth and survival of the lycopod embryo rather than as an organ adaptively modified to metabolically nurture and protect these ecologically vulnerable components of the life cycle.

This view should not detract from the other remarkable convergences of the calamites and lepidodendrids. Both adopted an arborescent growth habit, reaching heights comparable to those of many very tall living woody dicot and conifer tree species. Both independently evolved wood produced by a vascular cambium. Both evolved leaves. Both evolved conelike reproductive structures in which sporangia were economically aggregated and propitiously arranged aerodynamically in a manner reminiscent of the seed and pollen cones of pine and spruce. Both evolved heterospory, and therefore both evolved unisexual, endosporic gametophytes. Both evolved a means to disperse their spores either by wind or by water. Both converged in the same kind of general environment, characterized by a readily available and dependable supply of water. And both suffered much the same evolutionary fate as their habitats dwindled in geographic extent and became drier during the close of the Carboniferous and the inauguration of the Permian period. The calamites and lepidodendrids were once the monarchs of the Paleozoic, towering above their contemporaries and growing and dispersing to form great forests that stretched over much of the globe. In addition to being ecological dominants, the calamites and lepidodendrids were also rich in species numbers. Despite this ecological and taxonomic success, global changes in Earth's biological and physical environment had an ineluctable effect on them. The fossil record reports a gradual decrease in the size of successively evolving species within each lineage and also a reduction in the number of surviving species. Today the sphenophyte and lycopod lineages are represented only by herbaceous species that survive and grow in comparatively restricted habitats, which are rapidly disappearing owing to human activity. If this trend persists, then the last vestiges of once grand lineages will surely become extinct. If there is a biological lesson to be learned from this, it is that the endgame of evolution, regardless of the myriad adaptive modifications that have permitted organisms to survive and prosper in the past, is death. Herein lies the most sober lesson we may learn from the study of the fossil record.

Gene Homologies and Archetypes

Modern molecular biology has an opportunity to resolve some of the most perplexing questions of comparative morphology—for example, why the basic body plans of major plant and animal groups are so few and why these "archetypes" are so morphologically conservative. This is perhaps best illustrated by the exciting discovery that organisms sharing the same basic body plans possess homologous gene sequences that regulate similar developmental systems. This discovery stems from detailed genetic and molecular studies of homeotic mutations. Recall (from chapter 2) that these mutations result in the transformation of one body part into a different but otherwise normal-appearing body part. In some animals homeotic mutants are often dramatic. The homeotic mutation of the fruit fly called *antennapedia* causes leglike appendages to grow where antennae normally appear, and another mutation, called *tetraptera*, causes a set of wings to appear in the place of halteres. Molecular comparisons among these and other homeotic fruit fly loci show that these genes all share a short sequence of highly conserved DNA consisting of

180 base pairs, called the homeobox. This homologous DNA sequence has been discovered in the gene loci of many kinds of animals (e.g., fruit flies, frogs, chickens, mice, and humans) that have similar biological functions. In all cases the homeobox encodes for a protein (consisting of sixty amino acids) similar to bacterial repressor proteins.

Homeotic loci have also been discovered in the flowering plants (see chapter 2). These loci all share a homologous DNA base-pair sequence, which is also approximately 180 base pairs long, called the *MADS*-box. The first homeotic plant gene discovered was the DEFICIENS gene (*DEF*) in the snapdragon. The isolation of this gene revealed a homology with genes known to encode for the yeast and human transcription factors (Sommer et al. 1990). Thus far only two yeast genes, called *MCM1* and *ARG90*, are known to be homologous with *MADS*-box genes. *MCM1* is a ubiquitous transcription factor that serves as an activator or repressor in many aspects of yeast development (e.g., the gene is involved in the determination of the yeast mating type), and *ARG90* acts as a repressor in the control of arginine metabolism. Larger families of *MADS*-box genes are present in humans (Yu et al. 1992), among the best known of these is the serum response factor gene, a transcription factor binding to the serum response element, a short sequence in the promoters of many mammalian genes.

The homeotic genes in plants and animals encode for transcription factor proteins that regulate manifold developmental systems (see Davies and Schwarz-Sommer 1994). Because mutations at homeotic loci can shunt the development of meristematic or embryonic cell clusters from one developmental fate to another, they can result in the replacement of one body part with another in the angiosperm or animal body plans. Although their mutations are largely dysfunctional, homeotic loci shared by otherwise diverse plants or animals shed light on why the structural features of flowering plants or animals are so conservative. *MADS*-box genes in the flowering plants and homeobox genes in animals suggest that each of these large groups descended from a last common ancestor. In each case the genome of the ancestor contained the homeotic genes seen in modern descendants. And in each case the homeotic genes, shared as a consequence of common ancestry, continue to regulate developmental pathways in ways similar to those used in the last common ancestor to each group. In this sense homeotic loci are genetic synapomorphies—ancient genetic fragments or "uralt-genes," if you will. Nevertheless, within each large group the developmental regulatory roles of uralt-genes underwent modifications as other parts of the genome evolved. Even though the angiosperm and animal archetypes persist as major structural themes, the details of each have been developmentally modulated in ways that

have led to the morphological and anatomical diversification and specialization of body parts.

A potentially important concept emerging from the study of homeotic
genes is that the developmental regulatory systems that homologous gene
sequences evoke may serve as a basis for defining "angiosperm" or "animal." Although the structures that develop along the length of the floral
axis or the animal embryo can differ in appearance from one species to
another, the gene sequences and the developmental signals they inspire
are essentially the same in each case. The likelihood that these genes and
the developmental systems they regulate are all very ancient gives cre-
dence to the concept of genetic-developmental "type" (e.g., "zootype,"
see Slack, Holland, and Graham 1993). Here, the notion is that the homologous gene sequences thus far identified are shared ancestral genetic
"traits" that simultaneously evoke and define the archetypes identified on
traditional morphological grounds.

This provocative suggestion raises the issue of whether homologies
should be based on genetic or morphological criteria. Consider that the
vessel-like cells reported for a number of fern, lycopod, and gymnosperm
species have been traditionally interpreted as convergent adaptations
because their development differs in subtle but presumably significant
ways from that of angiosperm vessels and because these cell types occur
in diverse groups of tracheophytes that must have phyletically diverged in
the ancient past. Yet based on what we currently know about homeotic
genes, it is not inconceivable that the developmental regulatory systems
responsible for these cells are controlled by homologous gene sequences
derived from the last common ancestor shared by all tracheophytes. If so,
then all vessel-like cells would be homologous in terms of genetic correspondences (i.e, "vesseltype").

But to what depths are we willing to plumb the genetic "memory" of
lineages and accept homologous gene sequences as evidence of structural
homology? Recall the criteria for determining homology—two or more
character states are homologues if they are derived directly from the
same previous character state, if they are each part of the same character
state transformation series, or if the most recent common ancestor also
had the character state. Do comparatively short homologous gene sequences that appear to have shuttled within genomes for hundreds of
millions of years pass the test of these criteria for morphological homology? The answer will likely be debated by scientists for years. But one of
the great lessons of evolutionary biology is that organisms differ largely
because of combinatorial variations of comparatively few features rather
than as a result of an unlimited capacity to endlessly add new features to
those that came before. Thus the diversity of life is more a consequence

of permutation than of simple addition (see Bonner 1988; Wagner and Altenberg 1996). If morphological diversity is a consequence of genetic permutation—the shuffling of the genetic deck of cards—and not an endless addition of new genes, and if it is true that there is no gene for a structure per se because each gene operates in a larger genetic milieu, then homologies seen at the level of short gene sequences can serve as evidence for common descent but not as evidence that the structures they evoke are the same, however alike they may seem.

8 Tempos and Patterns

We may think of the evolution of genetic systems as a course of evolution which, although running parallel to and closely integrated with the evolution of form and function, is nevertheless separate enough to be studied by itself.

G. Ledyard Stebbins

For an evolutionary biologist to ignore extinction is probably as foolhardy as for a demographer to ignore mortality.

David M. Raup

How fast and in what direction have plants evolved? This is a question asked in many different avenues of biological research. Yet as the title of this chapter implies, there is no simple answer. Plants have evolved at different rates and in different directions, even within the genetic confines of a single lineage. Thus the tempos and modes of plant evolution are as diverse as life itself. Nevertheless partial answers can be given, each necessarily emphasizing a particular group of plants and each restricted to a particular time in a lineage's history. The objective of this chapter is to examine some of the rates and patterns of genotypic and phenotypic evolution in an effort to better understand evolution as a whole.

There are three general approaches to determining the tempos and patterns of evolution. The first is to draw molecular comparisons among living species so as to gauge the rate and direction of genotypic change. Biochemical techniques allow the sequencing of gene products, such as proteins, and permit comparisons among the variants of the same gene product. Provided the divergence times of species can be inferred from the fossil record, rates of molecular evolution can be calculated. With the advent of modern molecular biology and the ability to directly sequence genes, even more detailed comparisons can be made among species. However, in this chapter we shall find that a remarkable feature of the ordinary plant cell is that it is a consortium of three genomes (the chloroplast, mitochondrion, and nucleus genomes) and that each of them has evolved in different ways. Thus the rate and direction of evolutionary change as gauged by the plant genome depend on which genome we choose to study. The strength of the molecular approach is that it offers a detailed rendition of genotypic change; its limitation is that comparisons can be drawn almost exclusively among living species.

The second, perhaps more traditional approach to determining the rate and direction of evolution employs morphological or anatomical comparisons among species. The rate of phenotypic evolution is determined from the number of differences that have occurred since species diverged from one another as inferred from the fossil record. Rates of phenotypic evolution can be calculated for individual traits or sets of traits, just as they may be calculated for individual genes or gene products or for pooled data from many genes or gene products. The strength of the phenotypic approach is that it can be applied to both fossil and living species. It also emphasizes the phenotype as the vehicle of evolution. Its weakness is that phenotypic changes do not provide insights into the rate of genomic change, which is often of great importance. Ideally, the molecular and phenotypic approaches should be used complementarily, but this requires a very broad scope of analysis that can be impractical. Also, even though it is possible to reduce a genome to its various genes or gene products and to deconstruct an organism into its various individual morphological or anatomical parts, experimental data continue to verify that different molecular and morphological characters can evolve at different rates and in different directions even for the same organism. Thus no single molecule or morphological character can serve as the sole basis for measuring the rates or patterns of evolution. Likewise, the results even from pooled data may be biased depending on the kinds of molecules or phenotypic traits selected for study.

The challenge to deduce evolutionary tempos and patterns from either molecular or phenotypic data is not easily met because there is no steadfast relation between genotypic and phenotypic change. Alterations in the appearance, structure, or behavior of any kind of organism are undoubtedly due to genotypic changes. Nevertheless, significant reorganizations of a genome may have little or no effect on the appearance of the organism, whereas dramatic phenotypic changes may result from seemingly small genic or chromosomal alterations (see chapter 2 for examples). For this reason cladistic studies based on molecular data sometimes give phyletic hypotheses that differ significantly from those based on morphological or anatomical data. Clashing patterns of molecular and phenotypic changes within the same lineage are yet another expression of mosaic evolution, the ability of different characters to evolve at different rates and in different directions. Fortunately mosaic evolution, which is not unusual in plants, has boundaries. Just as a gene can change to varying degrees before its gene product is functionally impaired, there also are limits to how far phenotypic traits can be altered before the survival or reproductive success of the entire organism is jeopardized. The organism is not a mere collection of individual genes, molecules, or phenotypic characters: it must be comprehended as an integrated whole whose survival and reproductive success are dictated by highly coordinated genetic and developmental efforts. The scope of permissible variation and the degree of coordination may be difficult to determine because they depend on the type of organism, the environment in which it survives and reproduces, and the point in an organism's history at which we choose to examine it. Thus the patterns of mosaic evolution are typically studied ad hoc. Nevertheless, these patterns give great insights into the integration of the genome and phenotype.

The third approach to understanding evolution is to catalog and study the birth and death of species. The rates of speciation and extinction provide powerful measures of the evolutionary fitness of organisms, which is greater than the sum of the contributions to fitness made by individual genes, gene products, or phenotypic parts. Natural selection acts on the phenotype, not on individual genes or individual morphological or anatomical features. Because each species represents an exclusive combination of traits, each one is a unique evolutionary experiment in survival and reproductive success. Studying the birth and death of species—the historical fates of phenotypes in their entirety—allows us to study long-term evolutionary rates and patterns. Thus the study of paleontology requires neither apology nor further justification.

Tempos of Gene Products

The tempo of genotypic evolution can be indirectly inferred from molecular differences in gene products, such as proteins, or determined directly from differences in the molecular structure of DNA. Alterations in the genetic code for a protein result in protein polymorphisms, that is, proteins that share the same general molecular structure but differ in their amino acid sequences. These genetic alterations result in DNA polymorphisms. Although differences in amino acid sequences result from differences in DNA nucleotide sequences, protein polymorphisms provide only an incomplete measure of the changes in the DNA molecule. Some DNA mutations cannot be detected from differences in protein structure because there is great genetic redundancy. To understand genetic redundancy, recall how DNA encodes the information for protein synthesis (see fig. 1.7). Each amino acid in a protein is specified by a triplet sequence of DNA nucleotides called a codon. Transcription of the DNA codons takes the form of messenger RNA codons. The nucleotide triplet in a DNA codon must be altered by genetic mutation to substitute one amino acid for another in a protein. This kind of mutation is called a base-pair substitution because replacing a nucleotide in one strand of DNA requires replacing its corresponding partner on the companion DNA strand. Base-pair substitutions can adversely affect the function of the protein, or they may have little or no effect on the protein's ability to perform its metabolic task because synonymous codons abound. As noted in chapter 1, synonymous codons encode for the same amino acid but differ in their triplet sequences of nucleotides. This genetic redundancy results from the fact that there are 4^3 or 64 triplet permutations, yet there are only 20 kinds of amino acids (there are many more codons than there are different amino acids).

Base-pair substitutions giving rise to synonymous codons are *silent mutations* in the sense that they go undetected by protein polymorphisms. For this reason some changes in the genetic code cannot be inferred from changes in gene products. Polymorphisms of proteins also fail to reflect genetic mutations in noncoding portions of the DNA molecule. One of the many surprises about eukaryotic DNA is that intelligible, encoding gene sequences called exons are interrupted by noncoding DNA sequences, or "gibberish," called introns. During transcription, introns are excised and exons are spliced together to create a continuous sequence of mRNA codons (fig. 8.1). The function of introns, if any, is unknown. It is possible that noncoding portions of the DNA may be important to the structure of eukaryotic chromosomes (see below). Alternatively, introns may reflect previously coding portions of DNA that have lost meaning over the long course of evolutionary history. Regardless of the function(s) of noncoding DNA, it is clear that many eukaryotic genes are "split genes" and that base-pair mutations altering their intervening introns likely have little or no effect on the molecular structure of gene products like proteins. Thus, in addition to base-pair mutations resulting in synonymous codons, silent mutations of introns cannot be detected based on protein polymorphisms.

Experiments verify abundant silent mutations in both noncoding and coding portions of DNA. Not unexpectedly, the nucleotide sequences of introns are, on the average, much more variable than the nucleotide sequences of exons. Because base-pair mutations in noncoding DNA have no known effect on gene products, they can be randomly maintained or lost in the genome. DNA sequencing studies reveal differences in the variation of exon nucleotide sequences specifying different portions of the same protein. Some but not all of this variation can be accounted for by base-pair substitutions resulting in synonymous codons. The remaining variation is in part due to amino acid substitutions that do not alter the

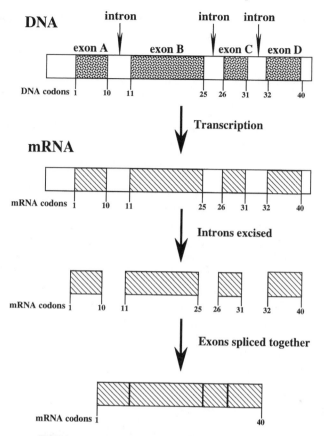

Figure 8.1. Introns and exons of a "split gene." Some eukaryotic genes consist of noncoding segments (introns) and coding segments (exons) of DNA. The introns and exons are transcribed onto the mRNA molecule during transcription. The introns are excised and the exons are spliced together before the mRNA exits the nucleus and mRNA translation occurs in the cytoplasm.

electrical charge or the activity of proteins. The amino acid sequence of a protein determines the three-dimensional active form of the protein. Provided amino acid substitutions do not alter the active form of a protein, these substitutions are functionally permissible and may be lost or fixed within the genome in a random way. How far a base-pair substitution is fixed or extinguished in a genome, therefore, is determined by the molecular constraints on the relation between the form and the function of a protein.

Molecular constraints on amino acid substitutions (and therefore genetic constraints on nucleotide substitutions) are apparent when we compare the tempos of change in different proteins or gene loci. For example, fibrinopeptide evolution occurs about 225 times faster than histone evolution (Dickerson 1971). This difference likely reflects how far the function of either of these two kinds of proteins is affected by amino acid substitutions. Fibrinopeptides have a comparatively generalized molecular structure and function little or not at all after they cleave from fibrinogen to yield the blood-clotting protein fibrin. The molecular constraints on fibrinopeptides imposed by amino acid sequences are comparatively small. Histones, on the other hand, bind to DNA along their entire length, thereby condensing the end-to-end length of the DNA molecule to one-seventh that of the unbound molecule and contributing to the nucleosomal organization of the eukaryotic chromosome (fig. 8.2). Changes almost anywhere in the amino acid sequence of a histone would likely alter binding with DNA, thereby affecting the organization of nucleosomes. For this reason the tempo of histone evolution is very slow (fig. 8.3). For example, among the five histone proteins, the amino acid sequence of the histone called H4 is virtually identical in the pea and the cow (De Lange et al. 1969; De Lange, Hooper, and Smith 1973). Clearly the molecular structure of H4 has changed little since plants and animals evolutionary diverged roughly 1.6 billion years ago. In contrast, fibrinopeptides have evolved faster and in apparently random and unpatterned ways (fig. 8.3).

The Chloroplast Genome (cpDNA)

As noted, the plant cell normally contains three genomes—the chloroplast genome (cpDNA), the mitochondrion genome (mtDNA), and the nuclear genome (nDNA). These three genomes interact with each other in numerous ways and have changed at very different rates and in different directions over evolutionary history. But in general terms the cpDNA and the mtDNA are the more conservative of the three genomes. Numerous comparisons among phyletically diverse species show that the genomes of these two organelles have undergone comparatively little molecular change over billions of years of evolution compared with the nDNA, which has evolved dramatically and may vary in size by several orders of magnitude even among closely related species of flowering plants.

Because of its smaller size and conservative evolution, the cpDNA provides a good starting point for comparing the three plant cell genomes. The chloroplast genome consists of a double-stranded circular DNA molecule that, unlike the noncircular DNA of the nucleus, is not bound to histones. The number and organization of cpDNA genes is very similar

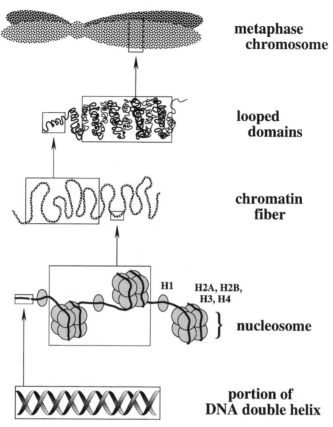

metaphase chromosome

looped domains

chromatin fiber

H1   H2A, H2B, H3, H4

} **nucleosome**

portion of DNA double helix

Figure 8.2. Histones and the nucleosomal structure of the eukaryotic chromosome at metaphase. From bottom to top: The double-stranded DNA molecule is wrapped around histone molecules that condense the great length of the DNA molecule. The nucleosomes are arranged in sequence along the chromatin fiber, which is further condensed in length by looped domains in the chromosome at metaphase.

even for phyletically diverse species like the nonvascular liverwort *Marchantia polymorpha* and the tobacco plant *Nicotiana tabacum*. Despite the vast phyletic gulf separating these two organisms, the chloroplast genomes of both plants carry very nearly the same complement of genes arranged in very nearly the same order (fig. 8.4). Both chloroplast genomes have two inverted repeat regions (denoted as IR_A and IR_B) separated by short and long single copy regions (SSC and LSC). The inverted repeat segments of the liverwort and tobacco cpDNA differ in size (10,058 base pairs versus 25,339 base pairs), but the principal difference in the gene sequence is an inversion of about 30,000 base pairs in the LSC region in the tobacco cpDNA. Both genomes contain roughly one

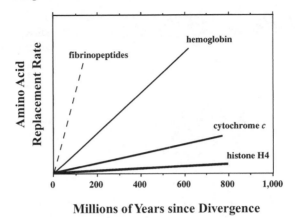

Millions of Years since Divergence

Figure 8.3. Comparison of the rates of amino acid substitution in four proteins. The re-placement rate is plotted against the divergence times of species inferred from the fossil record. The slope of the line drawn for each protein reflects the rate of amino acid substi-tution in terms of either the number of replacements per amino acid per year or the aver-age time required for one replacement per one hundred amino acid sites (a 1% change). Fibrinopeptides have the highest amino acid substitution rate; histone H4 has the lowest rate. Adapted from Dickerson 1971.

hundred different gene functions (Palmer 1985a, b, 1987). The liverwort cpDNA has genes for a complete set of four rRNA molecules and thirty-two tRNA molecules. It also has fifty-four genes encoding for known proteins (most of which are involved in transcription, translation, or photosynthesis) and at least twenty-eight more DNA regions (called open reading frames, or "orfs") that may code for other proteins that have not yet been identified. The tobacco cpDNA has fifty-two currently identified genes and thirty-eight unidentified open reading frames.

The size of the cpDNA varies somewhat, ranging from 85,000 to 195,000 base pairs, but is largely uniform among vascular plants, typi-cally falling between 120,000 and 160,000 base pairs (table 8.1). Some of this variation is due to differences in the length of the inverted repeat segments, which have "shrunken" or "expanded" to engulf genes from other parts of the cpDNA genome. The inverted repeat of tobacco cpDNA has incorporated the genes for two rRNA proteins and a tRNA that are adjacent to the repeat segment on the LSC region of the liverwort cpDNA. Species also differ in the number and size of introns, which are frequent in cpDNA genes. Some introns are much longer in some species than in others; other introns are invariant in size among species.

Another reason for differences in cpDNA size is that, among some plant lineages, many gene functions have been transferred from the cpDNA to the nuclear genome. Because many more genes are required for the

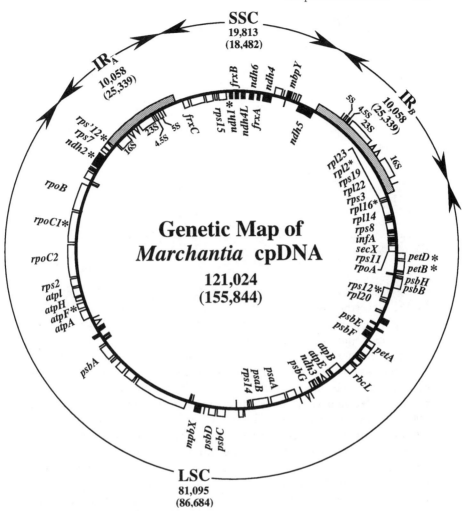

Figure 8.4. Genetic map of the chloroplast genome of the liverwort *Marchantia*. Base-pair numbers for regions of the *Marchantia* cpDNA are compared with those of tobacco (given in parentheses). The cpDNA is a circular molecule consisting of a short and a long single copy region (SSC and LSC, respectively) separated by two inverted repeat regions (IR$_A$ and IR$_B$). Asterisks indicate genes with introns. Adapted from Ohyama et al. 1986.

photosynthetic machinery than are currently found in the cpDNA, the transfer of genetic information from the cpDNA to the nDNA must have been considerable. For example, among the green algae and their descendants (the embryophytes), the genetic information required to construct the enzyme called Rubisco (an abbreviation for *ribulose-1,5-bis*phosphate

Table 8.1   Size of Some Chloroplast Genomes (cpDNA) in Thousands of Base Pairs (Kilobase Pairs, kbp)

Group	Species	Size
Green algae		
	Chlamydomonas reinhardii	195
	Codium fragile	85
Bryophytes		
	Marchantia polymorpha	121
Dicots		
	Epifagus virginiana (beechdrops)	71
	Nicotiana tabacum (tobacco)	155
	Spinacia oleracea (spinach)	145
Monocots		
	Triticum aestivum (wheat)	135
	Zea mays (corn)	139

Source: Data from a variety of sources.

carboxylase/oxygenase) is now shared between the cpDNA and the nDNA. Nevertheless, it must originally have resided in toto within the chloroplast genome, as it does in modern cyanobacteria. Rubisco is abundant in chloroplasts, making up about 15% of the total chloroplast protein. This enzyme is said to be the most abundant protein on earth because plants constitute over 90% of the world's living matter. Rubisco is the key enzyme in photosynthetic carbon dioxide fixation. It catalyzes the carboxylation of ribulose-1,5-bisphosphate (the first step in the Calvin cycle) and the oxygenation of ribulose-1,5-bisphosphate (the first step in photorespiration). Rubisco consists of two protein subunits that interact to form the active site of the enzyme. Amino acids on each of the two subunits are brought together by the three-dimensional folding of the two joined subunits to form the active site. Experiments verify that the cpDNA gene *rbcL* encodes for the larger of the two subunits of Rubisco. The smaller subunit of Rubisco is encoded by nDNA genes. Thus it is clear that it takes some coordination between the two genomes to assemble this important chloroplast protein.

No one knows precisely why portions of the cpDNA were transferred to the nuclear genome, as in the green algae and their descendants, but it seems reasonable to suppose that one selective advantage for gene transfer is that the nuclear genome acquired control over the metabolic activities of the chloroplast. According to the endosymbiotic theory (see chapter 3), the first chloroplasts evolved from photosynthetic prokaryotes that must initially have had self-sufficient genomes. The available evidence indicates that the ancestral host cell, which along with its endosymbionts eventually evolved into the green algae, was a heterotroph. This host genome could not have had genetic control over the photosynthetic machinery of its first photosynthetic endosymbionts. Such control could be achieved by the transferal of some genes of the chloroplast-like genome to the nuclear genome of the host cell, thereby establishing the important linkage between the two genomes required to coordinate the metabolic efforts of the cell as a whole. Another advantage of gene transfer may be "economy." Within each chloroplast, there exist multiple copies of the cpDNA that can account for a large percentage of the total DNA in some plants (e.g., the older chloroplasts of cells in rapidly expanding leaves can contain twenty-eight to thirty copies of cpDNA per plastid, and the cpDNA in leaves can account for 10%–20% of the total DNA). The transferal of some cpDNA genes (particularly large genes) to the nuclear DNA essentially reduces the total amount of DNA per cell.

In addition to transferal of genes, some cpDNA genes were also *lost* during the evolution of photosynthetic eukaryotes because photosynthesis and cytoplastic respiration share enzymes whose genes occur exclusively in the nuclear genome of all living eukaryotes (e.g., triose phosphate isomerase). Some cpDNA genes may also be lost when plants evolve a parasitic lifestyle, as is well illustrated by the flowering plant *Epifagus virginiana* (see fig. 8.9). The plastid genome of this parasite consists of 71,000 base pairs, containing only about forty-seven gene functions (roughly half the normal genetic complement). The cpDNA of *Epifagus* has lost most, if not all, of the thirty or more genes for photosynthesis and most of the large family of *ndh* genes, which are suspected to encode for the respiratory chain of the chloroplast (dePamphilis and Palmer 1990). Large portions of the *Epifagus* cpDNA must have been lost fairly recently, 5 to 50 million years ago, because the cpDNA of a related parasitic, albeit photosynthetic plant, *Striga asiatica,* has the typical complement of cpDNA genes for photosynthesis and plastid respiration. *Epifagus* and *Striga* belong to the same family (the Orobanchaceae in the order Scrophulariales), the oldest fossils of which are 5 million years old (the oldest fossil remains known for the Scrophulariales are 50 million years old). Because *Epifagus* and *Striga* shared an ancestor, the loss of numerous genes from the *Epifagus* cpDNA must have been comparatively rapid and probably was due to the relaxation of selection pressures on the requirement for photosynthesis once their common ancestor evolved a heterotrophic, parasitic existence. From a morphological perspective, it is interesting to note that parasitic flowering plants have highly reduced leaves in addition to a reduced cpDNA genome.

Variation in cpDNA gene evolution is apparent, even for the same loci among closely related plants. For example, the base-pair substitution rate of the *rbcL* locus is higher among grass species than other monocot species like palms (Wilson, Gaut, and Clegg 1990), and the nucleotide sequences of the *atpE* gene have changed more slowly in grasses than in dicots (Rodermal and Bogorad 1987). Although some genes evolve at similar rates in monocot species (e.g., rice and corn), the same genes evolve at very different rates in dicot species (Gaut, Muse, and Clegg 1993). These findings and others verify that cpDNA gene evolution is variable despite an overall staid and steady genome evolution (Bousquet et al. 1992; Gaut et al. 1992; Clegg et al. 1994).

The Mitochondrial Genome (mtDNA)

The organization and evolution of the plant mitochondrial genome contrast sharply with those of the cpDNA. The mtDNA generally consists of one large, double-stranded circular "master" chromosome attended by a constellation of smaller, circular DNA molecules. Because they encode for portions of the larger genome, the smaller satellite mtDNA molecules are believed to be derived from (and can recombine to form) the larger master chromosome. The size of the master chromosome and the number and size of satellite DNA molecules are extremely variable among species. Among angiosperms, the size of the mtDNA varies elevenfold (from 218,000 to 2,500,000 base pairs). This variation is attributable to differences in the number and size of direct repeat and inverted repeat base-pair sequences in the master chromosome in addition to differences in the number and size of the satellite DNA molecules. For example, the mtDNA master chromosome of corn has 570,000 base pairs (with five direct repeat sequences and one inverted repeat sequence). This chromosome is attended by six smaller satellite DNA molecules ranging from 47,000 to 503,000 base pairs in length. In the mustard (*Brassica campestris*), the master chromosome is only 218,000 base pairs long and is attended by only two other circular DNA molecules roughly 83,000 and 85,000 base pairs long (see Birky 1988).

Unlike the gene sequence of the chloroplast genome, which is fairly uniform among species, the limited number of gene maps for mtDNA indicates substantial variation in the arrangement of genes on the master chromosome (fig. 8.5). This variation is principally due to the physical exchange and rearrangement of homologous DNA segments among the various mitochondrial DNA molecules in a manner analogous to crossover in the nuclear chromosomes (see fig. 1.9). Crossover occurs with high frequency both within and among the DNA molecules of the mitochondrial genome.

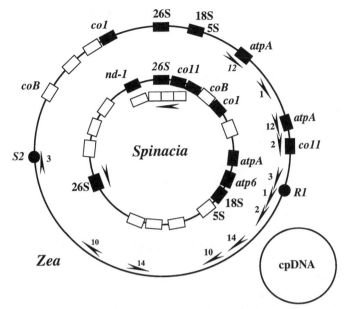

Figure 8.5. Comparison of the mitochondrial genomes of corn (*Zea*) and spinach (*Spinacia*). Genes are indicated by solid boxes, regions homologous with the cpDNA by open boxes. Arrows show repeated regions (numbers are kilobase pairs). Comparative size of the cpDNA is indicated by lower right circle. Adapted from Dawson et al. 1986 and Birky 1988.

Gene maps show that some mtDNA genes are homologous with those of the chloroplast genome, so portions of the chloroplast genome must have been incorporated into the mitochondrial genome as well as the nuclear genome. For example, homologous genes encoding for the large subunit of Rubisco and chloroplast rRNA and tRNA occur in the mitochondrial genome of corn (*Zea mays*). Although chloroplast genes may be transcribed in the mitochondrial DNA, their gene products likely are nonfunctional in mitochondria. In addition to chloroplast genes, a number of nuclear genes have also been found in mtDNA, indicating that some genetic material is promiscuous, shuttling among the three plant genomes in an apparently random way. Different kinds of organelles often come into contact within a cell, making possible the exchange of promiscuous DNA. Transposable elements ("jumping genes") similar to those found in nuclei may serve as the actual vehicle for genetic exchange. Jumping genes permit the movement of short DNA segments from one site to another on the same or different DNA molecules. Yet another mechanism for genetic transfer within and across different organelles consists of plasmids, small rings of DNA that carry accessory genes separated from bacterial chromosomes. Plasmids can associate with and dissociate from the mtDNA and may exchange genetic information between different kinds of organelles within and among plant cells. As yet, however, there is no good evidence for transposable elements in chloroplasts, nor is there good evidence for plasmids. Thus the mechanism(s) responsible for the transfer of genetic information among the three plant genomes remain unclear.

The Nuclear Genome (nDNA)

The eukaryotic nuclear genome consists of linear double-stranded DNA molecules bound to histones, organized into discrete nucleosomes within the chromosome (see fig. 8.2). Every eukaryotic species has a characteristic amount of nDNA. The amount in the haploid cell of a species is called the C-value and is expressed in units of base pairs (bp), picograms (pg), or molecular weight (daltons). One picogram equals approximately 10^9 bp or 6.4×10^{11} daltons. The range of C-values is greater among plants than among any other group of organisms (Lewin 1985). In flowering plants alone, C-values range from 3×10^8 to 1×10^{11} base pairs. The C-values of algae extend below 4×10^7 base pairs (fig. 8.6).

Some correspondence exists between the size of the nuclear genome and phenotypic complexity, but only when the broadest comparisons are drawn among unicellular and multicellular plants. It takes a minimum of 10^7 base pairs to encode for an alga, and a minimum of 10^8 to make a flowering plant. There is no correlation between the C-value and phenotypic complexity among multicellular organisms, however. For example, the C-value of corn is nearly identical to that of a human being. The lack of correlation between the size of the nuclear genome and the apparent complexity of the phenotype has been called the C-value paradox.

The C-value paradox results from the expectation that the size of nuclear genome will increase in proportion to organismic complexity provided the number of genes correlates with the amount of DNA in the nucleus (the C-value) and that the number of genes is proportional to organismal complexity. Leaving aside the ingenuous belief that "organismal complexity" can be defined, modern molecular biology has shown that the first of these two assumptions, that nuclear genome size correlates with gene number, is false. Among eukaryotic organisms, genes make up a relatively small percentage of the plant nuclear genome. For example, only about 1%–2% of the total haploid nuclear DNA content of the garden pea encodes for proteins (assuming that the "average gene" consists of 1,000 base pairs and that there are from 40,000 to 100,000 genes). A far larger percentage of the nuclear genome is composed of

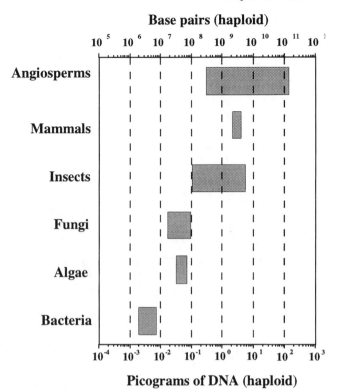

Base pairs (haploid)

Figure 8.6. The amount of DNA in the haploid nuclei of various organisms. The C-value paradox refers to the lack of correspondence in the amount of nuclear DNA and the morphological complexity within and across different groups of organisms.

repetitive DNA—reiterated nucleotide sequences that have no known coding function but that can occur in hundreds, thousands, or even millions of copies. In general terms, species with larger C-values contain larger amounts of repetitive DNA. Among flowering plants, up to 80% of the total nuclear genome can be repetitive DNA. Thus, the C-value paradox is partially explained by the lack of correlation between total genome size and the amount of repetitive nuclear DNA. Although "genetically inert," repetitive DNA may provide essential services. The centromeres of eukaryotic chromosomes are surrounded by repetitive DNA, whose sheer amount may provide a buffer against mutations that could adversely affect chromosome structure, pairing, and migration to cell poles during mitosis or meiosis. It is also possible that some repetitive DNA sequences are involved with the regulation of transcription or the repression or activation of different DNA regions. There is also good reason to believe that repetitive DNA influences the dynamics of the nuclear genome during meiosis. Recent studies of the fruit fly, *Drosophila melanogaster,* by Gary Karpen and his coworkers indicate that at least some of the chromosome pairings that take place during meiosis are initiated or maintained by heterochromatin, the type of DNA found on the centromeres and terminal regions of chromosomes consisting mainly of noncoding repetitive sequences (Karpen, Le, and Le 1996). The homologous chromosomes studied by these workers are those that do not exchange genetic information before separating (i.e., the chromosomes involved in achiasmatic disjunction). The degree and extent to which heterochromatin base-pair sequences in the regions of these homologous chromosomes overlap appear to influence whether disjunction or nondisjunction occurs. If the heterochromatin sequences on homologues overlap by 1,000 kilobases or more, achiasmatic disjunction occurs. Below this critical size, however, the frequency of nondisjunction increases linearly with a decrease in the size of overlapping sequences. How the heterochromatin draws homologous chromosomes together still remains a mystery. The heterochromatin may hold homologues together directly like a molecular zipper, or it may bind to proteins that in turn bind homologues together. Yet another big question is the prevalence of achiasmatic disjunction in other animal and plant species. Nevertheless, it is clear that noncoding repetitive DNA sequences are important to the normal behavior of chromosomes.

The C-value paradox also comes about because a substantial percentage of nonrepetitive nuclear DNA consists of intergenic spacers and introns that have no known coding function. As noted earlier, introns are nucleotide sequences within split genes. Intergenic spacers are nucleotide sequences between genes. Both either are not transcribed or, when transcribed into mRNA, are excised before translation (see fig. 8.1). Because the number of intergenic spacers and introns does not correlate well with the number of split genes or the amount of repetitive DNA, the size of the nuclear genome typically does not correlate with phenotypic complexity.

The only direct tests for the potential effects of the nuclear genome size on phenotypic features have involved crossing related species with the same chromosome numbers but different morphological characters and C-values. Comparisons between the C-values and morphological differences of the hybrid offspring and backcrossed progeny offer an opportunity to determine the relation between the size of the nuclear genome and phenotypic complexity. The few studies using this approach report no significant correlation between C-value and phenotypic complexity (Hutchinson, Rees, and Seal 1979; Bachmann, Chambers, and Price 1985). The only phenotypic character evincing a clear correlation with the C-value is the duration of the mitotic cycle (Van'T Hoff and Sparrow 1963). Presumably the greater the amount of nDNA in a cell, the more time is required for DNA replication. Other correlations have been reported (e.g., cell, tissue, and organ size) and suggested as having important adaptive significance (Bennett 1972), but no good evidence exists to support the belief that changes in the amount of nuclear DNA are directly guided by natural selection. Selection pressures on the duration of the mitotic cycle may place an upper limit on the size of the nuclear genome, and in some circumstances selection pressures for reduced generation times may dispose of plants with genomes that approach this size, but the absolute amount of nuclear DNA likely plays at best a modest evolutionary role.

Far more important to evolution than the amount of DNA is how coding DNA sequences (genes) can be rearranged within and across chromosomes and how genes increase in number and diverge in function. The nuclear genes have paramount importance because they provide the cell with the myriad molecular constituents necessary for growth, differentiation, and reproduction. Classical genetic studies have shown that genes linked to one another on chromosomes are frequently reorganized within the nucleus (see figs. 1.5 and 1.6). Segments of linked genes can be replicated within a single chromosome (duplication); replicated genes can subsequently diverge in their gene functions, thereby proving the potential for evolving novel phenotypic features. As I noted in chapter 1, chromosomes can break along their length, and segments of linked genes can detach and reattach in a number of ways. Segments can reattach to the same chromosome but in the inverse direction (inversion), or they can be permanently lost (deletion). Inversions can alter the appearance of a phenotype because gene expression is often influenced by neighboring genes (position effects). Deletions are frequently lethal. Segments of linked genes can also detach from one chromosome and reattach to a nonhomologous chromosome (translocation). The transferal of chromosome segments between nonhomologous chromosomes can be equal or unequal (reciprocal or nonreciprocal translocations). Rearrangements of chromosome segments can also occur when a chromosome replicates and forms two identical chromatids. Equal or unequal breakage of the chromatids can result in reciprocal or nonreciprocal crossover; although chromatid crossover is normally reciprocal, nonreciprocal crossovers can produce homologous chromosomes, one with a deletion and another with a duplicated segment of linked genes. Crossing over within introns may lead to exon duplication and exchange between genes. These events could expedite the evolution of new genes and gene products produced from modified nucleotide sequences. If so, then crossover within the introns of split genes may enhance the rate of change in the eukaryotic genome well beyond the capacity of simple mutation.

Transposable elements, or insertion elements, also provide an important mechanism for reshuffling genetic information. Transposable elements are DNA sequences that can be inserted into many different chromosome sites; some move around a genome on their own. The insertion of some transposable elements is under the influence of other DNA sequences. The first such "jumping genes" were detected and characterized in corn by Barbara McClintock, well before the structure of DNA was known (McClintock 1951). Jumping genes have been discovered in bacteria, yeast, invertebrates, and vertebrates as well as in different kinds of plants.

Gene duplication followed by divergence in gene function is an important mechanism for modifying or acquiring new phenotypic features. Among plants, the duplication of linked genes is commonly achieved by polyploidization (see chapter 2). The duplication of unlinked genes also seems to be common. Crossing over within introns undoubtedly has led to exon duplication. Despite modern molecular techniques, there are few insights into the rules or mechanisms for genomic reorganization (Tanksley and Pichersky 1984). Thus, although it is clear that linked and unlinked genes are constantly being rearranged within and among chromosomes, it is not certain that these rearrangements inspire gene duplication.

Finally, there is now good evidence that environmental conditions can trigger heritable genomic changes in plants (Cullis 1985). When flax (*Linum usitatissimum*) is grown at high temperatures and under certain soil and fertilizer conditions, it can phenotypically change into a large or a small form depending on the balance of nutrients supplied. These two forms are morphologically stable, and the difference in size is inherited by their progeny. Heritable differences in ribosomal RNA genes are associated with these two phenotypes and may be caused by unequal duplication of some DNA sequences, nonreciprocal crossing over, or some extrachromosomal event resulting in gene duplication (Cullis 1985). Results like these favor the genomic stress hypothesis—that environmental stresses may trigger genomic reorganizations leading to mutations capable of overcoming stressful conditions (McClintock 1984). One implication of the genomic stress hypothesis is that speciation may be more rapid in organisms living in unstable or repeatedly disturbed environments.

Rates of Genomic Evolution

Until the mid-1980s, much of what was known about rates of molecular evolution came from comparative studies of the DNA sequences of mammal genes and gene products. The picture that emerged from these DNA sequences was that the rate of mtDNA evolution was substantially faster than the rate of mammal nDNA evolution (Kimura 1983; Li, Luo, and Wu 1985). But this picture was recognized as incomplete because few data were available for estimating the rates of plant mtDNA and nDNA evolution. By the mid-1980s, DNA sequencing of the three plant genomes had reached the stage where tentative comparisons could be drawn among the relative rates of plant and mammal molecular evolution (see Wolfe, Li, and Sharp 1987; Birky 1988). Early studies showed that the plant mitochondrial genome was larger and more variable than its animal counterpart and that the plant nuclear genome appeared to change faster than mammal nDNA (Palmer 1985a, b). Even with the larger databases, however, direct comparisons between the rates of plant and mammal molecular evolution must be approached cautiously. Most plant DNA sequences are for monocot and dicot species that have been subjected to intense agricultural and horticultural selection, so our knowledge about changes in the plant nuclear genome is taxonomically skewed in favor of flowering plants whose evolution has been influenced by human activity. Also, rates of base-pair substitutions are calculated according to the divergence time of species, which may be in error. Specifically, rates of "silent" or synonymous base-pair substitutions (R_s) and rates of amino acid–changing or nonsynonymous (R_n) base-pair substitutions are calculated by dividing the number of substitutions per gene site (K_s and K_n, respectively) by twice the estimated divergence time T. That is, $R = K/2T$. The divergence time is typically given in billions of years, and so R is usually reported in units of the number of base-pair substitutions per gene site per 10^9 years. Because the divergence time is multiplied by a factor of two, the effects of an error in the estimated divergence time are compounded. Provided the same estimated divergence time is used across species comparisons, an error in the divergence time will have no effect on comparisons among the three plant genomes (cpDNA, mtDNA, and nDNA) or the two animal genomes (mtDNA and nDNA), but errors will unquestionably alter the impression of the relative rates of plant and animal mtDNA or nDNA evolution, because species of plants and animals have diverged at very different times. The divergence time typically taken for the divergence of monocots and dicots is 100 or 180 million years; a divergence time of 80 million years is usually taken for the mean divergence of major orders of mammals (e.g., Wolfe, Li, and Sharp 1987). Because the estimated divergence times for the monocot-dicot split and the divergence of mammal orders differ by a minimum of 20 million years (100 versus 80 Myr), the rates of animal mtDNA and nDNA evolution will necessarily be higher than those of plant mtDNA and nDNA even if the number of synonymous and nonsynonymous base-pair substitutions per gene site is the same for flowering plants and animals.

Direct comparisons drawn among the rates of plant and animal genomic change are also confounded because the nuclear genome differs from that of the chloroplast and the mitochondrion in two important ways. First, much of population and evolutionary genetic theory for nuclear genes is not directly applicable to organelle genes, which do not obey Mendel's "laws of inheritance." Second, it is dangerous to infer that plant and animal organelles are subject to the same selection pressures. In contrast to the nuclear genome, chloroplasts and mitochondria typically have uniparental inheritance. Among flowering plants, chloroplasts and mitochondria are transmitted preferentially or exclusively by the female gamete ("maternal inheritance"). Among gymnosperms, chloroplasts may have preferential "paternal inheritance," whereas mitochondria may be maternally transmitted. In most circumstances, therefore, plant cells that have two alleles for the same cpDNA or mtDNA gene (owing to mutation or biparental transmission) tend to undergo rapid vegetative segregation of the two alleles; so in contrast to the common heterozygosity of nuclear genes, most plant cells are homozygous for chloroplast and mitochondria alleles.

Another concern with drawing comparisons among the rates of evolution for plant and animal genomes is that rates may be computed using different genes. Experiments verify that different genes mutate at different rates, just as the same gene may mutate at different rates even among closely related organisms. Consequently, comparisons among the rates of genome evolution will yield equivocal results unless rates are computed from the same population of genes. Finally, it should be noted that amino acid and nucleotide substitution rates computed based on protein and DNA polymorphisms reflect the rates at which mutations are fixed in populations and are not a direct measure of mutation rates. Many more mutations may occur and go undetected because they are lethal and do not appear in the population.

With these caveats in mind, comparisons indicate that the number of base-pair substitutions per gene site and the estimated rate of substitution per site per year differ among the three plant genomes as well as between the two mammal genomes (table 8.2). For each plant or mammal genome, the rate of synonymous base-pair substitutions (R_s) is higher than the rate of nonsynonymous base-pair substitutions (R_n). Taking the quotient of the rates of synonymous and nonsynonymous base-pair substitutions, the average R_s/R_n for the plant mtDNA, cpDNA, and nDNA are 5.5, 11.6 and 12.5. That R_s is much greater than R_n for each of the

Table 8.2   Comparisons of the Number of Synonymous (K_s) and Nonsynonymous Base-Pair Substitutions (K_n) per Site and the Estimated Rate of Synonymous (R_s) and Nonsynonymous (R_n) Substitutions per 10^9 Years for Plant and Mammal Organelle Genomes

	cpDNA	mtDNA	nDNA
Angiosperms			
K_s	0.58	0.21	1.61
K_n	0.05	0.04	0.13
R_s	2.9	1.1	8.1
R_n	0.25	0.20	0.65
Mammals			
K_s	—	5.5–15	0.74
K_n	—	0.19–3.2	0.14
R_s	—	34.5–94.5	4.7
R_n	—	1.2–20	0.88

Note: Plant comparison is between monocots and dicots (data from Wolfe, Li, and Sharp 1987); mammal nuclear and mitochondrial genome comparison is for major orders (data for nuclear genome from Li, Luo, and Wu 1985; data for mitochondrion genome from Brown and Simpson 1982; Brown 1983; and Miyata et al. 1982).

three plant genomes is expected because "silent" base-pair substitutions do not alter the molecular structure of gene products, and so they are expected to be under less selection pressure than amino acid–changing (nonsynonymous) base-pair substitutions. The rate of silent base-pair substitutions in the chloroplast genome is almost three times that of the plant mtDNA. Taken at face value, this finding is somewhat surprising (the chloroplast genome is typically viewed as highly conservative) until we realize that the rates of amino acid–changing base-pair substitutions for the chloroplast genome is much more conservative than that of the plant nuclear genome.

Even though direct comparisons between the rates of plant and animal organelle evolution are highly problematic, they are nonetheless thought-provoking. For example, the rates of nonsilent base-pair substitutions in the plant and mammal nuclear genomes are similar enough (0.65×10^{-9} Myr and 0.88×10^{-9} Myr) to challenge the notion that the plant nuclear genome has evolved much faster than the animal nuclear genome. True, the rate of silent base-pair substitutions in the plant nuclear genome is nearly twice that in the mammal nuclear genome (8.1×10^{-9} Myr and 4.7×10^{-9} Myr), but because silent base-pair substitutions do not change gene products, they have no known evolutionary consequence. When seen in the light of nonsynonymous base-pair substitutions, the tempos of plant and mammal nuclear genomic change are similar.

Another interesting plant-animal comparison is the rate of change of the mitochondrial genome (table 8.2). Angiosperm mitochondria have substantially lower rates of synonymous or nonsynonymous base-pair substitutions than do mammal mitochondria. The available data for mammals show that mtDNA evolves at least five times as fast as nDNA. The reverse is true for the flowering plants that have been studied (Palmer 1985a, b; Wolfe, Li, and Sharp 1987), even though both mitochondrial genomes contain very similar sets of genes. The reasons for this difference are unknown. The mtDNA genome of angiosperms is larger and more variable in size than mammal mtDNA and may have evolved more along the path of gene-sequence rearrangement than along that of base-pair substitutions. Angiosperm and mammal mitochondria differ enzymatically. Unlike mammal mitochondria, angiosperm mitochondria membranes have two pathways for the reduction of oxygen, one of which is insensitive to cyanide, and two pathways for oxidizing NADH, one of which is found on the external membrane of the organelle (Taiz and Zeiger 1991). Base-pair substitutions effecting changes of mitochondrial gene products may be limited in some as yet unknown way by the presence of enzymatic pathways in angiosperm mitochondria not found in mammal mitochondria. In the end, however, the reason for the difference in the angiosperm and mammal mitochondrial genomic rates of change is unknown.

Base-pair substitutions are only one way the genetic structure of an organism can change. Alterations in chromosome number or organization, which are abundant in plants, are equally important in considering the tempo of genomic evolution, yet they are entirely neglected when R is calculated. Thus rates of base-pair substitution undoubtedly give an incomplete picture of the total changes occurring in the nuclear genome.

The Molecular Clock

There are two opposing views regarding the tempo of genomic evolution. One view is that most base-pair mutations are subject to molecular constraint; the other is that most substitutions are neutral in their effects. The selectionist view, which holds that most genes and gene products are subject to selection pressure acting at the level of molecular organization, is consistent with experiments showing that different gene loci and proteins can evolve at different rates, even among closely related species (see below). Amino acid and nucleotide substitutions can show patterns within and among lineages that undoubtedly reflect how far gene products and genes can be molecularly altered before their function becomes impaired. In turn, these patterns suggest that different base-pair substitutions confer different benefits to species existing in different environments. This selectionist view of molecular evolution does not discount random base-pair substitutions. It does, however, argue that randomly occurring substitutions will be fixed or eliminated in populations based on their effects on relative fitness at the molecular level of cellular organization.

The contrasting point of view is that selection forces are typically absent at the molecular level of organization, and so most base-pair substitutions in most organisms are neutral in their effect. This view holds that most base-pair substitutions are expected to be fixed or extinguished in diverse populations in an unpatterned, apparently random way. First proposed by Motoo Kimura in 1968, the theory of neutral molecular evolution assumes that rates of molecular change are constant because the random processes of mutation and fixation swamp the effects of natural selection. This perspective leads to the conclusion that the rate of molecular evolution is more or less constant within and among lineages, particularly when rates of amino acid or allelic substitution are averaged over millions of years. Kimura's theory of neutral evolution sharply distinguishes between molecular and phenotypic evolution. Many of those who subscribe to a "neutralist" perspective agree that phenotypic evolution is influenced more by natural selection than by mutation and chance.

The debate between molecular "neutralists" and "selectionists" is not a matter of whether some base-pair substitutions are fixed or eliminated from populations randomly or in patterned ways. The opposing points of view posit that both normally occur. Rather, the debate focuses on the relative frequency of neutral base-pair substitutions. Those who adhere to the neutralist view hold that most base-pair substitutions have little or no effect; molecular selectionists hold that most adversely affect the functions of gene products. The debate remains controversial and unresolved because few genes have been sequenced, and in only a few species, so the question of "relative frequency" thus far remains unanswered.

The concept of a molecular evolutionary clock was first proposed by Emile Zuckerkandl and Linus Pauling (1965), but it was not given full attention until the theory of neutral evolution was advanced and promoted by Motoo Kimura (1968). Data from several kinds of proteins support the theory of neutral molecular evolution and therefore the concept of a steady-state molecular evolutionary clock. For example, although the number of amino acid differences in the α-globin chain of hemoglobin progressively increases in direct proportion to the divergence time between vertebrate species, if any two vertebrate species are compared, the number of amino acid substitutions is very nearly the same. About seventy to eighty amino acid differences distinguish the α-globin chain in the shark from that of the carp, the newt, or a number of diverse mammal species (Kimura 1982). The approximate uniformity of amino acid substitutions in hemoglobin among species belonging to diverse groups of vertebrates is more in keeping with the theory of neutral evolution than with a selectionist theory of molecular evolution. Amino acid substitutions in proteins such as hemoglobin appear to occur rapidly and in a random rather than a patterned way, and the high rate of DNA mutation inferred from the pace of these amino acid substitutions could account for the great diversity in hemoglobin structure observed among vertebrate species. But the apparently constant rate of molecular change combined with an apparent absence of pattern suggests that a great deal of this diversity is randomly fixed or lost within populations.

That different gene products and genes have different evolutionary molecular clocks is evident from prior discussions, and the molecular clock cannot be based on a single gene or gene product. The whole of genomic evolution is greater than the sum of its parts, but if the molecular clock is a "stochastic clock"—the sum of the rates of random changes in all the genes in a genome—then the molecular clock may be determined by pooling data from as wide an assortment as possible of proteins or genes. Even so, the extent to which the stochastic molecular clock is more or less constant or erratic must be evaluated. Two tests have been used, the calibration-dependent test and the relative-rate test. The calibration-dependent test, first used by Langley and Fitch (1974), plots the number of amino acid or nucleotide substitutions determined for different species against the time species have evolutionarily diverged. Divergence times are inferred from the fossil record. A linear relation for averaged values of substitutions is expected if the tempo of the molecular clock is constant. In their seminal study, Langley and Fitch (1974) compared the amino acid sequences of four kinds of proteins from eighteen species to determine an average tempo of nucleotide substitutions. The molecular clocks of the four individual proteins varied, but this variation was largely eliminated when the data for the separate clocks were pooled and average rates were determined over a long period of evolutionary time.

When the molecular clock of the cpDNA is tested in this fashion for synonymous base-pair substitutions, a seemingly opposite situation is illustrated (fig. 8.7). Here synonymous base-pair substitutions in the chloroplast genome plotted against the estimated divergence times for the angiosperm/bryophyte split, the monocot/dicot split, and the split between corn and wheat give a nonlinear rather than a linear trend, as would be expected if the molecular clock were reasonably constant. Only three data points are plotted, so there is not sufficient information to determine the constancy or inconstancy of the cpDNA molecular clock (virtually any type of curve may be drawn to connect three data points). Yet some persist in drawing inferences based on data of no better quality.

The principal limitations of the calibration-dependent test are that even averaged rates of molecular evolution can be biased depending on the

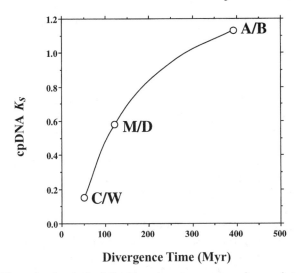

Figure 8.7. The molecular clock of the chloroplast genome. A nonlinear relation is found when the number of synonymous base-pair substitutions is plotted against the estimated divergence times for angiosperms/bryophytes (A/B), monocots/dicots (M/D), and corn/wheat (C/W). A linear relation is expected for a constant molecular clock.

genes or gene products studied, and that averaged rates have considerable (and unmeasureable) error whenever fossil-based divergence times are incorrectly estimated. Also, divergence times for many living species and higher taxa are not available from the fossil record, so there are practical limits to how and when the calibration-dependent test can be applied. The alternative relative-rate test, first used by Sarich and Wilson (1973), does not rely on absolute species divergence times. Instead, it compares any three species or higher taxa of which two (called ingroups) are known to be more closely related than either is to the third group (called the outgroup). If amino acid or nucleotide sequences are the same between the outgroup and each of the two ingroup species, then the two ingroup species must have evolved at the same average rate. The relative-rate test has been used in numerous instances, but it suffers two important restrictions. First, it sheds no light on the absolute rate of molecular evolution, and second, it fails to detect changes in evolutionary rates whenever the tempo of molecular evolution varies proportionally among the species compared.

The concept of the molecular clock assumes that the rate of molecular change is steady over chronological time and is indifferent to the time required to complete an organism's life cycle (the generation time) as well as to the environment in which the organism exists. These assumptions are somewhat doubtful. Studies indicate that the pace of genomic evolution is influenced by the time required to complete the sexual life cycle. For example, the nucleotide substitution rate for the *rbcL* locus in the chloroplast genome is higher for annual than perennial species (Bousquet et al. 1992). Likewise, chromosomal evolution has proceeded faster for herbaceous species with short generation times than for shrubby or arborescent species with longer life cycles (Levin and Wilson 1976). These and other results, which are inconsistent with the belief that the molecular clock is unaffected by generation time, suggest that the shorter the time required to complete the life cycle of a species, the greater the potential for genomic evolution. If each generation represents an opportunity for selection or random genetic mutation to occur, then the molecular clock must tick at a rate partially defined by the generation time; and because different species have different generation times, the molecular clock likely varies among species. Other factors also likely influence the rate of genomic evolution and lead to differences in the molecular clock, perhaps even among related species. Annual and short-lived perennial herbs tend to occupy transient or disturbed habitats. Many such species are self-fertilizing and have small neighborhood populations that favor interpopulational differentiation by selection and random drift (Wright 1943, 1951). Likewise, the genomic stress hypothesis of McClintock (1984) suggests that stressful environments may induce heritable and adaptive genetic changes and so accelerate the molecular clock. In contrast, long-lived perennial herbs, shrubs, and tree species tend to occupy equitable and largely predictable environments. Most such species tend to be outcrossers and to have large neighborhood population sizes, features that may collectively slow the molecular clock.

The debate between molecular "neutralists" and "adaptationists," as well as the debate over the even or uneven tempo of the molecular clock, will continue because it is currently based on a limited number of genes and gene products from comparatively few species. Depending on the gene or gene product and on the type of organism considered, studies verify that molecular change can be either very fast or exceptionally slow. Pooling data from many different genes or gene products undoubtedly gives a better view of the pace of evolution, but this approach is also limited because it obscures the importance of variation among the rates of change of different genes. Also, tests currently used to assess the constancy of the tempo of molecular evolution either depend on species divergence times or assume the rate has remained constant within a lineage. The fossil record clearly shows that the second assumption is incorrect. Many lineages phenotypically burst upon the evolutionary scene when they first appear, then gradually decline in their rate of morphological or anatomical modification and elaboration. Assuming that phenotypic evolution reflects genotypic evolution, however crudely, such "boom and bust" trends in the fossil record challenge the belief that the molecular clock is always staid and steady even within the same lineage.

The Birth and Death of a Species

In chapter 2 the birth of a species was portrayed in two contrasting ways: either as an evolutionary phenomenon involving the gradual accumulation of selectively favorable phenotypic traits owing to selection pressures acting on genic or chromosomal variation in a population (phyletic gradualism), or as a rapid evolutionary phenomenon instigated by the random processes of mutation and genetic drift (genetic revolution). The second of these portraits attempts to explain the general absence of transitional or intermediate forms connecting ancestral to descendant species, the phenotypic stasis often evident between the time a species first appears and dies, and long-term patterns of speciation that are often punctuated by bursts of speciation.

The portraits for gradual versus rapid speciation are paralleled by two contrasting models for the way a species may die, either gradually as the inclusive population dwindles because genetic variation fails to inspire adaptive responses to persistent selection pressures, or suddenly owing to the extirpation of all individuals in the population by an environmental crisis beyond prior experience. The gradual death of a species by natural selection was the view advocated by Charles Darwin, who further believed that the principal selection pressure was interspecific rivalry for the same resources and space. The contrasting portrait of rapid extinction is motivated by two observations. First, the ecological niches formerly occupied by dead species are often not filled by new species until millions of years later. Second, the fossil record contains "mass extinctions," each characterized by the geologically sudden disappearance of most species, many of which formerly were geographically widespread and ecologically successful (see table 8.3). Neither of these observations resonates with the Darwinian view of gradual extinction by interspecific competition, which predicts immediate ecological replacement of dead species by their contemporary competitors and a more or less uniform rate of extinction reflecting the persistent pressures of competition.

Whether species typically evolve into new ones and eventually die slowly as a consequence of natural selection or whether they typically evolve and die rapidly because of random genetic or physical events is still a matter of debate. The authors of the modern synthesis have showed that genetic mechanisms can easily account for rapid as well as slow speciation or extinction, while the fossil record, which is the final arbiter of the "relative frequency" of rapid versus slow birth and death, has not been temporally resolved far enough to settle the issue (what appears "sudden" to a paleontologist may be a very slow genetic or physical process to a geneticist). What can be said at this time is that some environmental changes have been slow and others have been rapid over the long history of life and that the suppression and death of species by a global environmental crisis are as much an expression of natural selection as is interspecific competition. Note, however, that mass extinctions account for only 4% of all the extinctions in the fossil record (see Raup 1994), and so the geologically sudden extirpation of formerly successful species does not appear to be the most frequent portrait of extinction.

The debate over the rates and genetic modes of speciation or extinction likely will continue for years to come. In itself the tempo of speciation and extinction deduced from the fossil record reveals virtually nothing about the genetic modes of speciation or the proximate causes of extinction, although the fossil record can be used to examine a variety of hypotheses based on data from living organisms. In chapter 2 Sewall Wright's metaphor for adaptive evolution, the fitness landscape, was used to model "rapid" or "slow" speciation. Genetic or phenotypic divergence from an ancestral species was depicted as a series of steps propelled by the combined influence of natural selection and the random processes of mutation and genetic drift. When natural selection dominates, the walk proceeds from regions of low fitness to regions of higher fitness on the landscape, and subpopulations move up adaptive peaks, where they may evolve into phenotypic entities sufficiently different to acquire species status. Conversely, when random forces dominate, the mean fitness of a subpopulation may decrease by chance, miring the subpopulation in a fitness valley where its effective size can dwindle to the point that any additional stress engenders extinction. During a genetic revolution, genic or chromosomal variations may be rapidly fixed in a very small population by the combined effects of natural selection and random fluctuations in gene frequencies. The low fitness of heterozygotes may prevent significant gene flow among subpopulations and so foster continued genetic divergence and reproductive isolation (the conditions traditionally believed to inspire speciation). But genetic revolutions are equally likely to lead to the death of a small population, just as a catastrophic change in the biological or physical environment may result in the death of a formerly successful species.

A number of biological and physical factors have the potential to eliminate species either slowly or suddenly. Plague, fire, volcanism, glaciation, contact with an exotic species, or the extinction of a species with which a symbiotic relationship has evolved can reduce the effective population size of plant or animal species to the point that any additional stress will extinguish the last vestiges of a once ecologically robust species. The sudden declines of the American elm, *Ulmus americana,* and American chestnut, *Castanea dentata,* serve as tragic examples of the effect of plant pathogens on formerly widespread plant species. The extinction of plant species by glaciation is well documented for Europe and North America (Pielou 1991). The introduction of exotic herbivores often has devastating effects on plants. For example, although island ecosystems and their native species are not intrinsically more biologically fragile than their mainland counterparts (Mueller-Dombois, Bridges, and Carson 1981), human disturbance and the introduction of pigs and goats has accelerated the destruction of native Hawaiian species. The fencing of plant populations in Hawaii national parks has saved many species from extinction. Nevertheless, of the roughly 1,100 native Hawaiian plants, 423 (38%) are considered extinct or threatened to some degree. Small populations of many of the latter cling precariously to life on steep cliffs where, though largely inaccessible to introduced herbivores, they are nevertheless prone to catastrophic extinction by violent storms and rockslides. The endemic species *Brighamia insignis* was essentially saved from extinction by biologists who successfully cultivated plants in botanical gardens before wild stands of this Hawaiian plant were decimated by a series of violent storms.

Overspecialization can also underwrite extinction. Some species of orchid have evolved so intimate a dependency on specific insect pollinators that the elimination of the insect would speed the extinction of the orchid (fig. 8.8). Likewise, circumstantial evidence strongly suggests that the extinction of the dodo, *Raphus cucullatus,* by humans may account for the dwindling population of the once robust forests of the tambalacoque tree, *Calvaria major,* on the island of Mauritius in the western Indian Ocean. The seeds of this tree have failed to germinate during the past three hundred years despite the abundant production of fruits and seeds. Only a dozen trees occur in natural stands, all dating back to about the time the dodo went extinct in 1680. Apparently, in order to germinate, the seeds of the tambalacoque tree need to pass through the digestive tract of large animals like the dodo, which abrades the stony seed wall (Owadally 1979). The animals now living on Mauritius appear to find these seeds unattractive as food or have digestive tracts incapable of wearing away seed endocarps, and so new seedlings are no longer being naturally added to the tree population. Undoubtedly other factors, such as deforestation, have contributed to the decline of the tambalacoque tree,

Style and
anthers

Drakonorchis sp.

*Angraecum
sesquipedale*

Floral
spur

but the extinction of the dodo may have initiated the process of extinction, which is still going on.

Yet another example of overspecialization is the evolution of parasitic plant species whose survival depends exclusively on the survival of the host species. Mention was made earlier of the parasitic beechdrops, *Epifagus*. This angiosperm is an indirect parasite of the beech tree, *Fagus*. (Beechdrops feed directly on mycorrhizae, fungi that live in symbiotic harmony with the roots of many tree species.) Should beech trees become extinct, the beechdrops would surely perish. Other parasitic angiosperms are much less particular about their hosts. The Indian pipe, *Monotropa*, also feeds directly off mycorrhizae (fig. 8.9). But unlike the beechdrops, any mycorrhizae in symbiotic association with virtually any tree species will do. Even though the Indian pipe is a parasite and therefore highly specialized, its persistence as a species is far less dependent on the survival of particular kinds of trees. In this sense the Indian pipe is less prone to extinction than the beechdrops.

Experience and theory show that biological or physical stresses have an additive effect. A population reduced by disease or environmental disturbance is prone to further reduction in size and ultimate extinction when exposed to additional stresses (Lynch and Lande 1993; Bürger and Lynch 1995; Gomulkiewicz and Holt 1995). Reduced population size can have disastrous genetic and phenotypic consequences. Plant and animal breeders have long known that crosses between closely related individuals produce offspring with low relative fitness (Darwin 1876), an effect called inbreeding depression (fig. 8.10). In general, inbreeding depression is inversely proportional to the number of reproducing individuals in a population. Numerous biotic or abiotic factors may cause the erosion of a population until inbreeding depression provides the final push toward extinction.

Two genetic hypotheses attempt to explain inbreeding depression: directional dominance and overdominance. The first hypothesis assumes that populations of outcrossing species consist of a large number of genetically dissimilar individuals, that deleterious genes tend to be recessive, and that these morbid recessive genes are revealed in the homo-

Figure 8.8. Orchid floral morphology adapted to particular insect pollinators. The flower of the dragon orchid (*Drakonorchis* sp.) releases a chemical substance mimicking the pheromone produced by a female thymid wasp, thereby attracting the male to the hinged column of the flower (top left). The weight of the male wasp trips the hinged column, catapulting its head into the cup-shaped style and anthers and effecting pollination (top right). The flower of the Madagascar orchid *Angraecum sesquipedale* provides nectar deep within its floral spur and is pollinated only by a rare hawk moth with a tongue (proboscis) long enough to reach the reward for pollination (bottom).

Figure 8.9. The parasitic flowering plants *Epifagus virginiana* (left) and *Monotropa uniflora* (right). Parasitism has evolved in a broad spectrum of flowering plant families. *Epifagus* belongs to the family Orobanchaceae; *Monotropa* belongs to the family Ericaceae.

Figure 8.10. Inbreeding depression in the mustard *Brassica campestris*. Starting with the parent generation (left), plants of succeeding generations were self-fertilized (from left to right). Reduction in plant size and numbers of flowers and fruits per plant is evident even in the first self-fertilized generation. Plants generously provided by Edward D. Cobb (Cornell University).

zygous state. Because the frequency of homozygotes in a population increases when closely related individuals mate, inbreeding increases the number of those expressing rare deleterious recessive genes. The overdominance hypothesis argues that there are loci at which the heterozgote is superior to either homozygote and that inbreeding decreases the frequency of heterozygotes for these loci. In most cases the directional dominance and overdominance hypotheses lead to the same expectations. For example, according to both, outcrossing should restore vigor.

Studies of domesticated plants like corn show that the effects of inbreeding depression can be observed after only a few generations (Jones 1939; Allard 1960). But the deleterious effects of inbreeding depression owing to a reduction in population size may take time to appear because the increase in the level of homozygosity is expected to be faster for neutral loci than for deleterious loci and because deleterious recessive alleles may initially be intensively purged from a population. The lag time will depend on a number of factors such as the effective population size, the nature and intensity of selection pressures, and the rate of any environmental changes. Inbreeding depression also depends on the breeding history and the mating system of the population. Continued inbreeding of a population with a long history of inbreeding may result in little additional depression because the population already has a high frequency of homozygotes among which deleterious genes have been purged. Crosses between inbred populations may elicit little immediate beneficial effect unless the frequency of deleterious recessive alleles has already been reduced by natural selection. Sustained selection against deleterious alleles will either purge a population's genome of morbid homozygotes or drive the population to extinction. Genomic purging is likely to occur in large populations of self-crossing organisms or in populations with a long history of inbreeding regardless of their size (Lande and Schemske 1985). In contrast, inbreeding depression is expected to be pronounced for populations of outcrossing organisms or for populations that have recently become small. And there is good evidence that its effects will be felt early in the life cycle (see Husband and Schemske 1995).

Because the potential to produce new variant combinations of genes is greater when plants cross-pollinate and because inbreeding depression may occur when plants fail to outcross, many flowering plants have evolved self-incompatibility (SI) systems that arrest the development of pollen produced by "self." SI is defined as "the inability of a fertile hermaphroditic seed plant to produce zygotes after self-pollination" (de Nettancourt 1977). The effectiveness of SI in promoting outbreeding is believed to be an important factor contributing to the evolutionary success of angiosperms (Whitehouse 1951). SI is genetically controlled, and in many cases the control is by a single locus (the S locus) that has many alleles. It is estimated that SI systems occur in more than half of all angiosperm species distributed in fifteen phylogenetically diverse families, although it has been difficult to unequivocally demonstrate it in many cases (Charlesworth 1985). Thus it is clear that SI either evolved early in the radiation of angiosperms or arose independently many times during angiosperm history (Mau et al. 1991), and that it is more common in families containing many species than in small families.

There are two types of SI (see Newbigin, Anderson, and Clarke 1993; Nasrallah and Nasrallah 1993). Gametophytic SI occurs when SI is determined by the genotype of the pollen grain (the microgametophyte); sporophytic SI occurs when the survival and growth of the pollen grain on the stigma and style of a flower are determined by the genotype of the pollen-producing plant (the sporophyte) (fig. 8.11).

Despite the potentially deleterious affects of inbreeding depression and the expectation that outcrossing is genetically advantageous, many plant species nevertheless consist of self-fertilizing individuals. For these species selfing may have evolved simply because it is better to reproduce in this fashion than not to reproduce at all. But genetic models show that the frequency of a gene causing self-fertilization in theory will increase in every generation unless some force, like inbreeding depression, opposes it (Fisher 1941). The reasoning is simple. A selfer will have, on average, three successful gametes—one in the form of an ovulate parent to its selfed seed, one as a pollen-bearing parent to its selfed seed, and one as a pollen-bearing parent to the outcrossed seed of other individuals. In contrast, an outcrosser will only have two successful gametes—one each as an ovulate parent and as a pollen-bearing parent. The "three-to-two average success rate" is an automatic selection advantage that may underwrite the evolution of self-fertilized species. The fate of any gene causing a change in the rate of self-fertilization is governed by two independent factors: the immediate fitness consequences of selfing versus outcrossing, and the functional and ecological factors that determine the relative success rates of selfed and outcrossed gametes in forming zygotes (Holsinger 1988). Of the numerous selective agents affecting the evolution of selfing, inbreeding depression is the only factor bearing on the immediate fitness consequences of selfing versus outcrossing. Other factors, like the cost of meiosis, influence the success rates of selfed and outcrossed gametes in forming zygotes (Lloyd 1992).

Birthrates and Death Rates

The actual birth and death of individual species have not been observed frequently enough or in sufficient detail by scientists studying living populations to resolve the debate over the "most frequent" genetic mode of speciation or extinction. Thus scientists will likely continue to debate the issue for many years. Much less uncertainty exists regarding the tempo of speciation and extinction, which can vary within as well as among groups of organisms and can change during the history of a group.

Gametophytic Self-Incompatibility

haploid genotypes of pollen (microgametophyte)

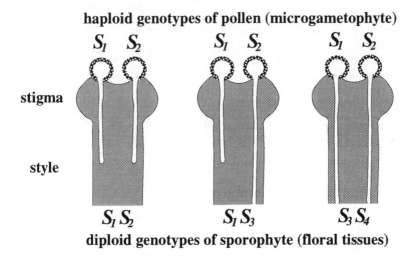

diploid genotypes of sporophyte (floral tissues)

Sporophytic Self-Incompatibility

haploid genotypes of pollen (microgametophyte)

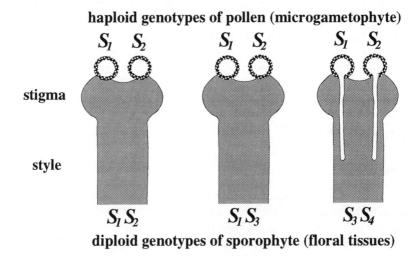

diploid genotypes of sporophyte (floral tissues)

The most obvious evidence for slow speciation comes from the study of living organisms that differ little or not at all from their antecedents over the course of tens or hundreds of millions of years ("living fossils") and from episodes of geologically rapid and abundant speciation, typically from one or a very few ancestral forms, resulting in the exploitation of a broad range of ecological niches (adaptive radiations). Living fossils belong to lineages characterized by little or no significant phenotypic change. They are often the only remaining representatives of groups that formerly were highly diversified. Taken in isolation, living fossils tell us nothing about the rate of speciation per se, but their exceptionally long duration in the fossil record and the dwindling number of related species speak volumes about phenotypic stasis and the speciation rate of their lineages. Perhaps the best-known living plant fossils are the maidenhair tree, *Ginkgo biloba,* the dawn redwood tree, *Metasequoia glyptostroboides,* and the Wollemi pine, *Wollemia nobilis,* all gymnosperms. The maidenhair tree is the only living representative of an ancient group of gymnosperms, the Ginkgophales (see fig. 4.15), that was once geographically widespread and most diverse during the Mesozoic era (Li 1956; Tralau 1968; fig. 8.12A). The history of the ginkgophytes can be traced in the geological column primarily because of their distinctive fan-shaped leaves with open dichotomous venation (Gifford and Foster 1989; Taylor and Taylor 1993). During the Jurassic, plants bearing very similar leaves once lived in such widely separated areas as England, western North America, Alaska, Australia, and Japan. The fossil record clearly shows that the ginkgophytes gradually declined in abundance and geographic distribution throughout the late Mesozoic and the Tertiary era until they were reduced to the single species *Ginkgo biloba,* geographically confined to Asia. Living maidenhair trees were cultivated in Japan and were

Figure 8.11. Self-incompatibility systems in flowering plants. Gametophytic self-incompatability system (top): Self-incompatibility in a single-gene gametophytic system where the genotype of the pollen-producing parent is S_1S_2. When the haploid genotype of a pollen grain matches either diploid sporophyte allele, the growth of the pollen tube through the style is arrested and fertilization does not occur (e.g., pollen with either the S_1 or S_2 allele fails to develop on a flower with the diploid genotype S_1S_2 but will develop on a flower with the diploid genotype S_3S_4). Sporophytic self-incompatibility system (bottom): Self-incompatibility in a single-gene sporophytic system where the genotype of the pollen-producing parent is S_1S_2. Pollen fails to develop on the stigma when the S allele of the haploid pollen genotype matches either allele of the diploid sporophyte genotype. (In these examples, the S_1 allele is dominant or codominant to the S_2 in the pollen and the S_1 allele is dominant or codominant to the S_3 in the style.) Adapted from Newbigin, Anderson, and Clarke 1993.

Figure 8.12. Renderings of "living fossils": (A) *Ginkgo biloba*, the maidenhair tree; (B) *Metasequoia glyptostroboides*, the dawn redwood tree; (C) *Wollemia nobilis*, the Wollemi pine, a relative of the Araucariaceae. Based on line drawings by David Mackay, Royal Botanic Gardens, Sydney, Australia.

discovered by European botanists in 1690. Specimens were brought back to Europe, where they were cultivated and distributed among botanical gardens. Currently *Ginkgo biloba* is securely ensconced as a valued ornamental tree species, gracing gardens and parks all over the globe. *Ginkgo* has the dubious distinction of being one of the few organisms saved from extinction by humans.

The dawn redwood tree, *Metasequoia glyptostroboides*, has a similar history (fig. 8.12B). The genus had long been known from fossil remains uncovered in various parts of western North America, where it survived until the Middle Miocene, and in Asia, where it occurred as late as the Pleistocene in Japan, after which no reliable fossil evidence for its existence is known. In 1944, however, Chinese botanists discovered a few hundred living trees in remote corners of the Szechwan Province of China. This small population contained the only survivors of a once geographically widespread and ecologically successful plant. Much like the maidenhair tree, the dawn redwood tree is now grown in numerous parks and gardens for its great beauty and botanical interest. The discovery of *Wollemia nobilis* in 1994 provides further testimony to the ability of some genera to survive millions of years without changing appreciably in internal structure or external appearance (Briggs 1995). *Wollemia* is a member of the Araucariaceae, a family of gymnosperms that includes *Araucaria* and *Agathis*. Indeed, by all accounts *Wollemia* stands morphologically between these two other living genera, but it bears a great similarity to a fossil genus described from Eocene rocks. *Wollemia nobilis* was found growing in the Wollemi National Park northwest of Sydney, Australia, in 1994 (fig. 8.12C). At the time of its discovery, only twenty-three adult trees and sixteen juveniles had been found, making this one of the world's rarest plants. Seeds are currently being propagated to increase the population of this rare species.

Just as studies of living organisms show that the rates of speciation can vary within and among groups, studies of living populations show that extinction rates can vary. The world's biota appears to be entering an era of species extirpation rivaling in scale even the most intense past episodes of extinction. Most of these extinctions are focused in tropical habitats as a consequence of intense deforestation and other human practices. Estimates based on the strong correlation between habitat area and the number of species within the area indicate that plant and animal species are disappearing with frightening speed. Thus the rate at which habitat area is constricting provides a good gauge of the rate of species extinctions. Even the most conservative estimates of habitat destruction by human activity attest that the biodiversity of tropical forests is in grave danger. Analyses of data from thirteen nations encompassing 18% of all tropical moist forests show that roughly 110,000 km^2 of land is defor-ested per year (Sommer 1976). When other forms of habitat conversion are considered along with deforestation, estimates of the rate of habitat regression range from 119,000 km^2/yr to 200,000 km^2/yr (Myers 1980). This amounts to the elimination of twenty-three to thirty-eight hectares of habitat, or fifty-seven to ninety-four acres, every minute! Biologists estimate that if the destruction of tropical forests continues at this pace, over 66% of all tropical plant species will die out by the end of this cen-tury. The estimate for the extinction of tropical bird species is even higher (Simberloff 1986). Our species appears to be responsible for extinction rates whose scale rivals any that has occurred since life first emerged 3.6 billion years ago (Myers 1980).

The Fossil Record of Speciation and Extinction

Studies of living plants as well as animals verify that speciation and extinction rates can vary within and among lineages, yet these studies provide insight into only a minute fraction of the taxonomic and mor-phological variety that once populated the Earth. They also offer a very truncated view of long-term changes in the tempos of speciation and extinction. In contrast, the fossil record provides abundant information on the rates of speciation and extinction for literally thousands of species over hundreds of millions of years. Compilations of the first and last appearances of species in the geological column and statistical analyses of these "births and deaths" verify that the intensities of speciation and extinction have varied widely over a continuum of low to high. Episodes of high speciation rates are often but not invariably identified with adap-tive radiations, while episodes of high extinction rates are called "mass extinctions." These intense episodes differ profoundly in magnitude from "background" speciation and extinction rates, which taken at face value reflect the normal course of evolutionary affairs.

Paleontologists have learned a great deal about the timing, magnitude, and taxonomic biases of previous episodes of rapid diversification and intense extinction, but the implications of the patterns of speciation and extinction are still not fully understood. Much less is known about the patterns of plant speciation and extinction. Nevertheless, the fossil record for terrestrial vascular plants shows that the pattern of intense episodes of plant extinctions does not correspond well temporally with the pattern of mass extinctions among either marine invertebrates or terrestrial ver-tebrates (Knoll 1984; Traverse 1988). The apparent discrepancy between the patterns of plant and animal extinction suggests that our understanding of the overall history and proximate causes of mass extinction are far from complete. Here I will review some of what is known about these long-term patterns.

The method typically used to calculate speciation and extinction rates is to survey the paleontological literature worldwide and tabulate the first and last occurrences of each fossil species ever discovered. This is so daunting a task that many workers prefer to tabulate the appearance and disappearance of higher taxa (genera or families), which are far fewer and less likely to be missed when the vast paleontological literature is surveyed. But the origination or extinction of higher taxa must coincide a priori with the birth or death of its smallest taxonomic member, the species. Thus it is possible for a higher taxon to go to extinction well after most of its subtaxa die out. Likewise, the birth or death of an ecologically important species would go unrecognized from the speciation or extinction pattern of higher taxa like families. For these reasons species compilations provide the best database for studying patterns of speciation and extinction. Once the fossil record is compiled, species may be sorted into higher taxonomic units (genera, families, and so forth) to determine how speciation or extinction rates vary within or across particular lineages. Species may also be sorted into morphological, anatomical, or reproductive grades of organization (tracheophytes, pteridophytes, etc.) to examine whether certain life history traits correlate with speciation or extinction rates.

Geological stages are the smallest useful time periods that can be consistently recognized for all fossil groups. However, the geological stages are unequal in duration and are typically on the order of several million to ten million years each. Thus speciation and extinction rates are *average* rates for each stage and reflect averages computed for different lengths of geological time. An important limitation on averaged rates is that they cannot be used to determine whether speciation or extinction events are clustered in finer time periods. Paleontologists therefore have access to a very coarse-grained view of the tempos of speciation and extinction. Another limitation is uncertainty about the absolute time scale, resulting from experimental and geologic errors in calibrating and correlating the ages of rock units worldwide. This limitation is not a severe problem, however; reliable geologic time scales are currently available and set a uniform standard for calculating and comparing the speciation and extinction rates of different groups of organisms (see Harland et al. 1990).

A more important issue is the metric for speciation and extinction. Rates can be expressed in several ways—as the number of species appearing or disappearing per stage duration or as the number of births or deaths relative to the standing diversity of taxa per stage duration. Other metrics are possible, but the latter is generally good demographic practice because it "normalizes" speciation and extinction rates with respect to the number of species that can potentially give rise to new species or that can die out. Nonetheless, each metric for speciation or extinction has its limitations, and none is clearly superior.

With all its potential pitfalls, uncertainties, and limitations, the fossil record for vascular land plants (tracheophytes) verifies that the rates of speciation and extinction have varied widely over the past 400 million years. The highest speciation rates occurred during Devonian and Mississippian times, when vascular plants taxonomically diversified into all but the angiosperm clade (fig. 8.13). The Devonian may be a special case, however, because extinction rates become inflated whenever the standing diversity is low, as is the case for this time period (i.e., the Devonian extinction rates may be artifacts). But episodes of high speciation also occurred during Permian, Triassic, and Tertiary times, when standing diversity was comparatively high. A recurrent pattern of initially high and variable rates of speciation followed by declining and less variable speciation rates is typically seen when vascular plant species are sorted into the three grades of reproduction (pteridophytes, gymnosperms, and angiosperms). Lyrically speaking, this pattern begins with a vibrant crescendo whose reproductive theme is attenuated as a diminuendo ending in a basso profundo, the fundamental harmonic (see fig. 8.13). The mechanism(s) responsible for this pattern remain unknown.

In theory, each of the three reproductive grades could have been inspired by a genetic revolution that literally created a new, totally unoccupied fitness landscape containing many adaptive peaks, each offering an opportunity for reproductive isolation and subsequent speciation. As the new fitness landscape became occupied, the subsequent genotypic changes required either to depose older species or to create new fitness peaks would become increasingly more unlikely or difficult to achieve, and so the rate of speciation would steadily decline. The baseline for the speciation rate may reflect the normally slow replacement of old species by new ones either by means of interspecific rivalry or through random species extinction owing to changes in the physical environment. Alternatively, the pattern may simply be a taxonomic artifact resulting from an inflated view of the real number of species during the early phase of the evolution of a grade. All we can say with any confidence is that the pattern is an observation awaiting explanation.

Clearly, one of the factors that may be responsible for patterns of speciation is the pattern of extinction. The death of formerly successful

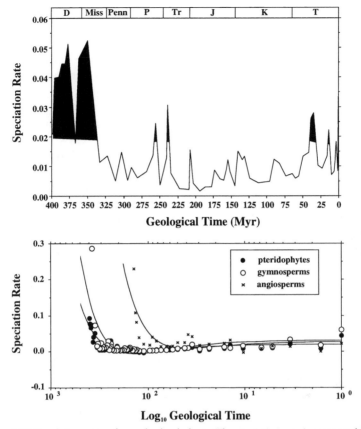

Figure 8.13. Speciation rates of vascular land plants. The speciation rate is computed as the number of new species appearing per geological stage time period per standing species diversity in each stage. When plotted against geological time, some stages have exceptionally high speciation rates (black regions in upper graph). When plotted against log geological time, the speciation rates are initially very high and then decrease to a low baseline level for each of the three major reproductive grades (lower graph).

species affords opportunities for other species to fill vacated niches. Recall that the traditional Darwinian view is that extinction is a gradual process driven by rivalry among species for the same resources and space. Darwin (1972) believed that less well adapted species are slowly but inevitably pushed to extinction by better-adapted ones competing for the same livelihood. According to this view, extinction is active in the sense that rivalry among species is expected to gradually reduce the population size of less well adapted species. At some point in its lifetime the inclusive population of a less well adapted species will drop below a critical mass, and then effects like inbreeding depression may supply the final nudge toward extinction. Darwin was well aware of sudden disappearances of species from the fossil record, but he was convinced that these abrupt departures were "artifacts" due to unrecognized gaps in the geological column. Like his mentor the geologist Charles Lyell, Darwin was rather contemptuous of the notion that extinctions were the result of environmental catastrophes. He firmly believed that the death of species was due to biological, not physical causes.

Darwin's view may explain background rates of extinctions—the comparatively low extinction rates that characterize much of the history of life. But the gradual replacement of old species by new does not comply with the lag in geological time between the extinction of major groups and their subsequent ecological replacement by others. Nor does it comply with mass extinction. If the cause of extinction is invariably interspecific rivalry, as Darwin believed, then no significant span of geological time should separate the extinction of a group of organisms from the filling of their ecological roles by competitors. In this sense interspecific competition abhors an ecological vacuum. Yet the fossil record shows that significant intervals often separate episodes of intense extinction and the filling of vacated niches by other groups of organisms (Simpson 1944). Ichthyosaurs became extinct millions of years before their role in the biosphere was replaced by marine mammals. Bats and birds diversified millions of years after the extinction of pterodactyls. And dinosaurs died out well before terrestrial mammals assumed similar roles in the terrestrial landscape. These and other examples show that extinction is not invariably the result of interspecific competition as Darwin believed. Rivalry over the same resources and space necessarily involves coexisting species. Logically, interspecific competition must have immediate and direct effects on the number of competing individuals.

A reasonable explanation for the geological time lag between extinction and ecological replacement is that the proximate cause of mass extinctions is an abrupt and radical departure from previous environmental conditions. Once a formerly successful group is by chance extinguished, another kind of organism, perhaps even a formerly less successful one, has the opportunity to adaptively radiate and fill the vacated ecological space. Because adaptive modifications take time to evolve, the replacement of one group by another will not be immediate, even in geological terms, and a temporal lag will be evident in the fossil record of taxonomic succession. Note that Darwin's theory of natural selection permits the physical environment to sort among species as ruthlessly as does biological competition for limited resources and space. At a fundamental level, Darwin's theory of natural selection is indifferent to whether competition among species or physical emergency drives the sorting of species or higher taxa. In this sense the salient difference between the two modes of extinction (those caused by physical stimuli versus those resulting from species rivalry) is not whether natural selection occurs but the time organisms are given to adapt to new conditions. During mass extinctions, the time scale is too brief to give even formerly well adapted organisms the opportunity to cope with events. During background extinctions the time scale is long enough for organisms to adaptively respond to changing conditions if it is within their genetic capacity to do so.

The fossil record for the mass extinctions of marine invertebrates and terrestrial vertebrates is clear. Marine invertebrates have experienced five mass extinctions (Raup and Sepkoski 1982; Jablonski 1986, 1991; Raup 1994). Terrestrial vertebrates have experienced two mass extinctions and a number of "minor" mass extinctions (Benton 1989, 1993). During each mass extinction, most animal species are extirpated in a geologically "brief" period (table 8.3). Many of these species were formerly very successful in terms of both geographical distribution and prior longevity in the fossil record. Attempts to find shared traits among the animals affected have largely gone unrewarded. For marine invertebrates, few physiological, morphological, or ecological threads appear to bind together the animals that died out (Jablonski 1991). In contrast to mass extinction events, background extinctions often involve great selectivity for life history traits. Animals that are geographically widespread appear to be spared more often than those that have narrow geographic coverage. The current view, therefore, is that mass and background animal extinctions differ in both tempo and mode. Mass extinctions occur quickly and appear to be due to sudden and severe physical environmental crises well beyond the normal ability of most animals to adapt. Background animal extinctions appear to reflect the more normal state of affairs involving gradually applied environmental stresses, of either a biological or a physical nature, that have normally been experienced by animals.

The fossil records for mass extinctions of vascular land plants and animals differ significantly. The history of tracheophytes is characterized by nine single or clustered episodes of intense species extinction (fig. 8.14). Only two of these nine episodes coincide with the "minor" mass extinctions reported for terrestrial vertebrates, and none coincide temporally with the four great mass extinctions observed for the marine invertebrate fossil record since the Silurian (see table 8.3). Because the mass extinctions of land plants and animals do not correspond temporally, it seems

Table 8.3 Comparisons of Nine Episodes of Intense Vascular Land Plant Species Extinction (+) with Major (M) and Minor (m) Extinctions of Terrestrial Vertebrates and Marine Invertebrates.

Period	Stage or Epoch	Vascular Land Plants	Terrestrial Vertebrates	Marine Invertebrates
Tertiary	Messinian (5.2)–Tortonian (6.7)	–	m	–
	Burdigalian (16.3)	+	–	–
	Rupelian (29.3)	+	m	–
Cretaceous	Maestrichtian (65)	–	M	M
	Neocomian (131.8)	+	–	–
Jurassic	Kimmeridgian (152.1)–Oxfordian (154.7)	+	m	M
Triassic	Norian (209.5)–Carnian (223.4)	–	m	M
	Anisian (240)–Scythian (241)	+	m	–
Permian	Tatarian (245)–Kazanian (251)	–	M	M
	Artinskian (260)–Sakmarian (269)	–	m	–
	Asselian (281.5)			
Pennsylvanian	Gezelian (290)	+	–	–
	Moscovian (303)	–	m	–
Devonian	Famennian (362.5)	+	–	M
	Frasnian (367)	–	–	–
	Givetian (377.5)	+	–	–
	Siegenian (390.6)	+	–	–

Source: Data for vertebrates from Benton 1993; data for maine invertebrates from Jablonski 1991. Parenthetical dates for stage or epoch in Myr from Harland et al. 1990.

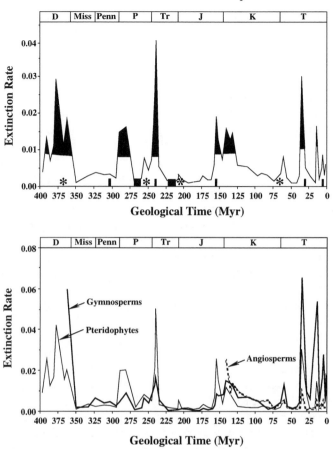

Figure 8.14. Extinction rates for vascular land plants. The extinction rate is computed as the number of species disappearing per geological stage time period per standing species diversity in each stage. Episodes of exceptionally high extinction rates occur (black regions in upper graph) but do not generally coincide with major or minor mass animal extinctions (denoted by asterisks and black boxes; see table 8.3). The intensity of species extinctions is not equivalent among the three plant reproductive grades (lower graph). Pteridophyte species tend to have higher extinction rates during the early Paleozoic, while gymnosperm species take the brunt of extinction during the Mesozoic and Tertiary.

likely that either the proximal cause of extinction or the mode of response to the cause is frequently different in land plants than in marine and land animals. Also, unlike the animal mass extinctions, which show no evidence of selectivity for specific life history traits, intense episodes of plant extinction appear to favor some reproductive grades over others (see fig. 8.14). For example, during the Paleozoic era and much of the Mesozoic, pteridophyte species tend to have much higher extinction rates than gymnosperms. The reverse is true for the intense episodes of tracheophyte extinctions during the Tertiary, when gymnosperm species took the brunt of extinction. The mass extinctions of Paleozoic pteridophytes may reflect a reduction in the availability of the moist environments required for their sexual reproduction. During the late Paleozoic, the Euramerican swamp environments in which most known pteridophyte species lived progressively dried out. Most seed plants require less moist environments for successful sexual reproduction, so Paleozoic gymnosperm species may have been much less affected by the global climatic changes leading to drier environments and the elimination of many pteridophyte species. The high extinction rate of Tertiary gymnosperm species is more difficult to explain in terms of the ecological requirements for sexual reproduction. The conventional explanation is that these species were competitively displaced by the ecologically more aggressive angiosperm species. But the pattern of high gymnosperm extinction rates coincides with those of pteridophyte and angiosperm extinction rates during this time, suggesting that physical factors in the environment may have changed and affected all plant species in much the same way. This hypothesis fits in with the fact that each of the episodes of high extinction rates in the Tertiary coincides with a time of global drying or warming. Indeed, it appears that each plant mass extinction coincides with a period of global climate change that differentially influenced the relative fitness of the three reproductive plant grades.

Well over 90% of all the species that ever lived are now dead. Thus the history of life is as much a saga of extirpation as of adaptive tenacity and success. But the fossil record shows that extinction has not taken the upper hand. The tremendous morphological and taxonomic diversity seen today reflects a slight surplus of species birth and survival over death accumulated over hundreds of millions of years (fig. 8.15). The terrestrial landscape was colonized by perhaps only a few small plant species existing in isolated populations. Today the surface of the Earth is draped in green, and the number of land plants is on the order of hundreds of thousands of species. The balance between speciation and extinction has shifted over time, and consequently the taxonomic composition of Earth's floras has dramatically changed over the past 450 million years (Niklas, Tiffney, and Knoll 1980, 1983). At one time Earth's forests were dominated by giant tree lycopods, horsetails, and ferns. As these forests passed into history, they left behind their fossil remains in the form of the great coal beds that fuel the fires of intellectual curiosity as well as those of commerce and industry. The early pteridophytic forests were replaced by

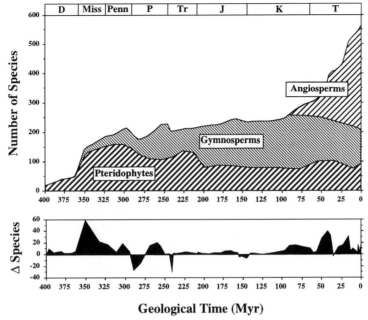

Figure 8.15. Taxonomic turnover during the Phanerozoic. Paleofloras were dominated by pteridophyte species during much of the early Paleozoic. These floras were replaced by ones dominated by gymnosperm species during the Mesozoic. The most recent floras are dominated by angiosperm species. The trend of increasing species number throughout the history of the vascular land plants is the result of a slight surplus of species accumulated over hundreds of millions of years (lower graph). Adapted from Niklas, Tiffney, and Knoll 1980.

forests dominated by cycad, maidenhair, and dawn redwood trees, which in turn gave way to the current stands of oak, maple, birch, and poplar. Evolution has no foresight—it is an unconscious process of descent with modification directed by the combined effects of random and nonrandom biological and physical forces. Thus the taxonomic composition and physical appearance of future forests cannot be predicted. What can be said with certainty is that some of today's descendants will become tomorrow's ancestors, despite the worst human efforts to extinguish many of evolution's fascinating organic products.

Literature Cited

Alberch, P. 1980. Ontogenesis and morphological differentiation. *Amer. Zool.* 20:653–67.

———. 1981. Convergence and parallelism in foot morphology in the Neotropical salamander *Bolitoglossa*. 1. Function. *Evolution* 35:84–100.

———. 1989. The logic of monsters: Evidence for internal constraints in development and evolution. In *Ontogenäse et evolution,* ed. D. B. Dommergues, J. L. Chaline, and B. Laurin, 21–57. Lyons: Geobios.

Allard, R. W. 1960. *Principles of plant breeding.* New York: John Wiley.

Anderson, E. 1949. *Introgressive hybridization.* New York: John Wiley.

Andersson, S., and R. G. Shaw. 1994. Phenotypic plasticity in *Crepis tectorum* (Asteraceae): Genetic correlations across light regimens. *Heredity* 72:113–25.

Andrews, H. H., Jr. 1952. Some American petrified calamitean stems. *Ann. Missouri Bot. Gard.* 39:189–218.

———. 1961. *Studies in paleobotany.* New York: John Wiley.

Angert, E. R., K. D. Clements, and N. R. Pace. 1993. The largest bacterium. *Nature* 362:239–41.

Arnold, J. 1993. Cytonuclear disequilibria in hybrid zones. *Ann. Rev. Ecol. Syst.* 24:521–54.

Arnold, J., and S. A. Hedges. 1995. Are natural hybrids fit or unfit relative to their parents? *Trends Ecol. Evol.* 10:67–71.

Arthur, W. 1984. *Mechanisms of morphological evolution.* Chichester: John Wiley.

Awramik, S. M. 1992. The oldest records of photosynthesis. *Photosyn. Res.* 33:75–89.

Ayala, F. J. 1988. Can "progress" be defined as a biological concept? In *Evolutionary progress,* ed. M. H. Nitecki, 75–96. Chicago: University of Chicago Press,

Babcock, E. B., and G. L. Stebbins. 1938. The American species of *Crepis*. *Carnegie Inst. Washing, Publ.* no. 504.

Bachmann, K. 1983. Evolutionary genetics and the genetic control of morphogenesis in flowering plants. *Evol. Biol.* 16:157–208.

Bachmann, K., K. L. Chambers, and H. J. Price. 1985. Genome size and natural selection: Observations and experiments in plants. In *The evolution of genome size,* ed. T. Cavalier-Smith, 267–76. New York: John Wiley.

Baker, H. G. 1959. Reproductive methods as factors in speciation in flowering plants. *Cold Spring Harbor Symp. Quant. Biol.* 24:177–91.

Bakker, R. T. 1977. Tetrapod mass extinctions—a model of the regulation of speciation rates and immigration by cycles of topographic diversity. In *Patterns of evolution as illustrated by the fossil record*, ed. A. Hallam, 439–68. Amsterdam: Elsevier.

Baldwin, B. G., and R. H. Robichaux. 1995. Historical biogeography and ecology of the Hawaiian silversword alliance (Asteraceae). In *Hawaiian biogeography*, ed. W. L. Wagner and V. A. Funk, 259–87. Washington, D.C.: Smithsonian Institution Press.

Banks, H. P. 1975. Early vascular land plants: Proof and conjecture. *BioScience* 25:730–37.

Barrington, D. S. 1990. Hybridization and allopolyploidy in Central American *Polystichum*: Cytological and isozyme documentation. *Ann. Missouri Bot. Gard.* 72:297–305.

Barton, N. H., and G. M. Hewitt. 1985. Analysis of hybrid zones. *Ann. Rev. Ecol. Syst.* 16:113–48.

Baum, D. A. 1992. Phylogenetic species concepts. *Trends Ecol. Evol.* 7:1–2.

Beale, S. I., and J. D. Weinstein. 1990. Tetrapyrrole metabolism in photosynthetic organisms. In *Biosynthesis of heme and chlorophylls*, ed. H. A. Dailey, 287–391. New York: McGraw-Hill.

Beck, C. B., R. Schmid, and G. W. Rothwell. 1982. Stelar morphology and the primary vascular system of seed plants. *Bot. Rev.* 48:691–816.

Bell, G. 1982. *The masterpiece of nature: The evolution and genetics of sexuality*. Berkeley: University of California Press.

———. 1994. The comparative biology of the alternation of generations. In *The evolution of haploid-diploid life cycles*, ed. M. Kerkpatrick, 1–26. Lectures on Mathematics in the Life Sciences, vol. 25. Providence: American Mathematical Society.

Bell, P. R. 1979. The contribution of the ferns to the understanding of life cycles of vascular plants. In *The experimental biology of ferns*, ed. A. F. Dyer, 57–85. London: Academic Press.

———. 1992. Apospory and apogamy: Implications for understanding the plant life cycle. *Int. J. Plant Sci.* 153:S123–36.

Bengtson, S., ed. 1994. *Early life on earth*. New York: Columbia University Press.

Bennett, M. D. 1972. Nuclear DNA content and minimum generation time in herbaceous plants. *Proc. R. Soc. Lond.*, ser. B, 191:109–35.

Benton, M. J. 1989. Mass extinctions among tetrapods and the quality of the fossil record. *Phil. Trans. R. Soc. Lond.*, ser. B, 325:369–86.

———, ed. 1993. *The fossil record*. 2d ed. London: Chapman and Hall.

Bernstein, H., G. S. Byers, and R. E. Michod. 1981. Evolution of sexual reproduction: Importance of DNA repair, complementation, and variation. *Amer. Nat.* 117:537–49.

Biao, D., M. V. Parthasarathy, K. J. Niklas, and R. Turgeon. 1988. A morphometric analysis of the phloem-unloading pathway in developing tobacco leaves. *Planta* 176:307–18.

Bierhorst, D. l971. *Morphology of vascular plants.* New York: Macmillan.

Birky, C. W., Jr. 1988. Evolution and variation in plant chloroplast and mito-chondrial genomes. In *Plant evolutionary biology,* ed. L. D. Gottlieb and S. K. Jain, 23–53. London: Chapman and Hall.

Blackstone, N. W. 1995. The units-of-selection perspective on the endosymbiont theory or the origin of the mitochondrion. *Evolution* 49:785–96.

Blankenship, R. E. 1992. Origin and early evolution of photosynthesis. *Photosynth. Res.* 33:91–111.

Bold, H. C. 1967. *Morphology of plants.* 2d ed. New York: Harper and Row.

Bold, H. C., and M. J. Wynne. 1978. *Introduction to the algae: Structure and reproduction.* Englewood Cliffs, N.J.: Prentice-Hall.

Bonner, J. T. 1988. *The evolution of complexity.* Princeton: Princeton University Press.

Bousquet, J., S. H. Strauss, A. H. Doerksen, and R. A. Price. 1992. Extensive variation in evolutionary rate of *rbcL* gene sequences among seed plants. *Proc. Natl. Acad. Sci. U.S.A.* 89:7844–48.

Bower, F. O. 1935. *Primitive land plants.* London: Macmillan.

Bowman, J. L., D. R. Smyth, and E. M. Meyerowitz. 1989. Genes directing flower development in *Arabidopsis. Plant Cell* 1:37–52.

Bradley, D., R. Carpenter, H. Sommer, N. Hartley, and E. Coen. 1993. Complementary floral homeotic phenotypes result from the opposite orientations of a transposon at the *plena* locus of *Antirrhinum. Cell* 72:85–95.

Bradshaw, A. D. 1972. Plant evolution in extreme environment. In *Ecological genetics and evolution,* ed. R. Creed, 20–50. Oxford: Blackwell.

Brent, R. P. 1973. *Algorithms for maximization without derivatives.* Englewod Cliffs, N.J.: Prentice-Hall.

Briggs, B. G. 1995. From 50 million years ago—the Wollemi pine. *Amer. Conifer Soc.* 12:8–9.

Brown, G. G., and M. V. Simpson. 1982. Novel features of animal mtDNA evolution as shown by sequences of two rate cytochrome oxidase subunit II genes. *Proc. Natl. Acad. Sci. U.S.A.* 79:3246–50.

Brown, W. L., Jr., and E. O. Wilson. 1956. Character displacement. *Syst. Zool.* 5:49–64.

Brown, W. M. 1983. Evolution of mitochondrial DNA. In *Evolution of genes and proteins,* ed. M. Nei and R. K. Koehn, 62–88. Sunderland, Mass.: Sinauer.

Bryant, D. A. 1987. The cyanobacterial photosynthetic apparatus: Comparison of those of higher plants and photosynthetic bacteria. In *Photosynthetic pico-plankton,* ed. T. Platt and W. K. W. Li, 423–500. Canadian Bulletin of Fisheries and Aquatic Science 214. Canada: Fishers and Oceans.

Bureau, T. E., and S. R. Wessler. 1994. *Stowaway:* A new family of inverted repeat elements associated with the genes of both monocotyledonous and dicotyledonous plants. *Plant Cell* 6:907–16.

Bürger, R., and M. Lynch. 1995. Evolution and extinction in a changing environment: A quantitative-genetic analysis. *Evolution* 49:151–63.

Cain, A. J. 1963. *Animal species and their evolution.* Rev. ed. London: Hutchinson University Library.

Cairns-Smith, A. G. 1982. *Genetic takeover and the mineral origins of life.* Cambridge: Cambridge University Press.

———. 1985. *Seven clues to the origin of life.* Cambridge: Cambridge, University Press.

Carlile, M. J. 1980. From prokaryote to eukaryote: Gains and losses. In *The eukaryotic microbial cell,* ed. G. W. Gooday, D. Lloyd, and A. P. J. Trinci, 1–40. Thirtieth symposium of the Society for General Microbiology. London: Cambridge University Press.

Carlquist, S. 1965. *Island life: A natural history of the islands of the world.* Garden City, N.Y.: Natural History Press.

———. 1975. *Ecological strategies of xylem evolution.* Berkeley: University of California Press.

———. 1980. *Hawaii: A natural history.* Honolulu: SB Printers.

———. 1995. Introduction to *Hawaiian biogeography,* ed. W. L. Wagner and V. A. Funk, 1–13. Washington, D.C.: Smithsonian Institution Press.

Carpenter, R., and E. S. Coen. 1990. Floral homeotic mutations produced by transposon-mutagenesis in *Antirrhinum majus. Genes Dev.* 4:1483–93.

Carson, H. L. 1983. Chromosomal sequences and interisland colonizations in Hawaiian *Drosophila. Genetics* 103:465–82.

———. 1987. The genetic system, the deme, and the origin of species. *Ann. Rev. Genet.* 21:405–23.

Cavalier-Smith, T. 1978. The evolutionary origin and phylogeny of microtubules, mitotic spindles and eukaryotic flagella. *BioSystems* 10:93–114.

Celakowsky, L. 1874. *Bedeutung des Generationswechsels der Pflanzen.* Prague.

Chabot, B. F., and J. F. Chabot. 1977. Effects of light and temperature on leaf anatomy and photosynthesis in *Fragaria vesca. Oecologia* 26:363–77.

Chaloner, W. G. 1967. Spores and land-plant evolution. *Rev. Palaeobot. Palynol.* 1:83–93.

Chaloner, W. G., and A. Sheerin. 1979. Devonian macrofloras. In *The Devonian system,* 145–61. Special Papers in Palaeontology. 23. London: Palaeontological Society.

Charlesworth, D. 1985. Distribution of dioecy and self-incompatibility in angiosperms. In *Evolution: Essays in honour of John Maynard Smith,* ed. P. J. Greenwood, P. H. Harvey, and M. Slatkin, 237–68. Cambridge: Cambridge University Press.

Church, A. H. 1919. Thalassiophyta and subaerial migration. *Oxford Bot. Mem.,* no 3:1–19.

Chyba, C. F., P. J. Thomas, L. Brookshaw, and C. Sagan. 1990. Cometary delivery of organic molecules to the early Earth. *Science* 249:366–73.

Clausen, J. 1951. *Stages in the evolution of plant species.* Ithaca: Cornell University Press.

Clausen, J., D. D. Keck, and W. W. Hiesey. 1940. *Experimental studies on the nature of species.* Publication 520. Washington, D.C.: Carnegie Institution.

Clegg, M. T., B. S. Gaut, G. H. Learn Jr., and B. R. Morton. 1994. Rates and patterns of chloroplast DNA evolution. *Proc. Natl. Acad. Sci. U.S.A.* 91:6795–801.

Cleveland, L. R. 1947. The origin and evolution of meiosis. *Science* 105:287–89.

Cohen, Y. 1984. Oxygenic photosynthesis, anoxygenic photosynthesis, and sulfate reduction in cyanobacterial mats. In *Current perspectives in microbial ecology,* ed. M. J. Klug and C. A. Reddy, 435–41. Washington, D.C.: American Society for Microbiology.

Cracraft, J. 1983. Species concepts and speciation analysis. *Current Ornithol.* 1:159–87.

Crandall-Stotler, B. 1980. Morphogenetic designs and a theory of bryophyte origins and divergence. *BioScience* 30:580–85.

Crepet, W. L. 1984. Advanced (constant) insect pollination mechanisms: Pattern of evolution and implications vis-à-vis angiosperm diversity. *Ann. Missouri. Bot. Gard.* 71:607–30.

Crepet, W. L., E. M. Friis, and K. C. Nixon. 1991. Fossil evidence for the evolution of biotic pollination. *Phil. Trans. R. Soc. Lond.,* ser. B, 333:187–95.

Cronquist, A. 1978. Once again, What is a species? *Beltsville Symp. Agri. Res.* 2:3–20.

Crosby, M. R. 1980. The diversity and relationships of mosses. In *The mosses of North America,* ed. R. J. Taylor and A. Leviton, 115–29. San Francisco: Pacific Division, AAAS.

Crow, J. F., and M. Kimura. 1965. Evolution in sexual and asexual populations. *Amer. Nat.* 99:439–50.

Cullis, C. A. 1985. Experimentally induced changes in genome size, In *The evolution of genome size,* ed. T. Cavalier-Smith, 197–209. New York: John Wiley.

Darlington, C. D. 1940. Taxonomic species and genetic systems. In *The new systematics,* ed. J. Huxley, 137–60. Oxford: Clarendon Press.

Darwin, C. 1876. *The effects of cross and self fertilisation in the vegetable kingdom.* London: Murray.

———. 1877. *The different forms of flowers on plants of the same species.* London: Murray.

———. 1972. *On the origin of species by means of natural selection.* 1859. 6th ed. London: Murray.

Davies, B., and Z. Schwarz-Sommer. 1994. Control of floral organ identity by homeotic MADS-box transcription factors. In *Results and problems in cell differentiation,* vol. 20, *Plant promoters and transcription factors,* ed. L. Nover, 235–58. Berlin: Springer-Verlag.

Davis, J. I., and P. S. Manos. 1991. Isozyme variation and species delimitation in the *Puccinellia nuttalliana* complex (Poaceae): An application of the phylogenetic species concept. *Syst. Bot.* 16:431–45.

Dawson, A. J., T. P. Hodge, P. G. Isaac, C. J. Leaver, and D. M. Lonsdale. 1986. Location of the genes for cytochrome oxidase subunits I and II, apocytochrome b, alpha-subunit of the F_1-ATPase and the ribosomal RNA genes on the mitochondrial genome of maize (*Zea mays* L.). *Curr. Genet.* 10:561–64.

De Beer, G. 1958. *Embryos and ancestors.* Oxford: Clarendon Press.

de Duve, C. 1991. *Blueprint of a cell: The nature and origin of life.* Burlington, N.C.: Neil Patterson.

De Lange, R.J., D. M. Fambrough, E. L. Smith, and J. Bonner. 1969. Calf and pea histone 4. II. Complete amino acid sequence of pea seedling histone 4; comparison with the homologous calf thymus histone. *J. Biol. Chem.* 224:5669–79.

De Lange, R. J., J. Hooper, and E. L. Smith. 1973. Histone 3. III. Sequence studies on the cyanogen bromide peptrides; complete amino acid sequence of calf thymus histone 2. *J. Biol. Chem.* 248:3261–74.

de Nettancourt, D. 1977. *Incompatibility in angiosperms*. Berlin: Springer-Verlag.

dePamphilis, C. W., and J. D. Palmer. 1990. Loss of photosynthetic and chlororespiratory genes from the plastid genome of a parasitic flowering plant. *Nature* 348:337–39.

Dickerson, R. E. 1971. The structure of cytochrome *c* and the rates of molecular evolution. *J. Mol. Evol.* 1:26–45.

DiMichele, W. A., J. I. Davis, and R. G. Olmstead. 1989. Origins of heterospory and the seed habit: The role of heterochrony. *Taxon:*38:1–11.

Dobzhansky, Th. 1937. *Genetics and the origin of species*. New York: Columbia University Press.

———. 1951. *Genetics and the origin of species*. 3d ed. New York: Columbia University Press.

Dobzhansky, Th., F. J. Ayala, G. L. Stebbins, and J. W. Valentine. 1977. *Evolution*. San Francisco: W. H. Freeman.

Dodge, J. D., and J. T. Vickerman. 1980. Mitosis and meiosis: Nuclear division mechanisms. In *The eukaryotic microbial cell*, ed. G. W. Gooday, D. Lloyd, and A. P. J. Trinci, 77–102. Thirtieth Symposium of the Society for General Microbiology. London: Cambridge University Press.

Doolittle, W. F., and J. R. Brown. 1994. Tempo, mode, the progenote, and the universal root. *Proc. Natl. Acad. Sci. U.S.A.* 91:6721–28.

Doyle, J. A., M. J. Donoghue, and E. A. Zimmer. 1994. Integration of morphological RNA data on the origin of angiosperms. *Ann. Missouri Bot. Gard.* 81:419–50.

Doyle, J. A., and L. J. Hickey. 1976. Pollen and leaves from the mid-Cretaceous Potomac Group and their bearing on early angiosperm evolution. In *The origin and early evolution of the angiosperms*, ed. C. B. Beck, 139–206. New York: Columbia University Press.

Duckett, J. G. 1970a. Sexual behavior of the genus *Equisetum*, subgenus *Equisetum*. *Bot. J. Linn. Soc.* 63:327–52.

———. 1970b. Spore size in the genus *Equisetum*. *New Phytol.* 69:333–46.

———. 1977. Towards an understanding of sex determination in *Equisetum*: An analysis of regeneration in gametophytes of the subgenus *Equisetum*. *Bot. J. Linn. Soc.* 74:215–42.

Duckett, J. G., and K. S. Renzaglia. 1988. Ultrastructure and development of plastids in the bryophytes. *Adv. Bryol.* 3:33–93.

Duysens, L. N. M. 1956. The flattening of the absorption spectrum of suspensions as compared to that of solutions. *Biochim. Biophys. Acta* 19:1–12.

Edwards, D., and U. Fanning. 1985. Evolution and environment in the late Silurian-early Devonian: The rise of the pteridophytes. *Phil. Trans. Roy. Soc. Lond.*, ser. B, 309:147–65.

Edwards, D., K. L. Davies, and L. Axe. 1992. A vascular conducting strand in the early land plant *Cooksonia*. *Nature* 357:683–85.

Edwards, D. S. 1980. Evidence for the sporophytic status of the Lower Devonian plant *Rhynia gwynne-vaughanii* Kidston. *Rev. Palaeobot. Palynol.* 29:177–88.

———. 1986. *Aglaophyton major*, a non-vascular land-plant from the Devonian Rhynie Chert. *Bot. J. Linn. Soc.* 93:173–204.

Eggert, D. A. 1961. The ontogeny of Carboniferous arborescent Lycoposida. *Palaeontographica*, ser. B, 108:43–92.

———. 1962. The ontogeny of Carboniferous arborescent Sphenopsida. *Palaeontographica*, ser. B, 110:99–127.

Ehrlich, P. R., and P. H. Raven. 1969. Differentiation of populations. *Science* 165:1228–32.

Eigen, M. 1971. Self-organization of matter and the evolution of biological macromolecules. *Naturwissenschaften* 10:465–523.

———. 1983. Self-replication and molecular evolution. In *Evolution from molecules to men*, ed. D. S. Bendall, 105–30. Cambridge: Cambridge University Press.

———. 1987. New concepts for dealing with the evolution of nucleic acids. *Cold Spring Harbor Symp. Quant. Biol.* 52:307–20.

———. 1992. *Steps towards life: A perspective on evolution*. Trans. Paul Woolley. Oxford: Oxford University Press.

Eigen, M., and P. Schuster. 1977. The hypercycle: A principle of natural self-organization. Part A: Emergence of the hypercycle. *Naturwissenschaften* 64:541–65.

Eldredge, N. 1985. *Unfinished synthesis*. Oxford: Oxford University Press.

Eldredge, N., and J. Cracraft. 1980. *Phylogenetic patterns and the evolutionary process*. New York: Columbia University Press.

Eldredge, N., and S. J. Gould. 1972. Punctuated equilibria. In *Models in paleobiology*, ed. J. M. Schopf, 82–115. San Francisco: Freeman, Cooper.

Endler, J. A. 1977. *Geographic variation, speciation, and clines*. Princeton: Princeton University Press.

Ereshefsky, M., ed. 1992. *The units of evolution: Essays on the nature of species*. Cambridge: MIT Press.

Ewens, W. 1979. *Mathematical population genetics*. New York: Springer-Verlag.

Faegri, K., and L. van der Pijl. 1979. *The principles of pollination ecology*. 3d ed. Oxford: Pergamon Press.

Fisher, D. C. 1991. Phylogenetic analysis and its application in evolutionary paleobiology. In *Analytical paleobiology*, ed. N. L. Gilinsky and P. W. Signot, 103–22. Short courses in paleontology 4. Knoxville: University of Tennessee.

Fisher, R. A. 1930. *The general theory of natural selection*. Oxford: Oxford University Press.

———. 1941. Average excess and average effect of a gene substitution. *Ann. Eugenics* 11:53–63.

———. 1958. *The genetical theory of natural selection*. 2d ed. New York: Dover.

Fitch, W. M., and K. Upper. 1987. The phylogeny of tRNA sequences provides evidence for ambiguity reduction in the origin of the genetic code. *Cold Spring Harbor Symp. Quant. Biol.* 52:759–67.

Fontdevila, A. 1992. Genetic instability and rapid speciation: Are they coupled? *Genetica* 86:247–58.

Ford, V. S., and L. D. Gottlieb. 1989. Morphological evolution in *Layia* (Compositae): Character recombination in hybrids between *L. discoidea* and *L. glandulosa*. *Syst. Bot.* 14:284–96.

———. 1990. Genetic studies of floral evolution in *Layia*. *Heredity* 64:29–44.

Fox, G. E., E. Stackebrandt, R. B. Hespell, J. Gibson, J. Maniloff, T. A. Dyer, R. S. Wolfe, W. E. Balch, R. S. Tanner, L. J. Magrum, L. B. Zablen, R. Blakemore, R. Gupta, L. Bonen, B. J. Lewis, D. A. Stahl, K. R. Luehrsen, K. N. Chen, and R. W. Woese. 1980. The phylogeny of prokaryotes. *Science* 209:457–63.

Frankel, R., and E. Galun. 1977. *Pollination mechanisms, reproduction and plant breeding*. Berlin: Springer-Verlag.

Franklin, I., and R. C. Lewontin. 1970. Is the gene the unit of selection? *Genetics* 65:707–17.

Friedman, W. E. 1994. The evolution of embryogeny in seed plants and the developmental origin and early history of endosperm. *Amer. J. Bot.* 81:1468–86.

Galen, C. 1996. Rates of floral evolution: Adaptation to bumblebee pollination in an alpine wildflower *Polemonium viscosum*. *Evolution* 50:120–25.

Ganong, W. F. 1901. The cardinal principles of morphology. *Bot. Gaz.* 31:426–34.

Garcia-Bellido, A. 1983. Comparative anatomy of cuticular patterns in the genus *Drosophila*. In *Development and evolution*, ed. B. C. Godwin, N. Holder, and C. C. Wylie, 227–55. Cambridge: Cambridge University Press.

Gates, D. M. 1965. Energy, plants, and ecology. *Ecology* 46:1–16.

Gaut, B. S., S. V. Muse, W. D. Clark, and M. T. Clegg. 1992. Relative rates of nucleotide substitution at the *rbcL* locus of monocotyledonous plants. *J. Mol. Evol.* 35:292–303.

Gaut, B. S., S. V. Muse, and M. T. Clegg. 1993. Relative rates of nucleotide substitution in the chloroplast genome. *Mol. Phylogenet. Evol.* 2:89–96.

Gensel, P. G. 1992. Phylogenetic relationships of the zosterophylls and lycopsids: Evidence from morphology, paleoecology, and cladistic methods of inference. *Ann. Missouri Bot. Gard.* 79:450–73

Gentry, A. H. 1974. Flowering phenology and diversity in tropical Bignoniaceae. *Biotropica* 6:64–68.

Gest, H., and J. L. Favinger. 1983. *Heliobacterium chlorum*, an anoxygenic brownish-green photosynthetic bacterium containing a "new" form of bacteriochlorophyll. *Arch. Microbiol.* 136:11–16.

Gest, H., and J. W. Schopf. 1983. Biochemical evolution of anaerobic energy conversion: The transition from fermentation to anoxygenic photosynthesis. In *Earth's earliest biosphere: Its origin and evolution*, ed. J. W. Schopf, 135–48. Princeton: Princeton University Press.

Gesteland, R. F., and J. F. Atkins, eds. 1993. *The RNA world*. Cold Spring Harbor, N.Y.: Cold Spring Harbor Laboratory Press.

Gifford, E. M., and A. S. Foster. 1989. *Morphology and evolution of vascular plants*. 3d ed. New York: W. H. Freeman.

Gill, P. E., W. Murray, and N. H. Wright. 1981. *Practical optimization.* London: Academic Press.

Gillespie, J. H. 1983. A simple stochastic gene substitution model. *Theor. Pop. Biol.* 23:202–15.

———. 1984. Molecular evolution over the mutational landscape. *Evolution* 38:1116–29.

Ginsburg, L. R. 1983. *Theory of natural selection and population growth.* Menlo Park, Calif.: Benjamin-Cummings.

Givnish, T. J. 1986. Introduction: On the use of optimality arguments. In *On the economy of plant form and function,* ed. T. J. Givnish, 3–9. Cambridge: Cambridge University Press.

Goldblatt, P., ed. 1981. Index to plant chromosome numbers, 1975–1978. *Missouri Bot. Garden, Monogr. Syst. Botany,* vol. 5.

Gomulkiewicz, R., and R. D. Holt. 1995. When does evolution by natural selection prevent extinction? *Evolution* 49:201–7.

Goodwin, R. H. 1937a. Notes on the distribution and hybrid origin of × *Solidago asperula. Rhodora* 39:22–28.

———. 1937b. The cyto-genetics of two species of *Solidago* and its bearing on their polymorphy in nature. *Amer. J. Bot.* 24:425–32.

Gottlieb, L. D. 1984. Genetic and morphological evolution in plants. *Amer. Nat.* 123:681–709.

Gould, S. J. 1966. Allometry and size in ontogeny and phylogeny. *Biol. Rev.* 41:587–640.

———. 1977. *Ontogeny and phylogeny.* Cambridge: Harvard University Press.

———. 1980. The evolutionary biology of constraint. *Daedalus* 109:39–52.

———. 1981. The meaning of punctuated equilibrium and its role in validating a hierarchical approach to macroevolution. In *Perspectives on evolution,* ed. R. Milkman, 83–104. Sunderland, Mass.: Sinauer.

———. 1990. Speciation and sorting as the source of evolutionary trends, or "Things are seldom what they seem." In *Evolutionary trends,* ed. K. J. McNamara, 3–27. Tucson: University of Arizona Press.

———. 1995. Tempo and mode in the macroevolutionary reconstruction of Darwinism. In *Tempo and mode in evolution: Genetics and paleontology fifty years after Simpson,* ed. W. M. Fitch and F. J. Ayala, 125–44. Washington D.C.: National Acadademy of Sciences.

Gould, S. J., and N. Eldredge. 1977. Punctuated equilibria: The tempo and mode of evolution reconsidered. *Paleobiology* 3:115–51.

Graham, L. E. 1993. *Origin of land plants.* New York: John Wiley.

Graham, L. E., J. M. Graham, W. A. Russin, and J. M. Chesnick. 1994. Occurrence and phylogenetic significance of glucose utilization by charophycean algae: Glucose enhancement of growth in *Coleochaeteorbicularis. Amer. J. Bot.* 81:423–32.

Granick, S. 1965. Evolution of heme and chlorophyll. In *Evolving genes and proteins,* ed. V. Bryson and H. J. Vogel, 67–88. New York: Academic Press.

Grant, V. 1949. Pollinating systems as isolating mechanisms. *Evolution* 3:82–97.

————. 1963. *The origin of adaptations.* New York: Columbia University Press.

————. 1971. *Plant speciation.* 2d ed. New York: Columbia University Press.

Gray, J. 1985. The microfossil record of early land plants: Advances in understanding of early terrestrialization, 1970–1984. In Evolution and environment in the late Silurian and early Devonian, ed. W. G. Chaloner and J. D. Lawson. *Phil. Trans. Roy. Soc. Lond.,* ser. B, 309:167–95.

Gray, J., and A. J. Boucot. 1977. Early vascular land plants: Proof and conjecture. *Lethaia* 10:145–74.

Gray, J., D. Massa, and A. J. Boucot. 1982. Cardocian land plant microfossils from Libya. *Geology* 10:197–201.

Gray, J., and W. Shear. 1992. Early life on land. *Amer. Sci.* 80:444–56.

Hagemann, W. 1975. Eine mögliche Strategie der vergleichenden Morphologie zur phylogenetischen Rekonstruktion. *Bot. Jahrb. Syst.* 96:107–24.

Haldane, J. B. S. 1954. The measurement of natural selection. In *Proceedings of the Ninth International Congress on Genetics,* 480–87.

Hallam, A. 1990. Biotic and abiotic factors in the evolution of early Mesozoic marine mollusks. In *Causes of evolution: A paleontological perspective,* ed. R. M. Ross and W. D. Allmon, 249–69. Chicago: University of Chicago Press.

Hamrick, J. L., and R. W. Allard. 1972. Microgeographical variation in allozyme frequencies in *Avena barbata. Proc. Natl. Acad. Sci. U.S.A.* 69:2100–104.

Hamrick, J. L., and D. Murawski. 1990. The breeding structure of tropical tree populations. *Plant Spec. Biol.* 5:157–65.

Harlan, H. V., and M. L. Martini. 1938. The effects of natural selection on a mixture of barley varieties. *J. Agr. Res.* 57:189–99.

Harland, W. B., R. L. Armstrong, A. V. Cox, L. E. Craig, A. G. Smith, and D. G. Smith. 1990. *A geological time scale.* Cambridge: Cambridge University Press.

Harper, J. L. 1961. Approaches to the study of plant competition. *Sym. Soc. Exp. Biol.* 15:1–39.

————. 1982. After description. In *The plant community as a working mechanism,* ed. E. I. Newman, 11–25. Oxford: Blackwell.

Harrison, R. G. 1990. Hybrid zones: Windows on evolutionary process. *Oxford Ser. Evol. Biol.* 7:69–128.

————, ed. 1993. *Hybrid zones and the evolutionary process.* New York: Oxford University Press.

Haughn, G. W., E. A. Schultz, and J. M. Martinez-Zapater. 1995. The regulation of flowering in *Arabidopsis thaliana:* Meristems, morphogenesis, and mutants. *Can. J. Bot.* 73:959–81.

Hawking, S. W. 1988. *A brief history of time.* New York: Bantam.

Hébant, C. l977. *The conducting tissues of bryophytes.* A. R. Gantner Verlag Kommanditgesellschaft. Vaduz, Ger.: J. Cramer.

Hennig, W. 1966. *Phylogenetic systematics.* Urbana: University of Illinois Press.

Hilu, K. W. 1983. The role of single-gene mutations in the evolution of flowering plants, In *Evolutionary biology,* vol. 16, ed. M. K. Hecht, B. Wallace, and G. T. Prance, 97–128. New York: Plenum.

Holsinger, K. E. 1988. The evolution of self-fertilization in plants: Lessons from population genetics. *Acta Oecol.* 9:95–102.

Honma, S., and M. J. Bukovac. 1966. Inheritance of gibberellin induced heterostyly in the tomato. *Euphytica* 15:362–64.

Horn, H. S. 1979. Adaptation from the perspective of optimality. In *Topics in plant population biology*, ed. O. T. Solbrig, S. Jain, G. B. Johnson, and P. H. Raven, 48–61. New York: Columbia University Press.

Horowitz, N. H. 1945. On the evolution of chemical sysntheses. *Proc. Natl. Acad. Sci. U.S.A.* 31:153–57.

Horowitz, N. H., and S. L. Miller. 1962. Current theories on the origin of life. *Fortschr. Chem. Org. Naturst.* 20:423–59.

Humphrey, N. 1987. "Scientific Shakespeare." *Guardian* (London), 26 August.

Huner, N. P. A. 1985. Morphological, anatomical, and molecular consequences of growth and development at low temperature in *Secale cereale* L. CV. Puma. *Amer. J. Bot.* 72:1290–1306.

Husband, B. C., and D. W. Schemske. 1995. Magnitude and timing of inbreeding depression in a diploid population of *Epilobium angustifolium* (Onagraceae). *Heredity* 75:206–15.

Hutchinson, J., H. Rees, and A. G. Seal. 1979. An assay of the activity of supplementary DNA in *Lolium*. *Heredity* 43:411–21.

Huxley, J. S. 1942. *Evolution: The modern synthesis.* New York: Harper.

Jablonski, D. 1986. Causes and consequences of mass extinctions. In *Dynamics of extinction*, ed. D. K. Elliott, 183–229. New York: John Wiley.

———. 1991. Extinctions: A paleontological perspective. *Science* 253:754–57.

Jackson, R. C., and C. T. Dimas. 1981. Experimental evidence for systematic placement of the *Haplopappus phyllocephalus* complex. *Syst. Bot.* 6:8–14.

Jenkins, C. D., and M. Kirkpatrick. 1994. Deleterious mutation and ecological selection in the evolution of life cycles. In *The evolution of haploid-diploid life cycles*, ed. M. Kirkpatrick, 53–68. Lectures on Mathematics in the Life Sciences 25. Providence: American Mathematical Society.

———. 1995. Deleterious mutation and the evolution of genetic life cycles. *Evolution* 49:512–20.

Jones, D. F. 1939. Continued inbreeding in maize. *Genetics* 24:462–73.

Joyce, G. F. 1991. The rise and fall of the RNA world. *New Biol.* 3:399–407.

Kaplan, D. R. 1977. Morphological status of the shoot systems of the Psilotaceae. *Brittonia* 29:30–53.

Karpechenko, G. D. 1927. Polyploid hybrids of *Raphanus sativas* L. × *Brassica oleracea* L. *Zeit. Ind. Abst. Vererbungslehre* 48:1–58.

Karpen, G. H., M.-H. Le, and H. E. Le. 1996. Centric heterchromatin and the efficiency of achiasmatic disjunction in *Drosophila* female meiosis. *Science* 273:118–22.

Kauffman, S. A. 1993. *The origins of order.* Oxford: Oxford University Press.

Kauffman, S. A., and S. A. Levin. 1987. Towards a general theory of adaptive walks on rugged landscapes. *J. Theoret. Biol.* 128:11–45.

Keck, D. D. 1935. Studies upon the taxonomy of the Madinae. *Madroño* 3:4–18.

Kenrick, P., and D. Edwards. 1988. The anatomy of Lower Devonian *Gosslingia breconensis* Heard based on pyritized axes, with some comments on the permineralization process. *Bot. J. Linn. Soc.* 97:95–123.

Kenrick, P., D. Edwards, and R. Dales. 1991. Novel ultrastructure in water-conducting cells of the Lower Devonian plant *Sennicaulis hippocrepiformis* Edwards. *Palaeontology* 34:751–66.

Kidston, R., and W. H. Lang. 1920. On Old Red Sandstone plants showing structure, from the Rhynie Chert Bed, Aberdeenshire. Part 2. Addition notes on *Rhynia Gwynne-Vaughanii*, Kidston and Lang; with descriptions of *Rhynia major*, n. sp., and *Hornea Lignieri*, n. g., n. sp. *Trans. Roy. Soc. Edinburgh* 52:603–27.

———. 1921. On Old Red Sandstone plants showing structure, from the Rhynie Chert Bed, Aberdeenshire. Part 4. Restorations of the vascular cryptogams, and discussion of their bearing on the genral morphology of the Pteridophyta and the origin of the organisation of land-plants. *Trans. Roy. Soc. Edinburgh* 52:831–54.

Kimura, M. 1968. Evolutionary rate at the molecular level. *Nature* 217:624–26.

———. 1983. *The neutral theory of evolution*. New York: Cambridge University Press.

———, ed. 1982. *Molecular evolution, protein polymorphisms, and the neutral theory*. Berlin: Springer-Verlag.

Kingsolver, J. G., and M. A. R. Koehl. 1985. Aerodynamics, thermoregulation, and the evolution of insect wings: Differential scaling and evolutionary change. *Evolution* 85:488–504.

Kirk, J. T. O. 1975. A theoretical analysis of the contribution of algal cells to the attenuation of light within natural water. 2. Spherical cells. *New Phytol.* 75:21–36.

———. 1976. A theoretical analysis of the contribution of algal cells to the attenuation of light within natural water. 3. Cylindrical and spheroidal cells. *New Phytol.* 76:341–58.

———. 1983. *Light and photosynthesis in aquatic ecosystems*. Cambridge: Cambridge University Press.

Kirkpatrick, M., ed. 1994. *The evolution of haploid-diploid life cycles*. Lectures on Mathematics in the Life Sciences 25. Providence: American Mathematical Society.

Klinger, T. 1993. The persistence of haplodiploidy in algae. *Trends Ecol. Evol.* 8:256–58.

Knoll, A. H. 1983. Biological interactions and Precambrian eukaryotes. In *Biotic interactions in Recent and fossil benthic communities*, ed. M. J. S. Tevesz and P. L. McCall, 251–83. New York: Plenum.

———. 1984. Patterns of extinction in the fossil record of vascular plants. In *Extinctions*, ed. M. Nitecki, 21–67. Chicago: University of Chicago Press.

———. 1992. The early evolution of eukaryotes: A geological perspective. *Science* 256:622–27.

Knoll, A. H., and K. J. Niklas. 1987. Adaptation, plant evolution, and the fossil record. *Rev. Paleobot. Palynol.* 50:127–49.

Knoll, A. H., K. J. Niklas, P. G. Gensel, and B. H. Tiffney. 1984. Character diversification and patterns of evolution in early vascular plants. *Paleobiology* 10:34–47.

Kondrashov, A. S. 1994. The asexual ploidy cycle and the origin of sex. *Nature* 370:213–16.

———. 1995. Contamination of the genome by very slightly deleterious mutations: Why have we not died 100 times over? *J. Theor. Biol.* 175:583–94.

Koornneef, M., J. van Eden, C. J. Hanhart, P. Stam, F. J. Braaksma, and W. J. Feenstra. 1983. Linkage map of *Arabidopsis thaliana. J. Hered.* 74:265–72.

Kvenvolden, K. A., J. Lawless, K. Pering, E. Peterson, J. Flores, C. Ponnamperuma, I. R. Kaplan, and C. B. Moore. 1970. Evidence for extraterrestrial amino acids and hydrocarbons in the Murchison meteorite. *Nature* 228:923–26.

Lande, R. 1985. Expected time for random genetic drift of a population between stable phenotypic states. *Proc. Natl. Acad. Sci. U.S.A.* 82:7641–45.

Lande, R., and D. W. Schemske. 1985. The evolution of self-fertilization and inbreeding depression in plants. I. Genetic models. *Evolution* 39:24–40.

Langley, C. H., and W. M. Fitch. 1974. An examination of the constancy of the rate of molecular evolution. *J. Mol. Evol.* 3:161–77.

Lee, R. E. 1989. *Phycology.* 2d ed. Cambridge: Cambridge University Press.

Lenski, R. E., and M. Travisano. 1994. Dynamics of adaptation and diversification: A 10,000-generation experiment with bacterial populations. *Proc. Natl. Acad. Sci. U.S.A.* 91:6808–14.

Levin, D. A. 1978. The origin of isolating mechanisms in flowering plants. *Evol. Biol.* 11:185–317.

Levin, D. A., and E. T. Brack. 1995. Natural selection against white petals in *Phlox. Evolution* 49:1017–22.

Levin, D. A., and A. C. Wilson. 1976. Rates of evolution in seed plants: Net increase in diversity of chromosome numbers and species numbers through time. *Proc. Natl. Acad. Sci. U.S.A.* 73:2086–90.

Levin, S. A. 1978. On the evolution of ecological parameters. In *Ecological genetics: The interface,* ed. P. F. Brussard, 3–26. New York: Springer-Verlag.

Levins, R. 1968. *Evolution in changing environments.* Monographs in Population Biology 2. Princeton: Princeton University Press.

Lewin, B. 1985. *Genes.* 2d ed. New York: Wiley.

Lewis, H., and W. L. Bloom. 1972. The loss of a species through breakdown of a chromosomal barrier. *Symp. Biol. Hungarica* 12:61–64.

Lewontin, R. C. 1974. *The genetic basis of evolutionary change.* New York: Columbia University Press.

———. 1977. Adaptation. *Sci. Amer.* 239:159–69.

Li, H.-L. 1956. A horticultural and botanical history of *Ginkgo. Bull. Morris Arboretum* 7:3–12.

Li, H., E. L. Taylor, and T. N. Taylor. 1996. Permian vessel elements. *Science* 271:188–89.

Li, W.-H., C.-C. Luo, and C.-I. Wu. 1985. Evolution of DNA sequences. In *Molecular evolutionary genetics,* ed. R. J. MacIntyre, 1–94. New York: Plenum.

Ligrone, R., and R. Gambardella. 1988. The sporophyte-gametophyte junction in bryophytes. *Adv. Bryol.* 3:225–74.

Lloyd, D. G. 1992. Self- and cross-fertilization in plants. 2. The selection of self-fertilization. *Int. J. Plant Sci.* 153:370–80.

Long, A. G. 1977. Lower Carboniferous pteridosperm cupules and the origin of angiosperms. *Trans. Roy. Soc. Edinburgh* 70:13–35.

Lynch, M., and R. Lande. 1993. Evolution and extinction in response to environmental change. In *Biotic interactions and global change,* ed. P. M. Kareiva, J. G. Kingsolver, and R. B. Huey, 234–50. Sunderland, Mass.: Sinauer.

Madigan, M. T. 1992. The family Heliobacteriaceae. In *The prokaryotes: A handbook on the biology of bacteria. Ecophysiology, isolation, idenification and applications,* ed. A. Balows, H. G. Trüper, M. Dworkin, K. H. Schleifer, and W. Harder, vol. 2, chap. 90, 1982–92. Berlin: Springer-Verlag.

Margulis, L. 1970. *Origin of eukaryotic cells.* New Haven: Yale University Press.

———. 1981. *Symbiosis in cell evolution.* San Francisco: Freeman.

Margulis, L., and D. Sagan. 1986. *Origins of sex: Three billion years of recombination.* New Haven: Yale University Press.

Mattox, K. R., and K. D. Stewart. 1984. Classification of the green algae: A concept based on comparative cytology. In *Systematics of the green algae,* ed. L. Irvine and M. J. John, 29–72. London: Academic Press.

Mau, S.-L., M. A. Anderson, M. Heisler, V. Haring, B. A. McClure, and A. E. Clarke. 1991. Molecular and evolutionary aspects of self-incompatibility in flowering plants. In *Molecular biology of plant development,* ed. G. Jenkins and W. Schuch, 245–69. Society of Experimental Biology Series 45. Cambridge: Company of Biologists.

Mauseth, J. D. 1988. *Plant anatomy.* New York: Benjamin Cummings.

Mauzerall, D. C. 1990. The photochemical origins of life and photoreaction of ferrous ion in the archaean oceans. *Orig. Life Evol. Biosph.* 20:293–302.

Maynard Smith, J. 1970. Natural selection and the concept of protein space. *Nature* 225:563–64.

———. 1977. Optimization theory in evolution. *Ann. Rev. Ecol. Syst.* 9:31–56.

———. 1978. *The evolution of sex.* Cambridge: Cambridge University Press.

Maynard Smith, J., R. Burian, S. Kauffman, P. Alberch, J. Campbell, B. Goodwin, R. Lande, D. Raup, and L. Wolpert. 1985. Developmental constraints and evolution. *Q. Rev. Biol.* 60:265–87.

Maynard Smith, J., and E. Szathmáry. 1995. *The major transitions in evolution.* Oxford: Freeman.

Mayr, E. 1942. *Systematics and the origin of species.* New York: Columbia University Press.

———. 1963. *Animal species and evolution.* Cambridge: Harvard University Press.

———. 1969. *Principles of systematic zoology.* New York: McGraw-Hill.

McClintock, B. 1951. Mutable loci in maize. *Carnegie Inst. Wash. Year Book* 49:157–67.

———. 1984. The significance of responses of the genome to challenge. *Science* 226:792–801.

McDonald, J. F. 1990. Macroevolution and retroviral elements. *BioScience* 40: 183–91.

McFadden, G., and P. Gilson. 1995. Something borrowed, something green: Lateral transfer of chloroplasts by secondary endosymbiosis. *TREE* 10:12–7.

McKay, D. S., E. K. Gibson Jr., K. L. Thomas-Keprta, H. Vali, C. S. Romanek, S. J. Clemett, X. D. F. Chillier, C. R. Maechling, and R. N. Zare. 1996. Search for past life on Mars: Possible relic biogenic activity in Martian meteorite ALH84001. *Science* 273:924–30.

Mercer-Smith, J. A., and D. C. Mauzerall. 1984. Photochemistry of porphyrins: A model for the origin of photosynthesis. *Photochem. Photobiol.* 39:397–405.

Meredith, D. D., K. W. Wong, R. W. Woodhead, and R. H. Wortman. 1973. *Design and planning of engineering systems.* Civil Engineering and Engineering Mechanics Series, ed. N. M. Newmark and W. J. Hall. Englewood Cliffs, N.J.: Prentice-Hall.

Mereschkowsky, C. 1905. Über Natur und Ursprung der Chromatophoren im Pflanzen Reiche. *Biol. Centr.* 25:593–604.

———. 1920. La plante considerée comme un complex symbiotique. *Bull. Soc. Sci. Nat. Ouest Fr.* 6:17–21.

Michod, R. E., and B. R. Levin, eds. 1988. *The evolution of sex: An examination of current ideas.* Sunderland, Mass.: Sinauer Press.

Miller, S. L. 1953. A production of amino acids under possible primitive Earth conditions. *Science* 117:528–29.

Miller, S. L., and H. C. Urey. 1959. Organic compound synthesis on the primitive Earth. *Science* 130:245–51.

Mirov, N. T. 1967. *The genus* Pinus. New York: Ronald Press.

Mishler, B. D., and S. P. Churchill. 1984. A cladistic approach to the phylogeny of the "bryophytes." *Brittonia* 36:406–24.

———. 1985. Transition to a land flora: Phylogenetic relationships of the green algae and bryophytes. *Cladistics* 1:305–28.

Mishler, B. D., L. A. Lewis, M. A. Buchheim, K. S. Rensaglia, D. J. Garbary, C. F. Delwiche, F. W. Zechman, T. S. Kantz, and R. L. Chapman. 1994. Phylogenetic relationships of the "green algae" and "bryophytes." *Ann. Missouri Bot. Gard.* 81:451–83.

Miyata, T., H. Hayashida, R. Kikuno, M. Hasegawa, M. Kobayashi, and K. Koike. 1982. Molecular clock of silent subsitution: At least six-fold preponderance of silent changes in mitochondrial genes over those in nuclear genes. *J. Mol. Evol.* 19:28–35.

Mogensen, H. L. 1995. The hows and whys of cytoplasmic inheritance in seed plants. *Amer. J. Bot.* 83:383–404.

Montgomery, W. L., and P. E. Pollak. 1988. *Epulopiscium fishelsoni* n. g., n. sp., a protist of uncertain taxonomic affinities from the gut of an herbivorous reef fish. *J. Protozool.* 35:565–69.

Mueller-Dombois, D., K. W. Bridges, and H. L. Carson, eds. 1981. *Island ecosystems: Biological organization in selected Hawaiian communities.* US/IBP Synthesis Series 15. Woods Hole, Mass.: Hutchinson Ross.

Muller, H. J. 1932. Some genetic aspects of sex. *Amer. Nat.* 66:118–38.

Myers, M. 1980. *Conversion of moist tropical forests.* Washington, D.C.: National Academy of Sciences.

Nasrallah, J. B., and M. E. Nasrallah. 1993. Pollen-stigma signaling in the sporophytic self-incompatability response. *Plant Cell* 5:1325–35.

Neal, D. B. 1983. Population genetic structure of the shelterwood regeneration system in southwest Oregon. Ph.D. diss., Oregon State University, Corvallis.

Newbigin, E., M. A. Anderson, and A. E. Clarke. 1993. Gametophytic self-incompatibility systems. *Plant Cell* 5:1315–24.

Newman, C. M., J. E. Cohen, and C. Kipnis. 1985. Neo-Darwinian evolution implies punctuated equilibria. *Nature* 315:400–401.

Niklas, K. J. 1981. Airflow patterns around some early seed plant ovules and cupules: Implications concerning efficiency in wind pollination. *Amer. J. Bot.* 68:635–50.

———. 1986. Computer-simulated plant evolution. *Sci. Amer.* 254:78–86.

———. 1988. Biophysical limitations on plant form and evolution, In *Plant evolutionary biology*, ed. L. D. Gottlieb and S. K. Jain, 185–220. London: Chapman and Hall.

———. 1992. *Plant biomechanics: An engineering approach to plant form and function.* Chicago: University of Chicago Press.

———. 1994. *Plant allometry: The scaling of form and process.* Chicago: University of Chicago Press.

———. 1995. Morphological evolution through complex domains of fitness. In *Tempo and mode in evolution: Genetics and paleontology fifty years after Simpson*, ed. W. M. Fitch and F. J. Ayala, 145–65. Washington, D.C.: National Academy Press.

Niklas, K. J., and V. Kerchner. 1984. Mechanical and photosynthetic constraints on the evolution of plant shape. *Paleobiology* 10:79–101.

Niklas, K. J., B. H. Tiffney, and A. H. Knoll. 1980. Apparent changes in the diversity of fossil plants: A preliminary assessment. In *Evolutionary biology*, vol. 12, ed. M. Hecht, W. Steere, and B. Wallace, 1–89. New York: Plenum.

———. 1983. Patterns in vascular land plant diversification. *Nature* 303:614–16.

Nisbet, E. G., J. R. Cann, and C. L. Van Dover. 1995. Origins of photosynthesis. *Nature* 373:479–80.

Nixon, K. C., W. L. Crepet, D. Stevenson, and E. M. Friis. 1994. A Reevaluation of seed plant phylogeny. *Ann. Missouri Bot. Gard.* 81:484–533.

Nixon, K. C., and Q. D. Wheeler. 1990. An amplification of the phylogenetic species concept. *Cladistics* 6:211–23.

Nobel, P. S. 1983. *Biophysical plant physiology and ecology.* New York: Freeman.

Nobs, M. A. 1963. Experimental studies on species relationships in *Ceanothus. Carneige Inst. Washington Pub.*, no. 623:1–94.

Odell, G. M., G. Oster, P. Alberch, and B. Burnside. 1981. The mechanical basis of morphogenesis. 1. Epithelial folding and invagination. *Dev. Biol.* 85:446–62.

Ohyama, K., H. Fukuzawa, T. Kohchi, H. Shirai, T. Sano, S. Sano, K. Umesono, Y. Shiki, M. Takeuchi, Z. Chang, S.-I. Aota, H. Inokuchi, and H. Ozeki. 1986.

Chloroplast gene organization deduced from complete sequence of liverwort *Marchanita polymorpha* chloroplast DNA. *Nature* 322:572–74.

Okubo, A., and S. A. Levin. 1989. A theoretical framework for data analysis of wind dispersal of seeds and pollen. *Ecology* 70:329–38.

Olson, J. M., and B. K. Pierson. 1987. Origin and evolution of photosyntheic reaction centers. *Orig. Life* 17:419–30.

Orgel, L. E. 1994. The origin of life on Earth. *Sci. Amer.* 271 (4): 77–83.

Oró, J. 1961. Comets and the formation of biochemical compounds on the primitive Earth. *Nature* 190:389.

Oster, G. F., G. Odell, and P. Alberch. 1980. Mechanics, morphogenesis and evolution. In *Lectures on mathematics in the life sciences,* ed. G. Oster, 165–255. Providence: American Mathematical Society.

Oster, G. F., and E. O. Wilson. 1979. *Caste and ecology in the social insects.* Princeton: Princeton University Press.

Owadally, A. W. 1979. The dodo and the tambalacoque tree. *Science* 203: 1363–64.

Oyzizu, H., B. Debrunner-Vossbrinck, L. Mandelco, J. A. Studier, and C. R. Woese. 1987. The green non-sulfur bacteria: A deep branching in the eubacterial line of descent. *Syst. Appl. Micro.* 9:47–53.

Palmer, J. D. 1985a. Evolution of chloroplast and mitochondrial DNA in plants and algae. In *Molecular evolutionary genetics,* ed. R. J. MacIntyre, 131–240. New York: Plenum.

———. 1985b. Comparative organization of chloroplast genomes. *Ann. Rev. Genet.* 19:325–54.

———. 1987. Chloroplast DNA evolution and biosystematic uses of chloroplast DNA variation. *Amer. Nat.* 130 (suppl.): S6–S29.

Paterniani, E. 1969. Selection for reproductive isolation between two populations of maize, *Zea mays. Evolution* 23:534–47.

Paterson, H. E. H. 1985. The recognition concept of species. In *Species and speciation,* ed. E. S. Vrba, 21–29. Transvaal Museum Monograph 4. Pretoria: Transvaal Museum.

Paterson, H. E. H., and M. Macnamara. 1984. The recognition concept of species. *S. Afr. J. Sci.* 80:312–18.

Pettitt, J. M. 1970. Heterospory and the origin of the seed habit. *Biol. Rev.* 45: 401–15.

Pettitt, J. M., and C. B. Beck. 1968. *Archaeosperma arnoldii:* A cupulate seed from the Upper Devonian of North America. *Contrib. Mus. Paleont. Univ. Michigan* 22:139–54.

Phillips, T. L. 1979. Reproduction of heterosporous arborescent lycopods in the Mississippian-Pennsylvanian of Euramerica. *Rev. Palaeobot. Palynol.* 27: 239–89.

Phillips, T. L., M. J. Avcin, and J. M. Schopf. 1975. Gametophytes and young sporophyte development in *Lepidodendron. Bot. Soc. Amer.* (Corvallis, Ore.), Abstr., 23.

Pickett-Heaps, J. D. 1975. *Green algae: Structure, reproduction and evolution in selected genera.* Sunderland, Mass.: Sinauer.

————. 1976. Cell division in eukaryotic algae. *BioScience* 26:445–50.

Pielou, E. C. 1991. *After the Ice Age: The return of life to glaciated North America*. Chicago: University of Chicago Press.

Prager, E. M., D. P. Fowler, and A. C. Wilson. 1976. Rate of evolution in conifers (Pinaceae). *Evolution* 30:637–49.

Pratt, L. M., T. L. Phillips, and J. M. Dennison. 1978. Evidence of non-vascular land plants from the early Silurian (Llandoverian) of Virginia, U.S.A. *Rev. Paleobot. Palynol.* 25:121–49.

Provine, W. B. 1986. *Sewall Wright and evolutionary biology*. Chicago: University of Chicago Press.

Raff, R. A., and G. A. Wray. 1989. Heterochrony: Developmental mechanisms and evolutionary results. *J. Evol. Biol.* 2:409–34.

Raikov, I. B. 1982. *The protozoan nucleus*. Heidelberg: Springer-Verlag.

Raper, J. R., and A. S. Flexer. 1970. The road to diploidy with emphasis on a detour. In *Organization and control in prokaryotic and eukaryotic cells: Twentienth symposium of the Society for General Microbiology*, ed. H. P. Charles amd B. C. K. G. Knight 401–32. Cambridge: Cambridge University Press.

Raup, D. M. 1994. The role of extinction in evolution. *Proc. Natl. Acad. Sci. U.S.A.* 91:6758–63.

Raup, D. M., and J. J. Sepkoski Jr. 1982. Mass extinctions in the marine fossil record. *Science* 215:1501–3.

Rausher, M. D. 1996. Genetic analysis of coevolution between plants and their natural enemies. *Trends Genet.* 12, 6:212–17.

Raven, J. A. 1984. Physiological correlates of the morphology of early vascular plants. *Bot. J. Linn. Soc.* 88:105–26.

————. 1985. Comparative physiology of plant and arthropod land adaptation. *Phil. Trans. Roy. Soc. Lond.,* ser. B, 309:273–388.

Raven, P. H. 1980. Hybridization and the nature of species in higher plants. *Canadian Bot. Assoc. Bull.* 13 (suppl.): 3–10.

————. 1984. Onagraceae as a model of plant evolution. In *Plant evolutionary biology,* ed. L. D. Gottlieb and S. K. Jain, 85–107. London: Chapman and Hall.

Remane, A. 1952. *Die Grundlagen des natürlichen Systems, der vergleichenden Anatomie und der Phylogenetik*. Leipzig: Akademische Verlagsgesellschaft.

Remy, W., P. G. Gensel, and H. Hass. 1993. The gametophyte generation of some early Devonian land plants. *Int. J. Plant Sci.* 154:35–58.

Remy, W., and H. Hass. 1991. Erganzende Beobachtungen an *Lyonophyton rhyniensis*. *Argumenta Palaeobot.* 8:1–27.

Renault, S., J. L. Bonnemain, L. Faye, and J. P. Caudillere. 1992. Physiological aspects of sugar exchange between the gametophyte and the sporophyte of *Polytrichum formosum*. *Plant Physiol.* 100:1815–22.

Rich, P. M. 1986. Mechanical architecture of arborescent rain forest palms. *Principes* 30:117–31.

Riedl, R. 1978. *Order in living things: A systems analysis of evolution*. New York: John Wiley.

Riesenberg, L. H. 1991. Homoploid reticulate evolution in *Helianthus:* Evidence from ribosomal genes. *Amer. J. Bot.* 78:1218–37.

———. 1995. The role of hybridization in evolution: Old wine in new skins. *Amer. J. Bot.* 82:944–53.

Riesenberg, L. H., and N. C. Ellstrand. 1993. What can molecular and morphological markers tell us about plant hybridization? *Crit. Rev. Plant Sci.* 12:213–41.

Riesenberg, L. H., B. Sinervo, C. R. Linder, M. C. Ungerer, and D. M. Arias. 1996. Role of gene interactions in hybrid speciation: Evidence from ancient and experimental hybrids. *Science* 272:741–44.

Robertson, M., and S. Miller. 1995. An efficient prebiotic synthesis of cytosine and uracil. *Nature* 375:772–74.

Rodermal, S. R., and L. Bogorad. 1987. Molecular evolution and nucleotide sequences of the maize plasmid genes for the subunit of CF1 (*atpA*) and the proteolipid subunit of CF0 (*atpH*). *Genetics* 116:127–39.

Rothwell, G. W. 1972. Evidence of pollen tubes in Paleozoic pteridosperms. *Science* 175:772–74.

Rothwell, G. W., and S. E. Scheckler. 1988. Biology of ancestral gymnosperms. In *Origin and evolution of gymnosperms,* ed. C. B. Beck, 85–134. New York: Columbia University Press.

Rothwell, G. W., S. E. Scheckler, and W. H. Gillespie. 1989. *Elkinsia* gen. nov., a late Devonian gymnosperm with cupulate ovules. *Bot. Gaz.* 150:170–89.

Rothwell, G. W., and R. Serbet. 1994. Lignophyte phylogeny and the evolution of spermatophytes: A numerical cladistic analysis. *Syst. Bot.* 19:443–82.

Roux, W. 1895. *Gesammelte Abhandlungen zur Entwicklungsmechanik der Organismen.* Leipzig: Engelmann.

Rowe, N. P. 1992. Winged late Devonian seeds. *Nature* 359:682.

Sakai, A. K., W. L. Wagner, D. M. Ferguson, and D. R. Herbst. 1995. Biogeographical and ecological correlations of dioecy in the Hawaiian flora. *Ecology* 76:2530–43.

Sarich, V. M., and A. C. Wilson. 1973. Generation time and genomic evolution in primates. *Science* 179:1144–47.

Scagel, R. F., R. J. Bandoni, J. R. Maze, G. E. Rouse, W. B. Schofield, and J. R. Stein. 1982. *Nonvascular plants: An evolutionary survey.* Belmont, Calif.: Wadsworth.

Scharloo, W. 1991. Canalization, genetic and developmental aspects. *Ann. Rev. Ecol. Sys.* 22:265–94.

Scheirer, D. C. 1980. Differentiation of bryophyte conducting tissues: Structure and histo-chemistry. *Bull. Torrey Bot. Club* 107:298–307.

Schidlowski, M. 1988. A 3800-million-year isotopic record of life from carbon in sedimentary rocks. *Nature* 333:313–18.

Schimper, A. F. W. 1883. "Über die Entwicklung der Chlorophyllkorner und Färbkornerm," part 1, *Bot. Zeit.* 41:105–14.

Schluter, D. 1996. Adaptive radiation along genetic lines of least resistance. *Evolution* 50:1766–74.

Schmalhausen, I. I. 1949. *Factors of evolution: The theory of stabilizing selection.* Chicago: University of Chicago Press.

Schmid, B. 1993. Phenotypic variation in plants. *Evol. Trends Plants* 6:45–60.

Schneider, E. L., and S. Carlquist. 1995. Vessel origins in Nymphaeaceae: *Euryale* and *Victoria. Bot. J. Linn. Soc.* 119:185–93.

Schopf, J. W. 1983. *Earth's earliest biosphere: Its origin and evolution.* Princeton: Princeton University Press.

Schoute, J. C. 1912. Über das Dickenwachstum der Palmen. *Ann. Jard. Bot. Buitenzorg,* ser. 2, 11:1–209.

Schuster, R. M. 1966. *The Hepaticae and Anthocerotae of North America.* Vol. 1. New York: Columbia University Press.

———. 1981. Paleoecology, origin, distribution through time, and evolution of Hepaticae and Anthocerotae. In *Paleobotany, paleoecology, and evolution,* ed. K. J. Niklas, 1:129–92. New York: Praeger.

Searcy, D. G., D. B. Stein, and K. B. Searcy. 1981. A mycoplasma-like archaebacterium possibly related to the nucleus and cytoplasm of eukaryotic cells. *Ann. N.Y. Acad. Sci.* 361:312–24.

Sepkoski, J. J. 1979. A kinetic model of Phanerozoic taxonomic diversity. 2. Early Phanerozoic families and multiple equilibria. *Paleobiology* 5:222–51.

Serbet, R., and G. W. Rothwell. 1992. Characterizing the most primitive seed ferns. I. A reconstruction of *Elksinia polymorpha. Int. J. Plant Sci.* 153:602-21.

Simberloff, D. 1986. Are we on the verge of a mass extinction in tropical rain forests? In *Dynamics of extinction,* ed. D. K. Elliott, 165–80. New York: John Wiley.

Simms, E. I., and M. D. Rausher. 1989. The evolution of resistance to herbivory in *Ipomoea purpurea.* II. Natural selection by insects and costs of resistance. *Evolution* 43:573–85.

Simpson, G. G. 1944. *Tempo and mode in evolution.* New York: Columbia University Press.

———. 1961. *Principles of animal taxonomy.* New York: Columbia University Press.

Singh, B. B., and A. N. Jha. 1978. Abnormal differentiation of floral parts in a mutant strain of soybean. *J. Hered.* 69:143–44.

Slack, J. M. W., P. W. H. Holland, and C. F. Graham. 1993. The zootype and the phylotypic stage. *Nature* 361:490–92.

Smith, G. M. 1950. *The fresh-water algae of the United States.* New York: McGraw-Hill.

———. 1955. *Cyrptogamic botany.* Vol. 1. New York: McGraw-Hill.

Sommer, A. 1976. Attempt at an assessment of the world's tropical moist forests. *Unasylva* 28:5–24.

Sommer, H., J.-P. Beltrán, P. Huijser, H. Pape, W.-E. Lönnig, H. Saedler, and Z. Schwarz-Sommer. 1990. *Deficiens,* a homeotic gene involved in the control of flower morphogenesis in *Antirrhunum majus:* The protein shows homology to transcription factors. *EMBO J.* 9:605–13.

Speck, T. S., and D. Vogellehner. l988. Biophysical examinations of the bending stability of various stele types and the upright axes of early "vascular" land plants. *Bot. Acta* 101:262–68.

Sprague, G. F. 1967. Plant breeding. *Ann. Rev. Genet.* 1:269–94.

Stanley, S. M. 1973. An explanation for Cope's rule. *Evolution* 27:1–26.

———. 1979. *Macroevolution: Pattern and process.* San Francisco: Freeman.

Stebbins, G. L. 1950. *Variation and evolution in plants.* New York: Columbia University Press.

———. 1959. The role of hybridization in evolution. *Proc. Amer. Phil. Soc.* 103:231–51.

———. 1960. The comparative evolution of genetic systems. In *Evolution after Darwin,* ed. S. Tax, 197–226. Chicago: University of Chicago Press.

Stebbins, G. L., and F. J. Ayala. 1985. The evolution of Darwinism. *Sci. Amer.* 253:72–82.

Stevens, A. A., and C. M. Rick. 1986. Genetics and breeding. In *The tomato crop,* ed. J. G. Atherton and J. Ruich, 35–109. London: Chapman and Hall.

Stewart, W. N., and G. W. Rothwell. 1993. *Paleobotany and the evolution of plants.* 2d ed. Cambridge: Cambridge University Press.

Stoeckenius, W., and R. A. Bogomolni. 1982. Bacteriorhodopsin and related pigments of halobacteria. *Ann. Rev. Biochem.* 51:587–616.

Strauss, E. G., J. H. Strauss, and A. J. Levine. 1996. Virus evolution. In *Fields virology,* 3d ed., vol. 1., ed. B. N. Fields, D. M. Knipe, and P. M. Howley, 153–71. Philadelphia: Lippincott-Raven.

Strutz, H. C., and L. K. Thomas. 1964. Hybridization and introgression between *Cowania* and *Purshia. Evolution* 18:183–95.

Sultan, S. E. 1987. Evolutionary implications of phenotypic plasticity in plants. *Evol. Ecol.* 21:127–78.

———. 1992. Phenotypic plasticity and the neo-Darwinian legacy. *Evol. Trends Plants* 6:61–71.

———. 1995. Phenotypic plasticity and plant adaptation. *Acta Bot. Neerl.* 44: 363–83.

Sultan, S. E., and F. A. Bazzaz. 1993. Phenotypic plasticity in *Polygonum persicaria.* I. Diversity and uniformity in genotypic norms of reaction to light. *Evolution* 47:1009–31.

Sung, Z. R., A. Belachew, B. Shunong, and R. Bertrand-Garcia. 1992. EMF, an *Arabidopsis* gene required for vegetative shoot development. *Science* 258: 1645–47.

Sussex, I. M. 1966. The origin and development of heterospory in vascular plants. In *Trends in plant morphogenesis,* ed. E. G. Cutter, 141–52. London: Longmans.

Suzuki, D. T., A. J. F. Griffiths, J. H. Miller, and R. C. Lewontin. 1989. *An introduction to genetic analysis.* 4th ed. New York: Freeman.

Szathmáry, E., and J. Maynard Smith. 1993. The origin of chromosomes. II. Molecular mechanisms. *J. Theor. Biol.* 164:447–54.

Taiz, L,. and E. Zeiger. 1991. *Plant physiology.* Redwood City, Calif.: Benjamin-Cummings.

Takhtajan, A. L. 1958. *Origins of angiospermous plants.* [Trans. from the original 1954 Russian version.] Washington, D.C.: American Institute of Biological Sciences.

———. 1959. *Essays on the evolutionary morphology of plants.* [Trans. from the original 1954 Russian version.] Washington, D.C.: American Institute of Biological Sciences.

Tanksley, S. D., and E. Pichersky. 1984. Organization and evolution of sequences in the plant nuclear genome. In *Plant evolutionary biology,* ed. L. D. Gottlieb and S. K. Jain, 55–83. London: Chapman and Hall.

Taylor, I. B. 1986. Biosystematics of the tomato. In *The tomato crop,* ed. J. G. Atherton and J. Ruich, 1–34. London: Chapman and Hall.

Taylor, T. N., and M. Millay. 1979. Pollination biology and reproduction in early seed plants. *Rev. Palaeobot. Palynol.* 27:329–55.

Taylor, T. N., and E. L. Taylor. 1993. *The biology and evolution of fossil plants.* Englewood Cliffs, N.J.: Prentice-Hall.

Templeton, A. R. 1980. The theory of speciation via the founder effect. *Genetics* 94:1011–38.

———. 1982. Adaptation and the integration of evolutionary forces. In *Perspectives on evolution,* ed. R. Milkman, 15–31. Sunderland, Mass.: Sinauer.

Thomas, R. D. K., and W.-E. Reif. 1993. The skeleton space: A finite set of organic designs. *Evolution* 47:341–60.

Thomas, R. J., D. S. Stanton, D. H. Longendorfer, and M. E. Farr. 1978. Physiological evaluation of the nutritional autonomy of a hornwort sporophyte. *Bot. Gaz.* 139:306–11.

Thomas, S. C., and F. A. Bazzaz. 1993. The genetic component in plant size hierarchies: Norms of reaction to density in a *Polygonum* species. *Amer. J. Bot.* 63:231–49.

Tiffney, B. H. 1981. Diversity and major events in the evolution of land plants. In *Paleobotany, paleoecology, and evolution,* ed. K. J. Niklas, 1:193–230. New York: Praeger.

Tiffney, B. H., and K. J. Niklas. 1985. Clonal growth in land plants: A paleobotanical perspective. In *Population biology and evolution of clonal organisms,* ed. J. B. C. Jackson, L. W. Buss, and R. E. Cook, 35–66. New Haven: Yale University Press.

Tobgy, H. A. 1943. A cytological study of *Crepis fuliginosa, C. neglecta,* and their F_1 hybrid, and its bearing on the mechanism of phylogenetic reduction in chromosome number. *J. Genet.* 45:67–111.

Tomlinson, D. B. 1961. *Anatomy of the monocotyledons.* Vol. 2. *Palmae.* Oxford: Claredon Press.

Tralau, H. 1968. Evolutionary trends in the genus *Ginkgo. Lethaia* 1:63–101.

Traverse, A. 1988. Plant evolution dances to a different beat. *Hist. Biol.* 1: 277–301.

Tryon, A. F., and B. Lugardon. 1991. *Spores of the Pteridophyta.* New York: Springer-Verlag.

Tryon, R. M., and A. F. Tryon. 1982. *Ferns and allied plants with special reference to tropical America.* New York: Springer-Verlag.

Tucker, J. M. 1953. Two new hybrid oaks from California. *Madroño* 12: 119–27.

Turesson, G. 1922. The genotypical response of the plant species to the habitat. *Hereditas* 3:211–350.

Uhl, C. H. 1992. Polyploidy, dysploidy, and chromosome pairing in *Echeveria* (Crassulaceae) and its hybrids. *Amer. J. Bot.* 79:556–66.

Van'T Hoff, J., and A. H. Sparrow. 1963. A relationship between DNA content, nuclear volume and minimum mitotic cycle time. *Proc. Natl. Acad. Sci. U.S.A.* 49:897–902.

Van Tienderen, P. H. 1990. Morphological variation in *Plantago lanceolata*: Limits of plasticity. *Evol. Trends Plants* 4:35–43.

———. 1991. Evolution of generalists and specialists in spatially heterogeneous environments. *Evolution* 45:1317–31.

Van Valen, L. 1975. Life, death, and energy of a tree. *BioTropica* 7:259–69.

Via, S., and R. Lande. 1985. Genotype-environment interaction and the evolution of phenotypic plasticity. *Evolution* 39:505–22.

Vogel, S. 1988. *Life's devices: The physical world of animals and plants.* Princeton: Princeton University Press.

Waddington, C. H. 1942. Canalization and development and the inheritance of acquired characters. *Nature* 150:563–65.

Wagner, G. P., and L. Altenberg. 1996. Complex adaptation and the evolution of evolvability. *Evolution* 50:967–76.

Wagner, W. H. 1954. Reticulate evolution in Appalachian aspleniums. *Evolution* 8:103–18.

Wagner, W. L., D. R. Herbst, and S. H. Sohmer. 1990. *Manual of the flowering plants of Hawai'i.* Vol. 1. Honolulu: University of Hawaii Press.

Wallace, B. 1981. *Basic population genetics.* New York: Columbia University Press.

Walter, M. R. 1983. Archean stromatolites: Evidence of the earth's earliest benthos. In *Earth's earliest biosphere: Its origin and evolution,* ed. J. W. Schopf, 187–213. Princeton: Princeton University Press.

Ward, D. M., R. Weller, J. Shiea, R. W. Castenholz, and Y. Cohen. 1989. Hot spring microbial mats: Anoxygenic and oxygenic mats of possible evolutionary significance. In *Microbial mats: Physiological ecology of benthic microbial communities,* ed. Y. Cohen and E. Rosenberg, 3–15. Washington, D.C.: American Society of Microbiology.

Waser, N. M., and M. V. Price. 1983. Pollinator behavior and natural selection for flower color in *Delphinium nelsonii. Nature* 302:422–24.

Webster, T. R. 1990. *Selaginella apoda* × *ludoviciana,* a synthesized hybrid spikemoss. *Amer. J. Bot.* 77 (suppl.): 108 (abstract).

Weis, A. E., and W. L. Gorman. 1990. Measuring selection on reaction norms: An explanation of the *Eurosta-Solidago* system. *Evolution* 44:820–31.

Weiss, R. L., J. R. Kukora, and J. Adams. 1975. The relationship between enzyme activity, cell geometry, and fitness in *Saccharomyces cerevisiae. Proc. Natl. Acad. Sci., USA* 72:794–98.

Wessler, S. R., G. Baran, M. Varagoa, and S. L. Dellaporta. 1986. Excision of *Ds* produces *waxy* proteins with a range of enzymatic activities. *EMBO J.* 5:2427–32.

Westermann, M., A. Ernest, S. Brass, P. Boeger, and W. Wehrmeyer. 1994. Ultrastructure of cell wall and photosynthetic apparatus of the phycobilisome-less *Synechocystis* sp. strain BO8402 and phycobilisome-containing derivative strain BO9201. *Arch. Microbiol.* 162:222–32.

Whatley, J. M., and F. R. Whatley. 1981. Chloroplast evolution. *New Phytol.* 87: 233–47.

Whitehouse, H. L. K. 1951. Multiple-allelomorph incompatibility of pollen and style in the evolution of angiosperms. *Ann. Bot.*, n.s., 14:198–216.

Wiley, E. O. 1978. The evolutionary species concept reconsidered. *Syst. Zool.* 27:17–26.

———. 1981. *Phylogenetics: The theory and practice of phylogenetic systematics.* New York: John Wiley.

Williams, G. C. 1975. *Sex and evolution.* Princeton: Princeton University Press.

Wilson, C., and J. W. Szostak. 1995. *In vitro* evolution of a self-alkylating ribozyme. *Nature* 374:777–82.

Wilson, E. O. 1980. Caste and division of labor in leaf-cutter ants (Hymenoptera: Formicidae: *Atta*). 2. The ergonomic optimization of leaf cutting. *Behav. Ecol. Sociobiol.* 7:157–65.

Wilson, M., B. Gaut, and M. T. Clegg. 1990. Chloroplast DNA evolves slowly in the palm family (Arecaceae). *Mol. Biol. Evol.* 7:303–14.

Wilson, P. 1992. On inferring hybridity from morphological intermediacy. *Taxon* 41:11–23.

Winn, A. A., and A. S. Evans. 1991. Variation among populations of *Prunella vulgaris* L. in plastic responses to light. *Funct. Ecol.* 5:562–71.

Woese, C. R. 1987. Bacterial evolution. *Microbiol. Rev.* 51:221–71.

Woese, C. R., and G. E. Fox. 1977. Phylogenetic structure of the prokaryotic domain the primary kingdoms. *Proc. Natl. Acad. Sci. U.S.A.* 74:5088–90.

Woese, C. R., O. Kandler, and M. L. Wheelis. 1990. Towards a natural system of organisms: Proposal for the domains of Archaea, Bacteria, and Eucarya. *Proc. Natl. Acad. Sci. U.S.A.* 87:4576–79.

Wolfe, K. H., W.-H. Li, and P. M. Sharp. 1987. Rates of nucleotide substitution vary greatly among plant mitochondrial, chloroplast and nuclear DNAs. *Proc. Natl. Acad. Sci. U.S.A.* 84:9054–58.

Wootton, R. J., and C. P. Ellington. 1991. Biomechanics and the origin of insect flight. In *Biomechanics in Evolution,* ed. J. M. V. Rayner and R. J. Wootton, 99–112. Cambridge: Cambridge University Press.

Wright, C. 1873. On the uses and origin of arrangements of leaves in plants. *Mem. Amer. Acad. Arts Sci.* 9:379–415.

Wright, S. 1931. Evolution in Mendelian populations. *Genetics* 16:97–159.

———. 1932. The role of mutation, inbreeding, crossbreeding, and selection in evolution. *Proc. Sixth Internat. Cong. Genet.* 1:356–66.

———. 1941. Analysis of local variability of flower color in *Linanthus parryae*. *Genetics* 28:139–56.

———. 1943. Isolation by distance. *Genetics* 28:114–38.

———. 1951. The genetical structure of populations. *Ann. Eugen.* 15:323–54.

Yang, C.-H., L.-J. Chen, and Z. R. Sung. 1995. Genetic regulation of shoot development in *Arabidopsis*: Role of the *EMF* genes. *Develop. Biol.* 169:421–35.

Yu, Y.-T., R. E. Breitbart, L. B. Smoot, L. Youngsook, V. Mahdavi, and B. Nadel-Ginard. 1992. Human myocyte-specific enhancer factor 2 comprises a group of tissue-restricted MADS box transcription factors. *Genes Dev.* 6:1783–98.

Zeyl, C., and G. Bell. 1996. Symbiotic DNA in eukaryotic genomes. *Trends Ecol. Evol.* 11:10–15.

Zimmermann, M. H., and P. B. Tomlinson. 1965. Anatomy of the palm *Rhapis excelsa*. 1. Mature vegetative axis. *J. Arnold Arbor.* 46:160–80.

———. 1967. Anatomy of the palm *Rhapis excelsa*. 4. Vascular development in apex of vegetative aerial axis and rhizome. *J. Arnold Arbor.* 48:122–42.

———. 1974. Vascular patterns in palm stems: Variations of the *Rhapis* principle. *J. Arnold Arbor.* 55:402–24.

Zimmerman, W. 1952. The main results of the "telome theory." *Palaeobotanist* 1:456–70.

Zuckerkandl, E., and L. Pauling. 1965. Evolutionary divergence and convergence in proteins. In *Evolving genes and proteins,* ed. V. Bryson and H. J. Vogel, 97–166. New York: Academic Press.

Author Index

Subject Index